THE DETONATION PHENOMENON

This book introduces the detonation phenomenon in explosives. It is ideal for engineers and graduate students with a background in thermodynamics and fluid mechanics. The material is mostly qualitative, aiming to illustrate the physical aspects of the phenomenon. Classical idealized theories of detonation waves are presented first. These permit detonation speed, gas properties ahead and behind the detonation wave, and the distribution of fluid properties within the detonation wave itself to be determined. Subsequent chapters describe in detail the real unstable structure of a detonation wave. One-, two-, and three-dimensional computer simulations are presented along with experimental results using various experimental techniques. The important effects of confinement and boundary conditions and their influence on the propagation of a detonation are also discussed. The final chapters cover the various ways detonation waves can be formed and provide a review of the outstanding problems and future directions in detonation research.

Dr. John H. S. Lee is a professor of mechanical engineering at McGill University. His research areas are in combustion, detonations and shock wave physics, and explosion dynamics; he has been carrying out fundamental and applied research in these areas for the past 40 years. He is a consultant and has served on numerous government and industrial advisory committees on explosion hazards and safety since the late 1960s. He is a recipient of the silver medal from the Combustion Institute (1980), the Dionizy Smolenski Medal from the Polish Academy of Sciences (1988), and the Nuna Manson gold medal (1991) for his outstanding contributions to the fundamentals and applied aspects of explosion and detonation phenomena. He is also a two-time recipient of the Faculty Outstanding Teaching Award (1989 and 1995), was elected an honorary professor of the Institute of Mechanics of the Chinese Academy of Sciences (2003), and received the Outstanding Alumni Award of the Polytechnic University of Hong Kong (2003). He is a Fellow of the Royal Society of Canada.

THE DETONATION PHENOMENON

JOHN H. S. LEE

McGill University

CAMBRIDGE
UNIVERSITY PRESS

32 Avenue of the Americas, New York NY 10013-2473, USA

Cambridge University Press is part of the University of Cambridge.

It furthers the University's mission by disseminating knowledge in the pursuit of education, learning and research at the highest international levels of excellence.

www.cambridge.org
Information on this title: www.cambridge.org/9781107430730

First published 2008
First paperback edition 2014

A catalogue record for this publication is available from the British Library

Library of Congress Cataloguing in Publication data

Lee, John H. S., 1938–
The detonation phenomenon / John H.S. Lee.
    p. cm.
Includes bibliographical references and index.
ISBN 978-0-521-89723-5 (hardback)
1. Detonation waves.   2. Detonation waves – Computer simulation.
3. Explosives.   I. Title.
QC168.85.D46L44   2008
662'.2 – dc22   2008000517

ISBN 978-0-521-89723-5 Hardback
ISBN 978-1-107-43073-0 Paperback

to Julie

Contents

Preface

Explosives are highly energetic substances with fast reaction rates and can be in gaseous, liquid, or solid form. Chemical reactions can propagate through the explosive at high supersonic speeds as a detonation wave: a compression shock with an abrupt increase in the thermodynamic state, initiating chemical reactions that turn the reactants into products. This book is devoted to a description of the detonation phenomenon, explaining the physical and chemical processes responsible for the self-sustained propagation of the detonation wave, the hydrodynamic theory that permits the detonation state to be determined, the influence of boundary conditions on the propagation of the detonation, and how detonations are initiated in the explosive.

The book is concerned only with detonation waves in gaseous explosives, because they are much better understood than detonations in condensed phase media. There are many similarities between detonations in gaseous and in condensed explosives, in that the detonation pressure of condensed explosives is much higher than the material strength of condensed explosives and the hydrodynamic theory of gaseous detonations is applicable also to condensed phase detonations. However, material properties such as heterogeneity, porosity, and crystalline structure can play important roles in the initiation (and hence sensitivity) of condensed explosives.

It is perhaps impossible to be entirely objective in writing a book, even a scientific one. The choice of the topics, the order of their presentation, and the emphasis placed on each topic, as well as the interpretation of theoretical and experimental observations, are bound to reflect the author's views. In an attempt to render the book readable for beginners, I have omitted topics that are mathematically complex (e.g., stability theory) or have given only a qualitative discussion of them.

I intend the book to be accessible to those with an undergraduate background in fluid mechanics and thermodynamics. Thus, this book is not a comprehensive treatment of detonation science, and many important topics have been omitted or given only superficial coverage. I have also refrained from giving a complete bibliography, for I feel that in the age of the internet, readers can readily search for relevant references on any topic they wish to pursue further.

I have not reviewed the extensive literature in the field and discussed the contributions of many of my colleagues for a balanced view of the subject. However, I have learned a great deal from all the researchers in the field, past and present; to all of them, I owe a great debt. In particular, I would like to mention that the detonation phenomenon is particularly fortunate in having had perhaps the world's best physicists and chemists working on the subject during the two world wars (Richard Becker, Werner Döring, G. I. Taylor, John von Neumann, George Kistiakowsky, John Kirkhood, Yakov Zeldovich, K. I. Schelkhin, and many others). It is from their wartime and postwar publications that I first learned the subject. I also have to acknowledge the strong influence of an international group of detonation researchers in my formative years in the early 1960s (e.g., Don White, Russell Duff, Gary Schott, Tony Oppenheim, Roger Strehlow, Numa Manson, Hugh Edwards, Rem Soloukhin, and Heinz Wagner). They took me as a junior colleague and mentored my development. T. Y. Toong is responsible for initiating me into scientific research. Even though I did not spend much time under his guidance, he set a lasting example of scholarship that I have always tried to follow throughout my career.

The person to whom I am most indebted in getting the manuscript finished and ready for publication is Jenny Chao. She spent endless hours proofreading and revising the various versions of the manuscript. Without her dedication, I doubt this book would ever have been ready for publication. Eddie Ng also contributed a lot to this book. I have drawn heavily from his Ph.D. dissertation in the chapters on the detonation structure, instability, and blast initiation. Both Jenny and Eddie deserve more than the usual "thank you," for it is their dedicated effort that turned my manuscript into this book. Steven Murray and Paul Thibault read the entire draft of the manuscript, pointed out errors, and gave valuable suggestions; Paul also spent many hours carefully proofreading the final proofs. Andrew Higgins and Matei Radulescu read selected chapters and offered useful comments. Jeff Bergthorson's thorough review of the copyedited manuscript resulted in a significant improvement of this book. It is not possible to incorporate everybody's suggestions, but I want to assure them that that is not because I chose to ignore their advice. There is a limit to how many times one can make changes to a manuscript.

I would also like to thank Christian Caron for encouraging me to start this effort, assuring me that a personal story of the phenomenon is acceptable for publication. The enthusiastic support and encouragement from Peter Gordon in the final process of revision and publication of the book is very much appreciated. I find that writing a book is a humbling experience.

Finally, I would like to acknowledge the essential indirect contributions of Julie, Julian, Leyenda, Heybye, and Pogo in providing a loving environment of fun and laughter outside of scientific work.

1   Introduction

It is of importance to first define deflagrations and detonations and give the characteristics that distinguish these two types of combustion waves. Since this book is concerned with a description of the detonation phenomenon, it is of value to first introduce the various topics that are concerned with detonations prior to their detailed description in later chapters. In this manner, a global perspective can be obtained and permit selective reading of the chapters for those who are already familiar with the subject.

In telling a story, it is natural to start from the beginning, and thus the presentation of the various topics follows more or less their historical development. However, no attempt is made here to discuss the extensive early literature. A historical chronology of detonation research covering the period from its first discovery in the late 1800s to the state of knowledge in the mid 1950s has been documented by Manson and co-workers (Bauer *et al.*, 1991; Manson & Dabora, 1993). An extensive bibliography of the early works is given in these two papers for those who want to pursue further the history of detonations. This chapter is in essence a qualitative summary of the material covered in this book.

1.1. DEFLAGRATIONS AND DETONATIONS

Upon ignition, a combustion wave propagates away from the ignition source. Combustion waves transform reactants into products, releasing the potential energy stored in the chemical bonds of the reactant molecules, which is then converted into internal (thermal) and kinetic energy of the combustion products. Large changes in the thermodynamic and gasdynamic states occur across the combustion wave as a result of the energy released. The gradient fields across the wave generate physical and chemical processes that result in the self-sustained propagation of the combustion wave.

Generally speaking, there are two types of self-propagating combustion waves: deflagrations and detonations. *Deflagration* waves propagate at relatively low subsonic velocities with respect to the reactants ahead of it. As subsonic waves, disturbances downstream can propagate upstream and influence the initial state of the reactants. Thus, the propagation speed of a deflagration wave depends not only on the properties and initial state of the explosive mixture, but also on the rear boundary condition behind the wave (e.g., from a closed-end or an open-end tube). A deflagration is an expansion wave where the pressure drops across the reaction front, and the combustion products are accelerated away from the wave in a direction opposite to its propagation. Depending on the rear boundary condition (e.g., for a closed-end tube the particle velocity is zero in the products), the expansion of the products causes a displacement of the reactants ahead of the reaction front. Thus, the reaction front propagates into reactants that are moving in the direction of propagation. The deflagration speed (with respect to the fixed laboratory coordinates) will then be the sum of the displacement flow velocity of the reactants and the velocity of the reaction front relative to the reactants (i.e., the burning velocity). Compression waves (or shocks) are also formed in front of the reaction front as a result of the displacement flow. Thus, a propagating deflagration wave usually consists of a precursor shock followed by the reaction front. The strength of the precursor shock depends on the displacement flow velocity, and hence on the rear boundary condition.

The mechanism by which the deflagration wave propagates into the reactants ahead of it is via diffusion of heat and mass. The steep temperature and chemical species concentration gradients across the reaction front result in the transport of heat and radical species from the reaction zone to the reactants ahead to effect ignition. Therefore, a deflagration is essentially a diffusion wave, and as such, it has a velocity proportional to the square root of the diffusivity and of the reaction rate (which governs the gradients). If the deflagration front were turbulent, we may, within a one-dimensional context, define a turbulent diffusivity to describe the transport processes. A *flame* is generally defined as a stationary deflagration wave (with respect to laboratory coordinates) stabilized on a burner with the reactants flowing toward it. However, the term *flame* is often also used for the reaction front even in a propagating deflagration wave.

A *detonation* wave is a supersonic combustion wave across which the thermodynamic states (e.g., pressure and temperature) increase sharply. It can be considered as a reacting shock wave where reactants transform into products, accompanied by an energy release across it. Because the wave is supersonic, the reactants ahead are not disturbed prior to the arrival of the detonation; hence they remain at their initial state. Because it is a compression shock wave, the density increases across the detonation, and the particle velocity of the products is in the same direction as that of the wave motion. The conservation of mass then requires either a piston or expansion waves to follow the detonation front. For a piston-supported detonation (known as a strong or overdriven detonation), the flow can be subsonic behind the detonation, since no expansion waves trail behind it. However, for a freely propagating detonation (without a supporting piston motion behind it), the expansion waves behind the detonation front will reduce the pressure and particle velocity to match the rear boundary condition. Since the flow is subsonic behind a strong detonation, any expansion wave will penetrate the reaction zone and attenuate the detonation. Thus, a freely propagating detonation must have either a sonic or a supersonic condition behind it. Detonations with a sonic condition behind them are called *Chapman–Jouguet* (CJ) detonations; those with a supersonic condition are called *weak* detonations. Weak detonations require special properties of the Hugoniot curve (i.e., curve representing the locus of equilibrium states of the detonation products for different detonation velocities) and are not commonly realized. Therefore, freely propagating detonations are generally CJ detonations with a sonic condition behind them.

Ignition of the reactants is effected by the adiabatic compression of the leading shock front that precedes the reaction zone of the detonation wave. An induction zone usually follows the leading shock where dissociation of the reactants and the generation of free radical species occur. The variation in the thermodynamic state in the induction zone is usually small. Following the induction zone, rapid recombination reactions occur with an accompanying temperature increase from the exothermic reactions. The pressure and the density drop through the reaction zone. Thus, the reaction zone of a detonation is similar to a deflagration wave, and a detonation wave is often considered to be a closely coupled shock–deflagration complex, except that ignition is due to adiabatic heating by the leading shock. The rapid pressure drop in the reaction zone, together with a further pressure decrease in the expansion waves that follow a freely propagating detonation, provides the forward thrust that supports the leading shock front. Thus, the classical mechanism of propagation of an unsupported detonation is autoignition by the leading shock front, which in turn is driven by the thrust from the expanding products in the rear.

Self-propagating deflagrations are intrinsically unstable, and there exist numerous instability mechanisms that render the reaction front turbulent, thereby increasing its propagation speed. Thus, self-propagating deflagrations accelerate, and when boundary conditions permit, they undergo an abrupt transition to detonations. Prior to transition to detonation, turbulent deflagrations can reach high supersonic speeds (relative to a fixed coordinate system). By *high-speed* deflagrations we usually mean these accelerating deflagrations during the transition period. When detonations propagate in very rough-walled tubes, their propagation speeds can be substantially less than the normal CJ velocity. These low-velocity detonations are referred to as "*quasi-detonations*." The velocity spectra of high-speed deflagrations and quasi-detonations overlap. The complex turbulent structure of these waves is similar, suggesting that their propagation mechanisms may also be similar. Thus, it is difficult to draw a sharp distinction between them.

The different types of combustion waves described in this section manifest themselves under different initial and boundary conditions. Their consideration is the subject of this book.

Figure 1.1. Marcelin Berthelot (1827–1907) and Paul Vieille (1854–1934).

1.2. DISCOVERY OF THE DETONATION PHENOMENON

It has been known since the fifteenth century that certain chemical compounds (e.g., mercury fulminate) undergo unusually violent chemical decomposition when subjected to mechanical impact or shock. However, it was not until the development of diagnostic tools, which permitted the rapid combustion phenomenon to be observed and the propagation velocity of the combustion wave to be measured, that we can say the detonation phenomenon was discovered. Abel (1869) was perhaps the first to measure the detonation velocity of explosive charges of guncotton. However, it was Berthelot and Vieille (Berthelot, 1881; Berthelot & Vieille, 1883) who systematically measured the detonation velocity in a variety of gaseous fuels (e.g., H_2, C_2H_4, C_2H_2) mixed with various oxidizers (e.g., O_2, NO, N_2O_4) and diluted with various amounts of inert nitrogen, thereby confirming the existence of detonations in gaseous explosive mixtures.

Mallard and Le Châtelier (1883) used a drum camera to observe the transition from deflagration to detonation, thus demonstrating the possibility of two modes of combustion in the same gaseous mixture. They also suggested that the chemical reactions in a detonation wave are initiated by the adiabatic compression of the detonation front. Therefore, in the late 1800s, supersonic detonation waves in gaseous explosive mixtures were conclusively demonstrated to be distinctly different from slowly propagating deflagration waves. The early pioneers (Berthelot and Vieille; Dixon, 1893, 1903) all recognized the role played by adiabatic shock compression in initiating the chemical reactions in a detonation wave.

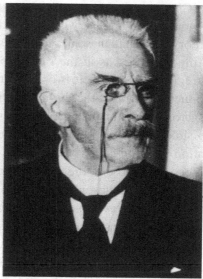

Figure 1.2. Ernest Mallard (1833–1899) and Henry Le Châtelier (1850–1936).

1.3. CHAPMAN–JOUGUET THEORY

A quantitative theory that predicts the detonation velocity of an explosive mixture was formulated by Chapman (1889) and Jouguet (1904, 1905) shortly after the discovery of the phenomenon.

Figure 1.3. Donald Leonard Chapman (1869–1958) and Ehrile Jouguet (1871–1943).

Both Chapman and Jouguet based their theory on the works of Rankine (1870) and Hugoniot (1887, 1889), who analyzed the conservation equations across a shock wave. For a detonation wave, the transformation of the reactants into products across the wave results in the release of chemical energy. Assuming equilibrium downstream of the wave, it is possible to determine the chemical composition of the products in terms of the thermodynamic state, and thus the chemical energy released across the detonation can be determined. Unlike a non-reacting shock wave, two possible solutions exist for a given detonation wave speed: the strong and weak detonation solutions. The pressure and density of the strong detonation solution are greater than those of the weak detonation solution. The flow downstream of a strong detonation is subsonic (relative to the wave), whereas for a weak detonation, it is supersonic. The two solutions converge when the detonation velocity is a minimum. No solution exists for detonation velocities below this minimum value. Since a continuous spectrum of detonation velocities above the minimum is possible for a given explosive mixture, the task of a detonation theory is to provide a criterion for the choice of the appropriate detonation velocity for an explosive mixture at given initial conditions.

Chapman's criterion is essentially to choose the minimum-velocity solution. The argument he provided was simply that for a given explosive mixture, experiments indicate that a unique detonation velocity is observed. Thus, the minimum-velocity solution must be the correct one. Jouguet, on the other hand, investigated the locus of the thermodynamic states for various detonation velocities (i.e., the Hugoniot curve). He determined the entropy variation along the Hugoniot curve and discovered a minimum. He further noted that the minimum entropy solution corresponds to sonic condition downstream of the detonation. Jouguet then postulated that the minimum-entropy solution (the sonic solution) is the appropriate one to choose. His collaborator, Crussard (1907), later showed that the minimum-velocity solution corresponds to the minimum-entropy solution and also gives sonic flow downstream of the wave. Thus, both Chapman and Jouguet provided a criterion (i.e., minimum velocity or minimum entropy) for the choice of the appropriate detonation velocity for a given explosive mixture, and this is now referred to as the CJ theory. Neither Chapman nor Jouguet provided physical or mathematical justification for their postulates.

It is of interest to note that Mikelson (1890) in Russia had earlier developed a similar theory for detonation. He also analyzed the conservation equations across a detonation and found the existence of two possible steady solutions that converge to a single solution when the detonation velocity is a minimum. Unfortunately, his doctoral dissertation, where his analyses were reported, was not known outside Russia. Although these three researchers had independently formulated a gasdynamic theory of detonation at about the same time, only Chapman's and Jouguet's names are associated with the theory.

Note that the CJ theory is incomplete until more rigorous physical or mathematical arguments are provided to justify the criterion for the selection of the solution.

A few subsequent researchers had based their arguments on entropy (Becker, 1917; 1922a, 1922b; Scorah, 1935). But Zeldovich (1940/1950) had refuted any thermodynamic argument by pointing out that the entropy increase across a shock wave alone does not imply that the shock wave will exist. The mechanism for generating the shock wave (e.g., piston motion behind it) is required. The justification for the CJ criterion used by the early investigators was based on the properties of the solution of the Rankine–Hugoniot equations across the detonation front (e.g., minimum velocity, minimum entropy, or sonic conditions). G.I. Taylor (1950), who studied the dynamics of the detonation products behind the front, was the first to point out that the boundary condition at the front must lead to a physically acceptable solution for the non-steady expansion flow of the combustion products behind the detonation. For planar detonations, the Riemann solution is compatible to the sonic conditions of a CJ detonation. However, for spherical detonations, a singularity in the form of an infinite expansion gradient is obtained when the CJ conditon is imposed. This has led to a controversy over the existence of steady CJ spherical detonations (e.g., Courant and Friedrichs, 1948; Jouguet, 1917; Zeldovich and Kompaneets, 1960). The strong detonation solution can be eliminated for freely propagating detonations, because the conservation of mass requires that an expansion wave must follow the detonation to reduce the density. Because the flow is subsonic behind the strong detonation, the expansion waves will penetrate the reaction zone and attenuate the detonation. However, the weak detonation solution is more difficult to eliminate. It was von Neumann (1942) who provided an interesting argument for rejecting the weak detonation solution by examining the structure of the detonation wave. He first assumed that intermediate Hugoniot curves can be constructed based on a given degree of completion of the chemical reactions. He then showed that if the intermediate Hugoniot curves do not intersect one another, then the weak detonation solution cannot be attained. However, if the chemical reactions are such that the intermediate Hugoniot curves do intersect, he showed that weak detonations are possible. Such detonations are referred to as *pathological* detonations and do exist for certain explosives with a temperature overshoot. It may be concluded that a gasdynamic theory based only on the Rankine–Hugoniot reactions across the front cannot justify the CJ criterion. Both the solution for the nonsteady flow of the detonation products and the nature of the chemical reactions within the structure must be considered in the selection of the appropriate solution of the Rankine–Hugoniot equations.

1.4. THE DETONATION STRUCTURE

The CJ theory completely bypasses the details of the detonation structure (i.e., the transition processes from reactants to products). It is essentially a consideration of the possible solutions of the steady one-dimensional conservation equations that link the upstream and downstream equilibrium states of the reactants and products, respectively. Without a description of the structure, the propagation mechanism of

Figure 1.4. Yakov B. Zeldovich (1914–1987), John von Neumann (1903–1957), and Werner Döring (1911–2006).

the detonation wave cannot be known. Although ignition via shock compression was known to the early pioneers who discovered the phenomenon, it was Zeldovich (1940), von Neumann (1942), and Döring (1943) who explicitly described the model of the detonation structure as comprising a leading shock front followed by a chemical reaction zone.

Due to the Second World War during the early 1940s, we may assume that these three researchers were unaware of each other's work. In his original paper, Zeldovich included heat and momentum losses within the structure to investigate their effects on the propagation of the detonation wave. An important consequence of the loss terms is that the integral curve encounters the sonic singularity prior to chemical equilibrium. Seeking a regular solution across the sonic singularity requires a unique value for the detonation velocity, and thus, the term *eigenvalue detonation* is often used in modern literature. With heat and momentum losses, the detonation velocity is less than the equilibrium CJ value. At some critical values for the loss terms, no steady solution can be obtained, which can be interpreted as the onset of the detonation limits observed experimentally. Heat and momentum losses to the walls are two-dimensional effects, and to model them as one-dimensional gives an incorrect description of their physical effects on the detonation structure. Nevertheless, Zeldovich's analysis led to an important mathematical criterion for determining the detonation solution, namely, regularity at the sonic singularity.

Von Neumann's analysis of the detailed transition processes in the detonation structure is an attempt to provide a more rigorous justification for the Chapman–Jouguet criterion, in particular the elimination of the weak detonation solution. He introduced a parameter n to denote the progress of the chemical reaction from the leading shock to the final products, with $0 \leq n \leq 1$. At each value of n, he assumed equilibrium states $(p(n), v(n))$ can be defined, permitting an intermediate Hugoniot curve (i.e., the locus of states that satisfy the conservation equations for a fixed value of n) to be constructed. Then, from the geometry of these intermediate Hugoniot curves, he demonstrated that weak detonations are not possible in general if the intermediate curves do not intersect one another. However, for certain reactions, where the Hugoniot curves do intersect, the detonation velocity obtained is higher than the equilibrium CJ value, and the solution itself lies on the weak detonation branch of the equilibrium Hugoniot curve, where $n = 1$. The importance of von Neumann's analysis is the demonstration of pathological detonations, which have velocities higher than the equilibrium CJ value. These pathological detonations are observed experimentally when there exists a temperature overshoot in the chemical reaction process toward equilibrium.

Werner Döring had studied under Richard Becker, who carried out important fundamental work on shock and detonation waves throughout the 1920s and 1930s. Becker had already conceived the idea that the detonation structure is in essence a shock wave where chemical transformation takes place. For this reason, Becker thought that heat conduction and viscosity effects could be important. As it turns out, chemical reactions occur much later downstream, and the leading shock can then be dissociated from the reaction zone.

Döring's analysis of the detonation structure is remarkably similar to that of von Neumann. He defined a reaction progress variable n (in terms of the concentrations of the reactants), which goes from 0 to 1 as the reaction proceeds toward equilibrium. He integrated the conservation equations across the reaction zone and obtained the profiles for the thermodynamic states within the detonation zone. In honor of the three researchers who carried out the analysis of the structure of the detonation, the model of a shock followed by chemical reactions is now referred to as the Zeldovich–von Neumann–Döring (ZND) model. The ZND model now provides the mechanism responsible for the propagation of the detonation wave, namely, ignition by adiabatic compression across the leading shock, which is in turn maintained by the thrust generated by the expansion of the gases in the reaction zone and in the products.

It should be noted that the CJ criterion that selects the minimum velocity solution is only a postulate and does not follow from the conservation laws across the detonation front. The minimum velocity solution implies that the Rayleigh line is tangent to the equilibrium Hugoniot curve, and therefore, the sonic condition is based on the equilibrium sound speed. In the alternate method where the ZND equations are integrated across the structure of the front, the criterion used in iterating for the desired detonation velocity is the regularity condition at the sonic singularity. The sonic condition is now based on the frozen sound speed. Although the solution still lies on the equilibrium Hugoniot curve, it is no longer the minimum-velocity (or tangency) solution and now lies on the weak branch of the equilibrium Hugoniot curve. There is no reason to expect that the two solutions are the same since the method and the criterion used to obtain them are different. The use of the CJ criterion is simpler since the details of the reaction zone are not involved. The detonation velocity can be found from computations using the equilibrium thermodynamic properties of the reacting mixture. On the other hand, integration across the ZND structure is rather involved and requires a knowledge of the detailed chemical kinetics of the reactions. However, solutions for *pathological detonations* can now be obtained. It is difficult to determine which solution corresponds to reality since the two detonation velocities differ slightly by only a few percent. Furthermore, the nonsteady three-dimensional cellular structure of real detonations and the influence of the boundary conditions on the propagation of the detonation wave probably have larger effects on the detonation velocity. In view of the relative ease in carrying out an equilibrium thermodynamic calculation, the CJ criterion that selects the minimum velocity solution based on tangency of the Rayleigh line to the equilibrium Hugoniot curve is generally used to find the detonation velocity of a given explosive mixture.

1.5. DYNAMICS OF THE DETONATION PRODUCTS

The analysis of the nonsteady flow of the detonation products is as important as the study of the conservation equations across the detonation front. Solutions for the flow behind planar and spherical detonations were first obtained by G.I. Taylor (1940/1950) and also independently by Zeldovich (1942). Taylor pointed out the important fact that a steady detonation is only possible if a solution for the nonsteady flow in the products can be found that can satisfy a steady state boundary condition at the CJ detonation front. For the planar case, the Riemann solution can be matched to the condition behind a CJ detonation. Thus, steady planar CJ detonations are possible. However, for diverging cylindrical and spherical detonations, it is found that there would exist a singularity in the form of an infinite expansion gradient behind the detonation if the sonic condition of a CJ wave were to be imposed. Such a singularity does not exist for strong or weak detonations. However, strong and weak detonations can be ruled out for other reasons. The infinite expansion singularity obtained behind the front raises a question as to the existence of steady cylindrical and spherical detonations. It is clear that if we were to consider the reaction zone thickness to be finite, then steady diverging CJ detonations cannot exist. This is due to the influence of curvature on the flow in the reaction zone, which leads to a detonation velocity less than the equilibrium CJ velocity. Since curvature varies with radius, the detonation velocity will change as it expands and will only reach the CJ value asymptotically at infinite radius. Lee *et al.* (1964) also pointed out that the direct initiation of spherical detonations requires a substantial amount of energy by the ignition source (Laffitte, 1923; Manson and Ferrie, 1952; Zeldovich *et al.*, 1957). If the initiation energy is considered, then a strong blast wave is generated at small radius and a CJ detonation would only be obtained asymptotically at infinite radius. Thus, both the consideration of a finite reaction zone thickness and the inclusion of the initiation energy led to the conclusion that steady CJ spherical detonations are not possible. Furthermore, the instability of the detonation front leads to a transient three-dimensional cellular structure, which differs from the one-dimensional structure assumed in the analysis of G.I. Taylor and Zeldovich. Therefore, the gasdynamic theory of detonation based on a consideration of just the conservation laws across the front (i.e., Rankine–Hugoniot equations) is incomplete. Both the nonsteady flow of the detonation products and the flow within the structure of the detonation must be considered for a complete theoretical description of the propagation of the detonation wave.

1.6. STABILITY OF THE DETONATION FRONT

For almost all explosive mixtures, detonation fronts are found to be intrinsically unstable and possess a transient three-dimensional structure, even though globally they still maintain a steady averaged propagation speed that is remarkably close to the ideal one-dimensional CJ velocity, even near the detonation limits. With the development of high-speed streak cameras in the 1920s and 1930s, these unsteady fluctuations in the detonation front began to be revealed. Near the detonation limits, the frequency of the fluctuation is low and the amplitude is large, and unsteady detonations could be easily observed. Campbell and Woodhead (1926) were the first to report the phenomenon of spinning detonations, and with some ingenious arrangements of the photographic slits of their streak cameras, Campbell and Finch (1928) were able to conclude that for spinning detonations, chemical reactions occur in an intense local region situated near the tube wall. As the detonation propagates, this reaction region rotates circumferentially, thus tracing out a helical path. Bright luminous bands that extend back into the detonation products can also be observed from the streak photographs of spinning detonations. Later studies by Bone and Fraser (1929, 1930) and Bone *et al.* (1935), with a faster streak camera designed by Fraser, revealed further details of the spinning detonation structure and permitted more accurate measurements of higher spin frequencies away from the limits.

Manson (1945) was the first to develop a theory to explain the phenomenon of spinning detonation. He recognized that the periodic luminous bands that extend well into the detonation products are due to compression from a transverse shock wave that rotates with the intense local reaction zone at the front. He also correctly assumed that this transverse shock is weak due to the high sound speed of the products and can be considered as an acoustic wave. Then, he associated the rotation of the transverse acoustic wave with the transverse vibrational mode of the cylindrical column of detonation products behind the front. Applying the solution for the acoustic wave equation given by Lord Rayleigh (1945), Manson was able to compute the transverse vibrational frequency, which was in excellent agreement with the experimental values measured by Bone and co-workers. Manson recognized that only the transverse waves of vibration are important and did not consider the radial and longitudinal modes, thus bypassing the complicated task of determining the boundary conditions of the perturbed detonation front and the rear boundary condition.

Manson's theory did not shed any light on the structure of the spinning detonation front. It was Shchelkin (1945a,b) who proposed that the intense localized reaction zone is associated with a "break" (a crease or fold) of the leading shock front. Since the shock waves on adjacent sides of the crease intersect at an angle, a third (transverse) shock is required to satisfy the conditions behind the intersecting shock waves. Thus, Shchelkin proposed a model of a triple-shock Mach configuration for the spinning detonation front. Zeldovich (1946) followed this with a detailed analysis of the triple-shock intersection and attempted to compute the angle of the path of the helix.

Thus, Manson, Shchelkin, and Zeldovich were the pioneers in the development of the theory for the unstable structure of the detonation front. It was not until the late 1950s and early 1960s that Voitsekhovskii *et al.* (1966) in the Soviet Union and Duff (1961), White (1961), and Schott (1965) in the United States carried out experiments that revealed the detailed structure of the unstable detonation front.

Away from the limits, the spin frequency increases and one may consider that higher transverse acoustic modes are excited, as described by Manson's acoustic theory. If one considers two sets of transverse waves, one rotating clockwise and the other counterclockwise, then the helical paths of these two sets of transverse waves intersect to produce a "diamond" or "fish scale" pattern. In fact, this pattern can be recorded on a smoked foil lining the inside of a detonation tube. The writing on a smoked foil by the passage of a triple-shock intersection across it was used by Mach and Sommers (1877) in the study of shock interaction from spark discharges. This technique was first applied to study the unstable detonation fronts by Denisov and Troshin (1959) and is now the standard technique used to measure the cell size of unstable detonation fronts.

Manson's acoustic theory does not predict the spin frequency (i.e., transverse wave spacing) for a given explosive mixture. However, if the frequency is measured, it is generally found to correspond to one of the transverse modes of Manson's acoustic theory, as was shown by Duff (1961). This is due to the fact that Manson's theory completely bypasses the details of the detonation front and the chemical kinetic rates of the reactions. It is the nonlinear coupling between the shock and the chemical reaction that determines the spectrum of unstable frequencies when the confinement of the tube wall no longer controls the unstable phenomenon (e.g., in spherical detonations). In spite of extensive studies on detonation stability theory in the past twenty-five years, a quantitative theory is still lacking. It should also be noted that advances in numerical simulation can reproduce the amazing details of the structure of two- and even three-dimensional spinning detonations, but proper analysis of the abundance of numerical information on the unstable detonation structure is still lacking.

1.7. INFLUENCE OF BOUNDARY CONDITIONS

The one-dimensional gasdynamic theory of detonation and the ZND model of the detonation structure do not consider the effect of boundary conditions on the propagation of the detonation wave. In reality, most detonations propagate in a confinement and are subjected to the influence of the confining walls. Even for the case of a spherical detonation, one may say that it is "self-confined" and that the spherical geometry introduces curvature, which is not unlike the effect of boundary layers on detonations propagating in tubes. The effect of tube diameter on the detonation velocity had been noted by early researchers as soon as velocity measurement techniques of sufficient accuracy were developed. It was also noted that excessive losses to the walls can lead to detonation limits.

Zeldovich (1940) was perhaps the first to consider the influence of boundaries on the propagation of the detonation waves by including heat and momentum losses in the conservation equations for the detonation structure. When loss terms are included, the detonation velocity is no longer obtained from equilibrium CJ theory, but by iteration, satisfying the condition of regularity when the sonic singularity is reached. It is also interesting to note that there exist multiple solutions to the steady one-dimensional conservation equations for the detonation structure for a given value of the momentum loss term. In reality, however, there should be only one solution. Hence, if one carries out a time-dependent calculation with given initial conditions, then only one steady-state solution will be obtained asymptotically. The problem of having multiple solutions using the steady flow equations will not be encountered. Losses to the tube walls are essentially a two-dimensional effect, and modeling these wall losses in a one-dimensional model introduces some unrealistic effects by distributing the wall effect over the entire cross section of the tube.

Fay (1959) developed a more correct model for the wall boundary-layer effect by noting that the negative displacement effect of the boundary layer (with respect to a coordinate system fixed to the shock) results in a divergence of the flow into the boundary layer and thus to wave curvature. He was able to correctly model the two-dimensional wall effect within the context of a one-dimensional model. Fay's model was later used by Dabora *et al.* (1965) and Murray and Lee (1985) to describe the velocity deficit of detonations in a tube with soft-yielding boundaries. With the inclusion of a curvature term in the conservation equations across the detonation front, multiple solutions are again obtained for a given value of the curvature term. However, if the nonsteady equations were used and the steady solution were obtained asymptotically, then only one solution will be obtained. It should be noted that the asymptotic solution can be unstable, and within the framework of a one-dimensional theory, a pulsating detonation can be obtained.

Perhaps the most important effect of the wall on the propagation of the detonation wave is the damping of the transverse waves of the detonation structure by an acoustically absorbing porous tube wall. It was shown that when the transverse waves are dampened out, and thus the unstable cellular structure is destroyed, self-sustained propagation of the detonation is not possible (Dupré *et al.*, 1988; Teodorczyk and Lee, 1995). This conclusively demonstrates the essential role played by instabilities on the self-sustained propagation of a detonation wave.

Abrupt changes in boundary conditions (confinement) can cause significant disruption of the propagation mechanism. When a confined planar detonation from a rigid tube suddenly exits into unconfined space, there exists a critical diameter of the tube below which the expansion waves from the corner penetrate to the tube axis and cause the wave to fail. When the tube diameter exceeds the critical value, the overall detonation wave curvature is not excessive when the rarefaction fan reaches the axis. This permits new cells to be generated, and the detonation then develops into a spherical wave.

Mitrofanov and Soloukhin (1964) were the first to note that the critical tube diameter corresponds to approximately thirteen detonation cell widths. In a two-dimensional geometry, it was found by Benedick *et al.* (1985) that the critical width of a two-dimensional channel, where a planar detonation can evolve into a cylindrical detonation upon exiting into unconfined space, is only 3λ (where λ is the detonation-cell width), instead of 6λ as would be expected on the basis of the difference in the curvature between cylindrical and spherical waves. This puzzle was later explained by Lee (1995), who pointed out that there are two failure mechanisms, depending on whether the detonation is stable or unstable. For a stable detonation, the structure is described by the ZND model and transverse waves play a negligible role in the propagation. For an unstable detonation, in contrast, transverse waves play a dominant role, and cell structure is essential to the propagation of the detonation. Lee argued that for an unstable detonation it is the inability to develop new cells via instability as the rarefaction waves penetrate into the detonation that governs failure. On the other hand, for a stable detonation it is the maximum curvature of the detonation attained by the quenching effect of the rarefaction fan that controls failure. Thus, for stable detonations, the critical channel should in fact correspond to 6λ. Conclusive experimental results for the critical channel width for stable detonations have not been obtained to date.

Experiments also indicate that sufficiently rough walls of the tube can have a drastic influence on the propagation of detonations. Shchelkin (1945b) and Guénoche (1949) had measured quasi-steady propagation speeds of detonations as low as 30% of the normal CJ detonation velocity of the mixture in tubes with a spiral wire coil inserted into the tube. The spiral coil generates turbulence as well as transverse shock waves that propagate into the reaction zone. The obstacles also generate hot spots due to shock reflections. All these tend to promote rapid combustion to maintain the quasi-steady propagation of these *quasi-detonations*. Thus, the combustion mechanisms are no longer distinct (diffusion or shock ignition) in these quasi-detonations, and it is hard to draw a clear distinction between turbulent deflagrations and detonations. Even for the case of unstable cellular detonations in smooth tubes, a clear distinction between shock ignition and turbulence cannot often be made. It appears that instability and cellular structure are nature's way of bringing together mechanisms to promote sufficiently high combustion rates that can maintain the propagation of a self-sustained combustion wave at the maximum speed compatible with the given boundary conditions.

Thus, contrary to ideal CJ and one-dimensional ZND theory, boundary conditions can not only influence the propagation of detonation waves, but can also dominate the combustion mechanisms to give, for a given mixture, a continuous spectrum of possible combustion wave speeds that is compatible with the given boundary conditions. A clear distinction between the deflagration and detonation modes of combustion can no longer be possible when boundary conditions exert a strong influence.

1.8. DEFLAGRATION-TO-DETONATION TRANSITION (DDT)

Unlike detonations, which propagate at a constant velocity once initiated, self-propagating deflagrations are intrinsically unstable and tend to accelerate continuously subsequent to ignition. Under the appropriate boundary conditions, a deflagration will accelerate to a high supersonic velocity and then undergo an abrupt transition to a detonation wave. An abrupt transition implies the transformation between two distinct states. The detonation state is well defined, and its propagation velocity corresponds to the CJ velocity of the explosive mixture. However, the deflagration state prior to the onset of detonation is not well defined, in general. In smooth tubes, the deflagration is observed to accelerate to some maximum velocity (relative to a fixed coordinate system) of the order of half the CJ detonation velocity of the mixture just prior to the onset of detonation. Furthermore, it is also observed that the duration of this predetonation regime at half the CJ detonation velocity can sometimes persist over many tube diameters. Therefore, there appears to be a metastable quasi-steady deflagration regime prior to the onset of detonation. Although the reaction zone and the precursor shock during this metastable regime both propagate at approximately the same velocity, they are not coupled as in a detonation wave, because the induction time for auto ignition due to adiabatic heating of the mixture by the precursor shock is orders of magnitude longer than that observed experimentally. Thus, the reaction zone propagates via a mechanism other than shock ignition. It should be noted that the reaction zone propagates into a mixture already set into motion by the precursor shock and that its propagation velocity relative to the reactants ahead of it is much less than its observed speed relative to the fixed coordinates. Turbulence appears to be the mechanism responsible for the self-propagation of the reaction zone in this metastable regime. The value of half the CJ detonation velocity is also quite close to the CJ deflagration speed of the mixture. Thus, it appears reasonable to assume that the deflagration will accelerate to its maximum velocity (the CJ deflagration velocity) prior to the transition to the detonation regime.

The flame acceleration phase of the DDT process involves the various flame instability mechanisms (Landau–Darrieus, thermal diffusion, etc.). The flame is also a density interface, which is unstable under acceleration and pressure wave interactions (Taylor instability, Richtmyer–Meshkov instability, Rayleigh instability, etc.). Since the displacement flow ahead of the deflagration is generated as a result of the specific volume increase across the flame, the reaction front propagates into the flow field it has generated. Pressure waves, generated as a result of fluctuations in the heat release rate in the reaction front, can also return to interact with the flame upon reflection from the boundaries. Thus, there are positive feedback mechanisms that render a self-propagating deflagration unstable and result in its continuous acceleration until the onset of detonation. A detailed description of the various flame acceleration mechanisms is given in a review paper by Lee and Moen (1980).

The onset of detonation (i.e., the formation of the detonation wave at the termination of the metastable regime) is not a unique phenomenon. Experiments indicate that there are various ways in which the genesis of detonation waves can take place; they are described by Urtiew and Oppenheim (1966). For example, the merging of precursor shock waves generated by the accelerating flame results in a high-temperature interface where autoignition and the subsequent formation of the detonation wave occur. Local hot spots, or explosion centers, are often observed to be formed in the turbulent reaction zone (presumably a result of turbulent mixing). The blast wave from an explosion center rapidly develops into a detonation *bubble*, which grows to catch up with the precursor shock and forms an overdriven detonation. The portion of the blast wave that propagates back into the combustion products is referred to as the *retonation* wave. When the spherical detonation bubble reflects from the tube wall, transverse pressure waves are generated. These transverse waves reflect back and forth from the tube wall in the products bounded by the detonation and retonation fronts. Another mode of the onset of detonation that is sometimes observed is a compression pulse, arising from the reaction zone, that amplifies and catches up with the precursor shock, thus forming the detonation. Also, the onset of detonation can often occur from the progressive amplification of transverse pressure waves that propagate through the reaction zone as they reflect from the tube wall. The pressure behind the precursor shock builds up, and eventually a detonation is formed. Thus, we note that there are numerous mechanisms from which the onset of detonation can result, and a general theory for the onset of detonation is not possible.

A mechanism to explain the abrupt formation of a detonation in the turbulent flame brush at the onset of detonation was proposed by Lee *et al.* (1978). The strength of the blast wave formed by an explosion center is generally weak, of the order of $M = 2$ at the most. However, a fully developed spherical detonation bubble appears to be formed almost spontaneously at these local explosion centers. Thus, there must exist a very effective shock amplification mechanism. Applying Rayleigh's criterion for the amplification of periodic oscillations coupled to a heat source, Lee proposed a mechanism for a traveling shock pulse. If the medium ahead of the pulse is preconditioned in such a manner that the chemical energy release can be synchronized with the pulse, then rapid amplification results. The mechanism is called the SWACER mechanism (for "shock wave amplification by coherent energy release"), and achieving this requires an induction-time gradient in the medium ahead of the shock pulse. With an induction-time gradient, the chemical reactions can be triggered by the arrival of the shock pulse, and the energy release synchronized with the propagating shock.

Since there are numerous aspects to DDT, it is unlikely that any general theory can be developed to describe the phenomenon. Thus, given an explosive mixture at given initial and boundary conditions, it is still not possible to predict if and when DDT occurs.

1.9. DIRECT INITIATION

Direct initiation refers to the spontaneous formation of a detonation by an ignition source without a predetonation deflagration regime. Thus, the flow field generated by the ignition source is responsible for the detonation formation processes. The mechanism of direct initiation is specific to the type of ignition source used. Direct initiation methods were developed initially to obtain spherical detonations, since the transition from deflagration to detonation in unconfined spherical geometry is extremely difficult if at all possible. This is due to the absence of almost all of the important flame acceleration mechanisms in an unconfined spherical geometry. In his attempt to initiate spherical detonation in CS_2–O_2 mixtures, Laffitte (1925) first used a planar detonation emerging from a tube at the center of a spherical vessel. He failed to initiate a spherical detonation with this method. He then used a powerful explosive charge of 1 mg of mercury fulminate and successfully obtained direct initiation of a spherical detonation wave. His failure to obtain a spherical detonation using a planar detonation from a tube is due to the fact that the tube diameter was too small. Later studies by Zeldovich *et al.* (1957) demonstrated that there exists a critical diameter below which the planar detonation is quenched as it emerges from the tube and no spherical detonation can be formed. The use of a powerful explosive charge (or spark) results in a strong blast wave, which subsequently decays to a spherical CJ detonation. Zeldovich *et al.* also found that there exists a critical blast energy below which the blast decays continuously to an acoustic wave, and no spherical detonation is formed. They also gave a criterion for the critical energy required for blast initiation by requiring that the decay rate of the blast must be such that when the blast has decayed to the CJ strength, the blast radius must be at least the induction-zone thickness of the detonation of the explosive mixture. This criterion led to the cubic dependence of the critical blast energy on the induction-zone thickness, and subsequent blast initiation theories all demonstrate a cubic dependence of the initiation energy on some characteristic length scale that describes the thickness of the detonation front (e.g., the detonation cell size).

An empirical correlation for the critical tube diameter for direct initiation was later found by Mitrofanov and Soloukhin (1964), who showed that for most mixtures, the critical tube diameter is about thirteen times the detonation cell size (i.e., $d_c = 13\lambda$). There are exceptions to this empirical correlation. For example, for C_2H_2–O_2 with a high concentration of argon dilution, the critical tube diameter can be as high as 40λ. The breakdown of the 13λ correlation is a consequence of the difference in the failure and reinitiation mechanisms of *stable* and *unstable* detonations, as defined by the role played by instability in the self-sustained propagation of the detonation. For stable detonations, ignition by the leading shock, as in the ZND model, is responsible for initiating the chemical reactions, whereas in unstable detonations, hot spots due to shock interactions and turbulence play the dominant role in the ignition and combustion processes in the detonation zone.

It should be noted that the use of a strong shock wave or a detonation in a more sensitive mixture to initiate a detonation in the test gas was well known to the early researchers on detonation. In relatively insensitive mixtures near the detonation limits where transition from deflagration cannot be obtained, the standard method used to initiate the detonation was to use a detonation driver, that is, a length of tube containing a more sensitive explosive mixture than the test gas. It was also found that if no diaphragm separates the driver and driven sections, detonations can be formed more readily when the detonation from the driver section transmits to the driven section. A diaphragm tends to destroy the detonation from the driver section, and only a shock wave will be transmitted downstream when the diaphragm ruptures. The detonation in the driven section will then have to be formed from the reactions initiated by the transmitted shock wave. Direct initiation can also be achieved via a reflected shock wave from the closed end of the detonation tube. The incident shock is insufficient to ignite the mixture, but upon reflection, the explosive mixture auto ignites and a detonation is formed, which then catches up with the re-flected shock. The detonation wave continues to propagate in the mixture processed earlier by the incident shock.

Perhaps the most interesting method of direct initiation is by photolysis and by turbulent mixing. In these methods, no shock wave is formed initially by the initia-tion source. The initiation source simply generates the critical chemical conditions required for the onset of detonation and bypasses the flame acceleration phase of DDT. The observation of large pressure spikes in their flash-photolysis apparatus first led Norrish, Porter, and Thrush (Norrish *et al.*, 1955; Thrush, 1955; Norrish, 1965) to conclude that detonations are formed directly subsequent to photodissoci-ation of the mixture irradiated by an intense UV pulse. Wadsworth (1961) later suc-cessfully initiated a converging cylindrical detonation directly by a UV pulse from a toroidal flash tube wrapped around a quartz cylinder containing the explosive mix-ture. Lee *et al.* (1978) took schlieren photographs of the photoinitiation process and confirmed that subsequent to a UV light pulse, no shock waves are formed in the explosive mixture initially. However, after a short induction period, the onset of detonation is observed to occur in the photodissociated mixture. By controlling the intensity and duration of the flash, they demonstrated that a concentration gradient of dissociated radical species is essential for detonation initiation. The concentra-tion gradient of dissociated species leads to a gradient in the induction time for auto ignition, and it is the sequential explosions in the induction-time gradient field that result in the formation of the detonation. Lee proposed the SWACER mechanism to describe this mode of direct initiation, and numerical simulations carried out by Yoshikawa (Lee *et al.*, 1978) confirmed the necessity of a critical induction-time gra-dient field.

An earlier numerical computation of the development of detonation in a tem-perature gradient field was also carried out by Zeldovich *et al.* (1970), whose aim was to investigate engine knock in internal combustion engines. They also demonstrated that, with a suitable temperature gradient field, very high-pressure spikes can be formed. However, they did not discuss the mechanism of pressure wave amplification due to the synchronization between the energy release and the propagating shock wave. An induction time gradient obtained by turbulent mixing leading to the direct initiation of a detonation wave was also demonstrated by Knystautas *et al.* (1978) and Murray *et al.* (1991). A number of studies of the SWACER mechanism have been carried out, and a comprehensive review is given by Bartenev and Gelfand (2000).

It is important to point out that in direct initiation, it is the initiation source itself that directly generates the critical conditions for the onset of detonation. However, the mechanism of the onset of detonation for the different methods of direct initiation appears to be similar to the case of DDT, where the critical conditions are achieved via flame acceleration. For example, in blast initiation under the critical energy condition, the blast first decays below the CJ detonation velocity, to about half the CJ value. A metastable quasi-steady regime then follows, at the end of which onset of detonation results. Thus, whether this critical metastable regime prior to the onset of detonation is obtained via flame acceleration or generated directly by the initiation source, the onset of detonation appears to involve the same mechanisms.

1.10. OUTSTANDING PROBLEMS

It is of interest to comment on the outstanding problems and on the direction of current and future research in detonation theory. Although the detonation structure is nonsteady and three dimensional, the averaged velocity in the direction of propagation is found to agree amazingly well with steady one-dimensional CJ theory, even for near-limit spinning detonations, where the three-dimensional fluctuations are significant. However, the hydrodynamic fluctuations should eventually decay, and an equilibrium plane should exist somewhere downstream. The application of steady one-dimensional conservation laws across the transition zone should still yield the CJ velocity if losses are negligible, for CJ theory considers only the initial and the final equilibrium Hugoniot curves (and bypasses the structure). The thickness of this transition zone (which is the effective thickness of the cellular detonation wave) is referred to as the *hydrodynamic thickness*, a term first introduced by Soloukhin (Lee *et al.*, 1969). As governed by CJ theory, this plane should also correspond to sonic conditions. Attempts have been made to measure this hydrodynamic thickness by Vasiliev *et al.* (1972) and more recently by Weber and Olivier (2004). In the latter experiments, the detachment of the bow shock from an obstacle in the flow at sonic conditions yields results that indicate that the hydrodynamic thickness is of the order of a few times the cell size. However, this experimental technique is not sufficiently accurate for definitive conclusions to be made. An alternate approach was attempted by Lee and Radulescu (2005) and Radulescu *et al.* (2007) by averaging the field data obtained in a numerical simulation of two-dimensional cellular detonations. The hydrodynamic thickness that is obtained (based on sonic conditions of the averaged properties) is found to be much larger than for the corresponding one-dimensional steady ZND structure. With such a hydrodynamic structure obtained, efforts can be devoted to models for the chemical and mechanical relaxation processes within the averaged one-dimensional structure. The hydrodynamic thickness represents a true characteristic length scale of the detonation in the direction of propagation, and this length scale is of great significance in the correlation of the various dynamic detonation parameters, which depend on a length scale that is governed by the non-equilibrium processes in the structure rather than on the equilibrium CJ detonation state. Thus, the investigation of the hydrodynamic thickness and the development of models for the relaxation process in the averaged one-dimensional structure of the detonation is a fruitful approach toward the next stage in the development of detonation theory.

The experimental fact that almost all self-sustained detonations are unstable implies that instabilities are essential for the self-sustained propagation of a detonation. The onset of detonation is evidenced by the spontaneous development of the cellular structure, and when the cellular structure is destroyed by damping out the transverse waves, a self-sustained detonation fails. This conclusively demonstrates that the cellular structure provides the mechanisms for the ignition and rapid combustion of the mixture in the reaction zone of the detonation wave.

In the one-dimensional ZND structure, the ignition mechanism is via adiabatic shock heating. From a physical point of view, shock heating can no doubt serve as a means of ignition via the thermal dissociation of the molecules to generate the radical species. However, in a one-dimensional model, the shock strength is based on the averaged value across the shock front. If the shocks intersect (as in a three-dimensional cellular front), then much higher local temperatures can be achieved at the Mach stems. Thus, ignition can greatly be facilitated in these local high-temperature regions of shock intersections and spread later to lower-temperature regions in the nonuniform reaction zone. Furthermore, very strong turbulent mixing in the reaction zone can be generated by vorticity produced in the three-dimensional shock interaction zones of a cellular detonation and in the shear layers of triple shock intersections, by baroclinic torque from interactions of pressure and density gradients, and by transverse-shock–vortex interactions. This permits the transport of free radicals from reacting mixtures as well as from products into the reactants to initiate chemical reactions without the need for thermal dissociation from shock heating.

A physical picture of a turbulent reaction zone in a detonation rather than the traditional view of the laminar ZND model should represent a more realistic interpretation of the actual detonation phenomenon. Accordingly, future research should be directed toward the development of *turbulence models* for the description of the reaction zone of unstable cellular detonations. However, these turbulence models will have to incorporate shock interactions as an integral part of the highly compressible turbulence in the reaction zone.

Experiments also indicate that, irrespective of whether the formation of deto-nation is via DDT or direct initiation, the final phase of the onset of detonation involves the rapid amplification of pressure waves to form an overdriven detona-tion. This rapid amplification (i.e., SWACER mechanism) is identified as a coher-ent coupling of the traveling shock wave with the chemical energy release. This coupling can be achieved in different ways, but a gradient in induction time that produces sequential explosions is an effective way to couple the energy release to the propagating shock in the proper phase relationship. The shockless direct ini-tiation achieved by an induction-time gradient field from photodissociation of the reactants by a light pulse demonstrated that this SWACER mechanism can lead to the onset of detonation. In steady oscillating systems, the SWACER mechanism is essentially Rayleigh's criterion for self-sustained oscillations via the coupling of the pressure waves with the chemical energy source under the proper phase relation-ship. Although there are a number of numerical simulations that demonstrate the possibility of the SWACER mechanism in an induction-time gradient field, no an-alytical theory has been developed to date that predicts the gradient necessary for direct initiation in a given explosive. This is an important problem that should be re-solved in order to obtain further progress in defining the requirements for the onset of detonation.

Although the study of detonation enjoyed early success from the development of the CJ theory, the phenomenon is far from being understood quantitatively, and the development of theories to predict the three-dimensional cellular structure remains a formidable task, being essentially a problem in high-speed compressible reacting flow.

Bibliography

Abel, F.A. 1869. *Phil. Trans. R. Soc.* 159:489–516. Also 1869. *C. R. Acad. Sci. Paris* 69:105–121.

Bartenev, A.M., and B.E. Gelfand. 2000. Spontaneous initiation of detonations. *Prog. Energy Combust. Sci.* 26:29–55.

Bauer, P., E.K. Dabora, and N. Manson. 1991. Chronology of early research detonation waves. In: 3–18. Washington, D.C.: AIAA.

Becker, R. 1917. *Z. Electrochem.* 23:40–49, 93–95, 304–309. See also 1922. *Z. Tech. Phys.*, pp. 152–159, 249–256; *Z. Phys.*, pp. 321–362.

Benedick, W., R. Knystautas, and J.H.S. Lee. 1985. In *Progress astronautics and aeronautics: dynamics of shock waves, explosions and detonations.* J.R. Bowen, N. Manson, A.K. Oppenheim, R.I. Soloukhin (Eds.) 546–555. NY: AIAA.

Berthelot, M. 1881. *C. R. Acad. Sci. Paris* 93:18–22.

Berthelot, M., and P. Vieille. 1883. *Ann. Chim. Phys. 5ème Sér.* 28:289.

Bone, W.A., and R.P. Fraser. 1929. *Phil. Trans. A.* 228:197–234.

Bone, W.A., and R.P. Fraser. 1930. *Phil. Trans. A.* 230:360–385.

Bone, W.A., R.P. Fraser, and W.H. Wheeler. 1935. *Phil. Trans. Soc. Lond. A.* 235:29–68.

Brinkley, S., and J. Richardson. 1953. In *4th Int. Symp. on Combustion*, 486–496.

Campbell, C., and A.C. Finch. 1928. *J. Chem. Soc.*, p. 2094.

Campbell, C., and D.W. Woodhead. 1926. *J. Chem. Soc.* 129:3010.

Chapman, D.L. 1889. *Phil. Mag.* 47:90–104.

Chue, R., J.F. Clarke, and J.H.S. Lee. 1993. *Proc. R. Soc. Lond. Ser. A* 441(1913):607–623.

Courant, R., and K.O. Friedrichs. 1948. *Supersonic Flow and Shock Waves*, Interscience Publishers.

Crussard, J.C. 1907. *C. R. Acad. Sci. Paris* 144:417–420.

Dabora, E.K., J.A. Nicholls, and R.B. Morrison. 1965. In *10th Int. Symp. on Combustion*, 817.

Denisov, Yu. N., and Ya. K. Troshin. 1959. *Dokl. Akad. Nauk SSSR* 125:110. See also: In *8th Int. Symp. on Combustion*, 600.

Dixon, H. 1893. *Phil. Trans. A* 184:97–188.

Dixon, H. 1903. *Phil. Trans. A* 200:315–351.

Döring, W. 1943. *Ann. Phys. 5e Folge* 43:421–436.

Duff, R., H.T. Knight, and J.P. Rink. 1958. *Phys. Fluids* 1:393–398.

Duff, R. 1961. *Phys. Fluids* 4(11):1427.

Dupré, G., O. Peraldi, J.H.S. Lee, and R. Knystautas. 1988. *Prog. Astronaout. and Aeronaut.* 114:248–263.

Erpenbeck, J.J. 1963. In *9th Int. Symp. on Combustion*, 442–453.

Erpenbeck, J.J. 1964. *Phys. Fluids* 7:684–696.

Erpenbeck, J.J. 1966. *Phys. Fluids* 9:1293–1306.

Fay, J. 1959. *Phys. Fluids* 2:283.

Fay, J. 1962. In *Progress in astronautics and rocketry*, vol. 6, 3–16. New York: Academic Press.

Fickett, W., and W.W. Wood. 1966. *Phys. Fluids* 9:903–916.

Guénoche, H. 1949. *Rev. Inst. Français Pétrole* 4:15–36, 48–69.

Hirschfelder, J.O., and C.F. Curtiss. 1958. *J. Chem. Phys.* 28:1130.

Hugoniot, H. 1887–1889. *J. Ecole Polytech.*, Cahiers 57, 58.

Jouguet, E. 1904. *C. R. Acad. Sci. Paris* 140:1211.

Jouguet, E. 1905–1906. *J. Math Pures Appl. 6th Seri.* 1:347, 2:5. See also 1917. *La Mécanique des Explosifs.* Paris: O. Doin.

Kirkwood, J.G., and W.W. Wood. 1954. *J. Chem. Phys.* 22:11, 1915–1919.

Kistiakowsky, G., and P. Kydd. 1955. *J. Chem. Phys.* 23(2):271–274.

Kistiakowsky, G., and E.B. Wilson. 1941. The hydrodynamic theory of detonation and shock waves. O.S.R.D. Rept. 114.

Knystautas, R., J.H.S. Lee, I. Moen, and H. Gg. Wagner. 1978. In *17th Int. Symp. on Combustion*, 1235.

Laffitte, P. 1923. Sur la propagation de l'onde explosive, CRAS: Paris 177:178–180.

Laffitte, P., 1925. *Ann. Phys. 10e Ser.* 4:487.

Lee, J.H.S., B.H.K. Lee, and I. Shanfield. 1964. In *10th Int. Symp. on Combustion*, 805–814.

Lee, J.H.S., R.I. Soloukhin, and A.K. Oppenheim. 1969. *Acta Astronaut.* 14:564–584.

Lee, J.H.S., R. Knystautas, and N. Yoshikawa. 1978. *Acta Astronaut.* 5:971–982.

Lee, J.H.S., and I. Moen. 1980. *Prog. Energy Combust. Sci.* 16:359–389.

Lee, J.H.S. 1995. *Dynamics of exothermicity,* 321–335. Gordon and Breach.

Lee, J.H.S. 2003. The universal role of turbulence in the propagation of strong shocks and detonation waves. In *High pressure shock compression of solids*, Vol. V, ed. Y. Horie, L. Davidson, and N. Thadhani, 121–144. New York: Springer Verlag.

Lee, J.H.S., and M.I. Radulescu. 2005. *Combust. Explos. Shock Waves* 41(6):745–765.

Mach, E., and J. Sommers. 1877. *Sitzungsber. Akad. Wien* 75.

Mallard, E., and H. Le Châtelier. 1883. *Ann. Mines* 8:274, 618.

Manson, N. 1945. *Ann. Mines Zémelivre*, 203.

Manson, N., and F. Ferrie. 1952. In *4th Int. Symp. on Combustion*, 486–494.

Manson, N., and E.K. Dabora. 1993. Chronology of research on detonation waves: 1920–1950. In *Dynamic aspect of detonations*, ed. A.L. Kuhl *et al.*, 3–39. Washington, D.C.: AIAA.

Mikelson, V.A. 1890. On the normal ignition velocity of explosive gaseous mixtures. Ph.D. dissertation, Moscow University.

Mitrofanov, V.V., and R.I. Soloukhin. 1964. *Dokl. Akad. Nauk SSSR* 159(5):1003–1007.

Mitrofanov, V.V. 1991. Modern view of gas detonation mechanism. In *Dynamic structure of detonations in gaseous and dispersed media*, ed. A.A. Borisov, 327–340. Kluwer Academic.

Meyer, J.W., and A.K. Oppenheim. 1971. *Combust. Flame* 14(1):13–20.

Murray, S.B., and J.H.S. Lee. 1985. *Progress in astronautics and aeronautics: dynamics of detonations and explosions: detonations*. 80–103. New York: AIAA.

Murray, S.B., I. Moen, P. Thibault, R. Knystautas, and J.H.S. Lee. 1991. *Progress in astronautics and aeronautics: dynamics of shockwaves, explosions and detonations*. 91–117. Washington, D.C.: AIAA.

Norrish, R.G., G. Porter, and B.A. Thrush. 1955. *Proc. R. Soc. Lond. A* 227:423.

Norrish, R.G. 1965. In *10th Int. Symp. on Combustion*. 1–18.

Pukhanachev, V.V. 1963. *Dokl. Akad. Nauk SSR (Phys. Sect.)*. 149:798–801.

Radulescu, M., G. Sharpe, C.K. Law, and J.H.S. Lee. 2007. *d. Fluid Mech*. 580:31–81.

Rakipova, Kh. A., Ya. Troshin, and K.I. Shchelkin. 1947. *Zh. Tekh. Fiz*. 17:1409–1410.

Rankine, W.J. 1870. *Phil. Trans*. 160:277–288.

Rayleigh, J.W.S. 1945. *Theory of sound*, Vol. II Dover Publications.

Romano, M., M. Radulescu, A. Higgins, J.H.S. Lee, W. Pitz, and C. Westbrook. 2003. *Proc. Combust. Inst*. 29(2):2833–2838.

Scorah, R.L. 1935. *V. Chem. Phys*. 3:425.

Schott, G. L. 1965. *Phys. Fluids* 8(1):850.

Shchelkin, K.I. 1940. *Zh. Eksp. Teor. Fiz*. 10:823–827.

Shchelkin, K.I. 1945a. *Dokl. Akad. Nauk SSSR* 47:482.

Shchelkin, K.I. 1945b. *Acta Phys. Chim. SSSR* 20:305–306.

Shepherd, J., and J.H.S. Lee. 1991. On the transition from deflagration to detonation. In *Major research topics in combustion*, ed. H.A. Kumar and R.G. Voigt, 439–490. Springer Verlag.

Sokolik, A.S. 1963. *Self ignition flame and detonation in gases*. Akad. Nauk SSSR; English translation, Jerusalem: I.P.S.T.

Taylor, G.I. 1950. *Proc. Soc. Lond. A* 200:235.

Teodorczyk, A., J.H.S. Lee, and R. Knystautas. 1991. In *23rd Int. Symp. Combustion*, 735–742.

Teodorczyk, A., and J.H.S. Lee. 1995. *Shock Waves* 4:225–236.

Thrush, B.A. 1955. *Proc. R. Soc. Lond. Ser. A* 233:147–151.

Urtiew, P., and A.K. Oppenheim. 1966. *Proc. R. Soc. Lond. Ser. A* 295:63–78.

Vasiliev, A.A., T.P. Gavrilenko, and M.E. Topchiyan. 1972. *Astron. Acta* 17:499–502.

Voitsekhovskii, B.V., V.V. Mitrofanov, and M. Ye. Topchiyan. 1966. The structure of a detonation front in gases. English transl., Wright Patterson AFB Rept. FTD-MT-64-527 (AD-633,821).

von Neumann, J. 1942. Theory of detonation waves. O.S.R.D. Rept. 549.

Wadsworth, J. 1961. *Nature* 190:623–624.

Weber, M., and H. Olivier. 2004. *Shock Waves* 13(5):351–365.

White, D.R. 1961. *Phys. Fluids* 4:465–480.

Wood, W.W., and F.R. Parker. 1958. *Phys. Fluids* 1:230–241.

Wood, W.W., and Z. Salzburg. 1960. *Phys. Fluids* 3:549–566.

Zaidel, R.M. 1961. *Dokl. Akad Nauk SSSR (Phys. Chem. Sect.)* 136:1142–1145.

Zaidel, R.M., and Ya. B. Zeldovich. 1963. *Zh. Prikl. Mekh. Tekh. Fiz.* 6:59–65.

Zeldovich, Ya. B. 1940. *Zh. Exp. Teor. Fiz.* 10(5):542–568. English translation, NACA TN No. 1261 (1950).

Zeldovich, Ya. B. 1942. *Zh. Eksp. Teor. Fiz.* 12:389–406.

Zeldovich, Ya. B. 1946. *Dokl. Akad. Nauk SSSR* 52:147.

Zeldovich, Ya. B., and A.I. Roslovsky. 1947. *Dokl. Akad. Nauk SSSR* 57:365–368.

Zeldovich, Ya. B., and A.A. Kompaneets. 1960. *Theory of Detonation*, New York: Academic. 284.

Zeldovich, Ya. B., S.M. Kogarko, and N.N. Simonov. 1957. *Sov. Phys. Tech. Phys.* 1:1689–1713.

Zeldovich, Ya. B., V.B. Librovich, G.M. Makhviladze, and G. I. Sivashinsky. 1970. *Astron. Acta* 15:313–321.

2 Gasdynamic Theory of Detonations and Deflagrations

2.1. INTRODUCTION

For given initial and boundary conditions, the possible combustion waves that can be realized are given by the solutions of the steady one-dimensional conservation equations across the wave. Since the three conservation equations (of mass, momentum, and energy) and the equation of state for the reactants and products constitute only four equations for the five unknown quantities (p_1, ρ_1, u_1, h_1, and the wave velocity u_0), an extra equation is required to close this set. For non-reacting gases, the solutions to the conservation equations were first investigated by Rankine (1870) and by Hugoniot (1887–1889), who derived the relationship between the upstream and downstream states in terms of a specified wave speed, pressure, or particle velocity downstream of the shock wave. Analysis of the solutions of the conservation equations also provided important information on the stability of shock waves and on the impossibility of rarefaction shocks in most common fluids. For reacting mixtures, similar analyses of the conservation laws were first carried out independently by Chapman (1889), Jouguet (1904), and Crussard (1907), but none of these investigators were aware of similar studies carried out by Mikelson (1890) in Russia. More thorough investigations of the properties of the solutions of these conservation laws were carried out later by Becker (1917, 1922a, 1922b), Zeldovich (1940, 1950), Döring (1943), Kistiakowsky and Wilson (1941), and von Neumann (1942). Their aim was to provide a more rigorous theoretical justification of the Chapman–Jouguet criterion.

In this chapter, we shall discuss the gasdynamic theory of steadily propagating combustion waves, because all steadily propagating detonations and deflagrations must satisfy these conservation equations. By investigating the existence and uniqueness of the solutions, we gain better insight into what kind of combustion waves can be realized under given initial and boundary conditions.

2.2. BASIC EQUATIONS

We shall first define the basic thermodynamic and fluid quantities. Let n_i denote the number of moles per unit volume (molar density) of the ith species in a gas mixture. The partial mass density of the ith species is then $\rho_i = n_i W_i$, where W_i is the molecular weight of the ith species. The density of the gas mixture is therefore $\rho = \sum \rho_i = \sum n_i W_i$. The mass fraction of the ith species is $X_i = \rho_i / \rho$, and the corresponding molar fraction is $Y_i = n_i / n$ where $n = \sum n_i$ is the total number of moles per unit volume. If $\vec{v}_i$ is the mean molecular velocity of the ith species, then the mass averaged velocity of the mixture (i.e., the flow velocity) is defined as

$$\vec{v} = \frac{\sum \rho_i \vec{v}_i}{\rho} = \sum X_i \vec{v}_i.$$

For a perfect gas mixture, the equation of state can be written as $p = \rho R T$, where $p = \sum p_i$ (p_i are the partial pressures), T is the temperature, and R is the gas constant of the mixture (given by $R = \frac{\bar{R}}{W}$ where $\bar{R}$ is the universal gas constant and W is the molecular weight of the mixture, $W = \sum Y_i W_i$). The enthalpy per unit mass of the ith species can be written as

$$h_i = h_{f_i}^\circ + \int_{298}^{T} c_{p_i} dT,$$

where $h_{f_i}^\circ$ is the enthalpy of formation of the ith species and c_{p_i} is its specific heat per unit mass. The enthalpy per unit mass of the mixture is therefore $h = \sum X_i h_i$. Similarly, we can write the specific heat of the mixture as $c_p = \sum X_i c_{p_i}$.

The basic conservation equations of mass, momentum, and energy for one-dimensional steady flow across a combustion wave with respect to a co-ordinate system fixed to the wave are given by

$$
\begin{array}{ccc}
p_0 & & p_1 \\
\rho_0 & \xrightarrow{u_0} \quad \xrightarrow{u_1} & \rho_1 \\
h_0 & & h_1
\end{array}
$$

$$\rho_0 u_0 = \rho_1 u_1, \tag{2.1}$$

$$p_0 + \rho_0 u_0^2 = p_1 + \rho_1 u_1^2, \tag{2.2}$$

$$h_0 + \frac{u_0^2}{2} = h_1 + \frac{u_1^2}{2}, \tag{2.3}$$

where the subscripts 0 and 1 denote reactant and product states, respectively.

In this chapter, all velocities will be measured relative to the combustion wave. Note that the combustion wave is not required to be a discontinuity. It is sufficient to require that the gradients of the state variables be negligible at the upstream and downstream boundaries of the control volume for Eqs. 2.1 to 2.3 to be valid. We may separate the enthalpy of formation from the so-called sensible enthalpy and rewrite Eq. 2.3 as

$$h_0 + q + \frac{u_0^2}{2} = h_1 + \frac{u_1^2}{2}, \tag{2.4}$$

where

$$q = \sum_i^{\text{reactants}} x_i h_{f_i}^\circ - \sum_j^{\text{products}} x_j h_{f_j}^\circ$$

is the difference between the enthalpies of formation of reactants and products. The enthalpy in Eq. 2.4 is now the sensible enthalpy of the mixture,

$$h = \int_{298}^T c_p \, dT,$$

where c_p is now the specific heat of the mixture (reactants or products).

For convenience, we choose the initial state 0 to correspond to the standard reference state in which the enthalpy of formation is defined (i.e., $p_0 = 1$ atm, $T_0 = 298$ K). Note that the chemical energy released is not generally known *a priori*, because neither the product species nor their concentrations are known initially. Furthermore, since the product species and their concentrations depend on the temperature of the products, they vary for different solutions of the conservation laws.

A more exact analysis of the conservation equations for combustion waves requires the simultaneous solution of the conservation equations and the chemical equilibrium equations for the unknown product species. There exist standard computer codes such as STANJAN (Reynolds, 1986) and CEA (McBride and Gordon, 1996) that facilitate these thermochemical equilibrium calculations. For convenience, in our subsequent analysis, we shall assume that q is known and remains a constant for all the different solutions of the conservation equations (i.e., detonations and deflagrations).

We also need to specify an equation of state, $h(p, \rho)$, for both reactants and products. We may assume that $h = c_p T$. Together with the equation of state for a perfect gas ($p = \rho R T$) and the relationships $c_p - c_v = R$ and $\gamma = c_p / c_v$, the caloric equation of state for the sensible enthalpy can also be written as

$$h = \frac{\gamma}{\gamma - 1} \frac{p}{\rho}. \tag{2.5}$$

If the initial state (i.e., p_0, ρ_0, h_0) were specified, we shall have four equations (Eqs. 2.1, 2.2, 2.4, and 2.5) for the five unknown quantities (p_1, ρ_1, u_1, h_1, and the combustion wave propagation speed u_0). An additional equation is therefore required to close the set of equations.

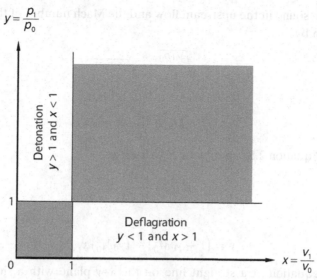

Figure 2.1. Domains of detonation and deflagration solutions in the p–v plane.

2.3. RAYLEIGH LINE AND HUGONIOT CURVE

The thermodynamic path along which the initial state proceeds to the final state across the combustion wave can be readily obtained from Eqs. 2.1 and 2.2:

$$\frac{p_1 - p_0}{v_0 - v_1} = \rho_0^2 u_0^2 = \rho_1^2 u_1^2 = \dot{m}^2, \tag{2.6}$$

where $v = 1/\rho$ is the specific volume and $\dot{m} = \rho u$ is the mass flux per unit area. From Eq. 2.6, we see that

$$\dot{m} = \sqrt{\frac{p_1 - p_0}{v_0 - v_1}}, \tag{2.7}$$

and thus, if $\dot{m}$ is real, we require that if $v_0 > v_1$ (or $\rho_0 < \rho_1$) then $p_1 > p_0$, and if $v_0 < v_1$ (or $\rho_0 > \rho_1$) then $p_1 < p_0$. Thus, Eq. 2.7 defines the regions in the p–v plane where real solutions exist. For $p_1 > p_0$, $v_0 > v_1$, we have the compression solution for detonation waves, and for $p_1 < p_0$, $v_0 < v_1$, we have the expansion solution for deflagration waves. If we define $x = v_1/v_0 = \rho_0/\rho_1$ and $y = p_1/p_0$, Eq. 2.7 can be rewritten as

$$\dot{m} = \sqrt{\left(\frac{y-1}{1-x}\right)\frac{p_0}{v_0}}. \tag{2.8}$$

In the p–v plane (or in the x–y plane), the detonation and deflagration regions are as shown in Fig. 2.1.

The speed in sound in the upstream flow and the Mach number of the combustion wave are given by

$$c_0 = \sqrt{\frac{\gamma_0 p_0}{\rho_0}} = \sqrt{\gamma_0 p_0 v_0}$$

and

$$M_0 = \frac{u_0}{c_0},$$

respectively. Equation 2.8 can now be rewritten as

$$\gamma_0 M_0^2 = \frac{y-1}{1-x},$$

or equivalently

$$y = (1 + \gamma_0 M_0^2) - (\gamma_0 M_0^2) x, \qquad (2.9)$$

which is the equation of a straight line on the x–y plane with a slope of $-\gamma_0 M_0^2$. Equation 2.9 defines the thermodynamic path in which the transition from state $(1, 1)$ to state (x, y) across the combustion wave takes place and is referred to as the Rayleigh line. From this equation, we also note that the speed of the combustion wave is proportional to the square root of the slope of the Rayleigh line. In addition, the slope of the Rayleigh line can also be written as

$$\left(\frac{dy}{dx}\right)_R = -\frac{y-1}{1-x}. \qquad (2.10)$$

Writing Eq. 2.6 as

$$u_0^2 = \frac{1}{\rho_0^2} \frac{p_1 - p_0}{v_0 - v_1}$$

or

$$u_1^2 = \frac{1}{\rho_1^2} \frac{p_1 - p_0}{v_0 - v_1},$$

the velocities in the energy equation (i.e., Eq. 2.4) can be eliminated to obtain

$$h_1 - (h_0 + q) = \frac{1}{2}(p_1 - p_0)(v_0 + v_1). \qquad (2.11)$$

This equation is for the Hugoniot curve, which represents the locus of downstream states for a given upstream state. We also may express the Hugoniot equation in terms of the internal energy instead of the enthalpy by using the definition $h = e + pv$, where e is the internal energy of the mixture. Equation 2.11 then becomes

$$e_1 - (e_0 + q) = \frac{1}{2}(p_1 + p_0)(v_0 - v_1). \qquad (2.12)$$

Note that in Eqs. 2.11 and 2.12, no assumptions have been made as yet regarding the form of the equation of state of the explosive medium. These equations are valid for gas, liquid, and solid media.

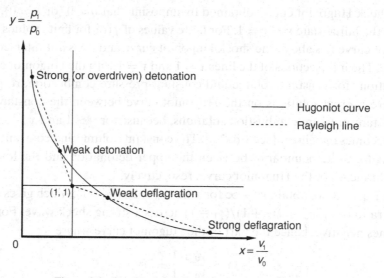

Figure 2.2. The Rayleigh line and the Hugoniot curve.

If we now assume a perfect gas with a caloric equation of state given by Equation 2.5, we can eliminate the enthalpy and express the Hugoniot curve in terms of p and v only. Equation 2.11 can then be written as

$$y = \frac{\frac{\gamma_0+1}{\gamma_0-1} - x + 2q'}{\frac{\gamma_1+1}{\gamma_1-1}x - 1},$$ (2.13)

where $q' = q/p_0v_0$. This equation can also be expressed in the more convenient form

$$(y + \alpha)(x - \alpha) = \beta,$$ (2.14)

where

$$\alpha = \frac{\gamma_1 - 1}{\gamma_1 + 1},$$

$$\beta = \frac{\gamma_1 - 1}{\gamma_1 + 1}\left(\frac{\gamma_0 + 1}{\gamma_0 - 1} - \frac{\gamma_1 - 1}{\gamma_1 + 1} + 2q'\right).$$

This last equation indicates that, in a perfect gas, the Hugoniot curve is in the form of a rectangular hyperbola. Since the solution of the conservation laws must satisfy both the Rayleigh line and the Hugoniot curve simultaneously, we see that the transition from reactants to products must follow the Rayleigh line from the initial state $x = y = 1$ to a final state (x, y), which lies on the Hugoniot curve as shown in Fig. 2.2.

Note that if we take q (or q') to be zero, Eq. 2.11 or 2.13 becomes the Hugoniot curve for a nonreacting shock wave. If we assume that chemical reactions occur progressively and that the chemical energy release can be written as λq (where $0 \leq \lambda \leq 1$ represents the progress of the reaction), we can define a family of Hugoniot curves that represents the locus of partially reacted states where there is a Hugoniot curve for each value of λ.

The shock Hugoniot curve, obtained by imposing that $q = 0$ (or $\lambda = 0$), will pass through the initial state $x = y = 1$. For finite values of q (or for finite values of λ), the Hugoniot curve lies above the shock Hugoniot curve and does not intersect the initial state. The intersections of the lines $x = 1$ and $y = 1$ with the Hugoniot curve give the solutions for constant-volume and constant-pressure combustion, respectively. There are no real solutions on the Hugoniot curve between the constant-volume and constant-pressure combustion solutions, because for $y > 1$ and $x > 1$, the mass flux $\dot{m}$ becomes imaginary (see Eq. 2.8). The constant-volume and constant-pressure solutions form the boundaries between the upper detonation and the lower deflagration branches of the Hugoniot curve, respectively.

From Eq. 2.13, we obtain $y \to \infty$ for $x \to (\gamma_1 - 1)/(\gamma_1 + 1)$, which gives a limiting density ratio of $\rho_1/\rho_0 \to (\gamma_1 + 1)/(\gamma_1 - 1)$ across a strong shock wave. For $x \to \infty$, y becomes negative, hence the part of the Hugoniot curve where

$$x > \frac{\gamma_0 + 1}{\gamma_0 - 1} + 2q$$

does not represent any physical solution.

The slope and curvature of the Hugoniot curve can be obtained by differentiating Eq. 2.14. On doing so, we get

$$\left(\frac{dy}{dx}\right)_{\mathrm{H}} = -\frac{y + \alpha}{x - \alpha} \tag{2.15}$$

and

$$\left(\frac{d^2y}{dx^2}\right)_{\mathrm{H}} = 2\frac{y + \alpha}{(x - \alpha)^2}. \tag{2.16}$$

Since $\gamma_1 > 0$ (i.e., $\alpha > 0$), we see that the curvature of the Hugoniot curve from Eq. 2.16 is always positive. Thus, the Hugoniot curve is concave upwards. For finite q', where the Hugoniot curve is above the initial state $x = y = 1$, we see that the Rayleigh line will intersect the Hugoniot curve at two points for both the upper detonation branch and the lower deflagration branch. The two solutions for the upper branch are called strong and weak detonations, whereas the lower branch gives weak and strong deflagrations. These solutions are represented in Fig. 2.2. Thus, for a given propagation velocity of the combustion wave (i.e., slope of the Rayleigh line), there correspond two possible solutions in (x, y). The two solutions merge when the Rayleigh line becomes tangent to the Hugoniot curve, giving a minimum velocity on the detonation branch and a maximum velocity on the deflagration branch of the Hugoniot curve. These two tangency solutions are referred to as the CJ solutions. There is no solution when the Rayleigh line does not intersect the Hugoniot curve.

2.4. THE TANGENCY (CHAPMAN–JOUGUET) SOLUTIONS

When the Rayleigh line is tangent to the Hugoniot curve, we obtain two solutions: a minimum-detonation-velocity solution and a maximum-deflagration-velocity

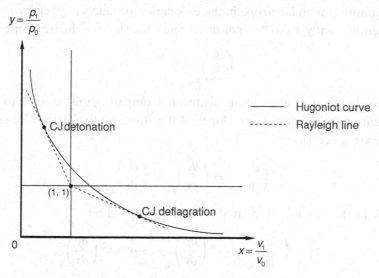

Figure 2.3. The tangency or CJ solutions.

solution. These tangency solutions are referred to as the CJ solutions (see Fig. 2.3). Since the tangency condition provides an extra criterion to close the conservation laws, the CJ solutions can be obtained from the conservation laws alone without the need to consider the detailed propagation mechanism of the combustion wave. However, whether these solutions are physically meaningful or not depends on their agreement with experimental observations or on additional physical consid- erations outside the conservation laws. For detonations, the upper tangency so- lution agrees remarkably well with experiments. For deflagrations, however, the maximum-velocity solution is generally not observed. The case of deflagrations re- quires further elaboration, because the reactants ahead of the wave are influenced by the dynamics of the combustion products, which in turn depend on the rear boundary conditions of the product flow. We will first discuss some general prop- erties of the CJ tangency solutions.

The CJ solutions (x^*, y^*) can be obtained by equating the slopes of the Rayleigh line and the Hugoniot curve, given by Eqs. 2.10 and 2.15, respectively. Thus,

$$y^* = \frac{-x^*(1-\alpha)}{1+\alpha-2x^*} = \frac{-x^*}{\gamma_1 - (\gamma_1+1)x^*}, \tag{2.17}$$

where the asterisk denotes the CJ points. Substituting the above into Eq. 2.10, we obtain the slope at the CJ points as

$$\left(\frac{dy}{dx}\right)_R^* = -\frac{y^*-1}{1-x^*} = \frac{\gamma_1}{\gamma_1 - (\gamma_1+1)x^*}. \tag{2.18}$$

Multiplying top and bottom by x and using Eq. 2.17, the slope of the Rayleigh line can be written as

$$\left(\frac{dy}{dx}\right)_R^* = \frac{\gamma_1 x^*}{[\gamma_1 - (\gamma_1+1)x^*]x^*} = -\frac{\gamma_1 y^*}{x^*} = \left(\frac{dy}{dx}\right)_H^*. \tag{2.19}$$

The equation for an isentrope in the detonation products is given by $pv^{\gamma_1} = $ constant or, equivalently, by $xy^{\gamma_1} = $ constant. Thus, the slope of the isentrope is

$$\left(\frac{dy}{dx}\right)_S = -\frac{\gamma_1 y}{x},\qquad(2.20)$$

where the subscript S refers to an isentrope. Comparing this result with Eq. 2.19, we see that at the CJ points, the slopes of the Rayleigh line, Hugoniot curve, and isentrope are all equal:

$$\left(\frac{dy}{dx}\right)_R^* = \left(\frac{dy}{dx}\right)_H^* = \left(\frac{dy}{dx}\right)_S^*.\qquad(2.21)$$

The slope of the isentrope is related to the sound speed by

$$c_1^2 = \left(\frac{dp}{d\rho}\right)_S = -v_1^2\left(\frac{dp}{dv}\right)_S = -p_0 v_0 x^2\left(\frac{dy}{dx}\right)_S,\qquad(2.22)$$

or

$$\left(\frac{dy}{dx}\right)_S = -\frac{c_1^2}{p_0 v_0 x^2}.\qquad(2.23)$$

From Eq. 2.6, we get

$$u_1^2 = \frac{1}{\rho_1^2}\frac{p_1 - p_0}{v_0 - v_1} = v_1^2\frac{p_1 - p_0}{v_0 - v_1} = p_0 v_0 x^2\frac{y-1}{1-x},\qquad(2.24)$$

and using Eq. 2.10, the slope of the Rayleigh line can be rewritten as

$$\left(\frac{dy}{dx}\right)_R = -\frac{u_1^2}{p_0 v_0 x^2}.\qquad(2.25)$$

At the CJ points, where the slope of the Rayleigh line is equal to the slope of the isentrope, Eqs. 2.22 and 2.25 give

$$\frac{u_1^*}{p_0 v_0 x_1^2} = \frac{c_1^*}{p_0 v_0 x_1^2},$$

or

$$\left(\frac{u_1^*}{c_1^*}\right)^2 = M_1^{*2} = 1.\qquad(2.26)$$

Therefore, the flow Mach number downstream of a CJ detonation or deflagration, M_1^*, is equal to unity. This sonic downstream flow condition was used by Jouguet to determine the desired detonation solution, whereas Chapman chose the minimum-velocity criterion for the correct detonation solution. From the above analysis, we see that Chapman's minimum-velocity criterion and Jouguet's sonic criterion are equivalent. However, there are cases where the sonic condition does not correspond to chemical equilibrium (e.g., pathological detonations).

2.5. ENTROPY VARIATION ALONG THE HUGONIOT CURVE

Entropy arguments were used by Becker (1917, 1922a, 1922b) and Scorah (1935) to provide a justification for the choice of the CJ solution. It is of interest to investigate the variation of entropy along the Hugoniot curve in order to determine the entropy change across a combustion wave for the various solutions.

We can first rewrite Eq. 2.12, for the Hugoniot curve, in a nondimensional form:

$$\bar{e} - (\bar{e}_0 + q') = \frac{1}{2}(y+1)(1-x),\qquad(2.27)$$

where $\bar{e} = e_1/p_0v_0$, $\bar{e}_0 = e_0/p_0v_0$, and $q' = q/p_0v_0$. Differentiating Eq. 2.27, we get the variation of the internal energy along the Hugoniot curve as

$$\left(\frac{d\bar{e}}{dx}\right)_H = \frac{1}{2}(1-x)\left(\frac{dy}{dx}\right)_H - \frac{1}{2}(y+1).\qquad(2.28)$$

From thermodynamics, we note that

$$de = T\,ds - p\,dv,$$

where s is the entropy. This equation can be rewritten in nondimensional variables as

$$d\bar{e} = xy\,d\bar{s} - y\,dx,$$

where $\bar{s} = s/R$. Specializing this equation along the Hugoniot curve, we obtain

$$\left(\frac{d\bar{e}}{dx}\right)_H = xy\left(\frac{d\bar{s}}{dx}\right)_H - y.\qquad(2.29)$$

Equating Eqs. 2.28 to 2.29 gives

$$\frac{1}{2}(1-x)\left(\frac{dy}{dx}\right)_H - \frac{1}{2}(y+1) = xy\left(\frac{d\bar{s}}{dx}\right)_H - y,$$

and solving for the entropy variation along the Hugoniot curve, we get

$$\left(\frac{d\bar{s}}{dx}\right)_H = \frac{1-x}{2xy}\left[\left(\frac{dy}{dx}\right)_H + \frac{y-1}{1-x}\right].\qquad(2.30)$$

Using Eq. 2.10 for the slope of the Rayleigh line, Eq. 2.30 can be written as

$$\left(\frac{d\bar{s}}{dx}\right)_H = \frac{1-x}{2xy}\left[\left(\frac{dy}{dx}\right)_H - \left(\frac{dy}{dx}\right)_R\right].\qquad(2.31)$$

At the CJ points where the slope of the Hugoniot curve is equal to the slope of the Rayleigh line, Eq. 2.31 becomes

$$\left(\frac{d\bar{s}}{dx}\right)_H^* = 0.$$

Thus, the entropy takes on extrema values at the CJ points. The curvature of the entropy curve can be obtained by differentiating Eq. 2.30:

$$\left(\frac{d^2\bar{s}}{dx^2}\right)_H = -\left[\frac{1}{2x^2y} + \frac{1-x}{2xy^2}\left(\frac{dy}{dx}\right)_H\right]\left[\left(\frac{dy}{dx}\right)_H + \frac{y-1}{1-x}\right]$$

$$+ \frac{1-x}{2xy}\left[\left(\frac{d^2y}{dx^2}\right)_H + \frac{1}{1-x}\left(\frac{dy}{dx}\right)_H + \frac{y-1}{(1-x)^2}\right]. \qquad (2.32)$$

Using Eq. 2.10 and the fact that the slopes of the Rayleigh line and Hugoniot curve are equal, we obtain the curvature of the entropy curves at the CJ points as

$$\left(\frac{d^2\bar{s}}{dx^2}\right)_H^* = \frac{1-x^*}{2x^*y^*}\left(\frac{d^2y}{dx^2}\right)_H^*. \qquad (2.33)$$

Since the curvature of the Hugoniot curve has been shown to be always positive, the curvature of the entropy curve then depends on whether $1 - x^*$ is positive or negative. For the detonation branch, where $x^* < 1$, the curvature of the entropy curve is positive, whereas for deflagrations, where $x^* > 1$, the entropy curvature is negative. Thus, the entropy is a minimum for CJ detonations and a maximum for CJ deflagrations. From the entropy curve, we also see that for strong detonations $(d\bar{s}/dx)_H < 0$, while for weak detonations $(d\bar{s}/dx)_H > 0$.

2.6. DOWNSTREAM FLOW CONDITIONS

Depending on whether the flow behind the combustion wave is subsonic or supersonic, the downstream boundary condition may or may not have an influence on the wave propagation speed. If the flow is subsonic behind the wave, then the back boundary condition must be satisfied by the solution of the conservation laws across the wave. In addition, if the wave speed is also subsonic, then perturbations can propagate upstream of the wave, and the upstream conditions will be altered. It is therefore of importance to determine the flow condition behind the combustion wave and determine if it is subsonic or supersonic.

Starting with the following general expression

$$dp(s, v) = \left(\frac{\partial p}{\partial s}\right)_v ds + \left(\frac{\partial p}{\partial v}\right)_s dv$$

and specializing the derivatives to be along the Hugoniot curve, we obtain

$$\left(\frac{dp}{dv}\right)_H = \left(\frac{\partial p}{\partial s}\right)_v \left(\frac{ds}{dv}\right)_H + \left(\frac{\partial p}{\partial v}\right)_s. \qquad (2.34)$$

Using the chain rule, we write

$$\left(\frac{\partial p}{\partial s}\right)_v = \left(\frac{\partial p}{\partial T}\right)_v \left(\frac{\partial T}{\partial e}\right)_v \left(\frac{\partial e}{\partial s}\right)_v.$$

Using the thermodynamic relationships

$$c_v = \left(\frac{\partial e}{\partial T}\right)_v, \qquad T = \left(\frac{\partial e}{\partial s}\right)_v, \quad \text{and} \quad \left(\frac{\partial p}{\partial T}\right)_v = \frac{R}{v} = \frac{p}{T},$$

Eq. 2.34 can be written as

$$\left(\frac{dp}{dv}\right)_H = \frac{p}{T}\left(\frac{ds}{dv}\right)_H + \left(\frac{dp}{dv}\right)_S.$$

In terms of nondimensional variables, the above equation becomes

$$\left(\frac{dy}{dx}\right)_H = (\gamma_1 - 1)y\left(\frac{d\bar{s}}{dx}\right)_H + \left(\frac{dy}{dx}\right)_S. \tag{2.35}$$

The above equation relates the slope of the Hugoniot curve to the slope of the isentrope and to the entropy change along the Hugoniot curve. Solving Eq. 2.31 for the slope of the Hugoniot curve and substituting it into Eq. 2.35 yields

$$(\gamma_1 - 1)y\left(\frac{d\bar{s}}{dx}\right)_H + \left(\frac{dy}{dx}\right)_S = \frac{2xy}{1-x}\left(\frac{d\bar{s}}{dx}\right)_H + \left(\frac{dy}{dx}\right)_R,$$

or

$$y(\gamma_1 + 1)\frac{\frac{\gamma_1-1}{\gamma_1+1} - x}{1 - x}\left(\frac{d\bar{s}}{dx}\right)_H = \left(\frac{dy}{dx}\right)_R - \left(\frac{dy}{dx}\right)_S. \tag{2.36}$$

Using the expressions obtained previously for the slopes of the isentrope and of the Rayleigh line (i.e., Eqs. 2.23 and 2.25), Eq. 2.36 becomes

$$y(\gamma_1 + 1)\frac{\frac{\gamma_1-1}{\gamma_1+1} - x}{1 - x}\left(\frac{d\bar{s}}{dx}\right)_H = \frac{c_1^2}{x^2 p_0 v_0}\frac{u_1^2}{x} = \frac{\gamma_1 y}{x}(1 - M_1^2),$$

or

$$\frac{\gamma_1 + 1}{\gamma_1}\left(\frac{d\bar{s}}{dx}\right)_H \left(\frac{\frac{\gamma_1-1}{\gamma_1+1} - x}{1 - x}\right)x = 1 - M_1^2. \tag{2.37}$$

For the case of a detonation where the value of x is bounded between $(\gamma_1 - 1)/(\gamma_1 + 1)$ and 1, we see that $1 - M_1^2$ has the opposite sign to the entropy derivative along the Hugoniot curve. Thus, for the case of a strong detonation, where $(d\bar{s}/dx)_H < 0$, $1 - M_1^2 > 0$, and thus $M_1 < 1$, the downstream flow is subsonic relative to the combustion wave. In the case of a weak detonation, where $(d\bar{s}/dx)_H > 0$, then $1 - M_1^2 < 0$ and thus $M_1 > 1$, and the downstream flow is supersonic. For the deflagration branch of the Hugoniot curve, the value of x is larger than 1, and the sign of $1 - M_1^2$ is therefore the same as the sign of the entropy derivative along the Hugoniot curve. From the variation of the entropy around the CJ deflagration point, we see from Eq. 2.37 that the flow is subsonic downstream of a weak deflagration wave and supersonic downstream of a strong deflagration wave.

When the downstream flow is subsonic, the back boundary conditions can influence the wave. Thus, strong detonations and weak deflagrations will depend on the downstream boundary conditions. For strong detonations, the back boundary influence can extend up to the wave front, but not beyond, because the detonation wave propagates at a supersonic speed. Specifying the back boundary conditions (e.g., the particle velocity or piston velocity) will allow the strong detonation solution to be obtained from the conservation laws. For weak deflagrations, the influence of the back boundary condition extends beyond the wave front to the reactants ahead of it. Therefore, the weak deflagration speed cannot be obtained from the conservation laws by specifying just the back boundary conditions. The mechanism of propagation must be specified in order to determine the wave speed relative to the moving reactants ahead of it.

For weak detonations and strong deflagrations where the flow is supersonic behind the wave, downstream boundary conditions cannot influence the propagation of the wave. Additional physical considerations must be used to determine if such waves can exist and, if so, what the requirements are to realize them in nature.

2.7. THE CHAPMAN–JOUGUET CRITERION

The conservation equations together with the equation of state are insufficient for the combustion wave speed to be determined for given initial and boundary conditions of an explosive mixture. The additional required relationship can be obtained from Chapman's and Jouguet's work and is generally referred to as the Chapman–Jouguet criterion. In Chapman's original study (1889), he found that when the Rayleigh line is tangent to the Hugoniot curve, the detonation velocity is a minimum. There are no solutions to the conservation equations for velocities less than this minimum value. Above this minimum velocity, the Rayleigh line intersects the Hugoniot curve, and there are two solutions for a given detonation velocity. Based on experimental observations that a unique velocity is obtained for a given explosive mixture at given initial conditions, Chapman then postulated that the correct solution must be the minimum-velocity or the tangency solution. The minimum-velocity solution agrees very well with experiments, thus justifying Chapman's choice.

Jouguet (1904), on the other hand, chose the solution that corresponds to the condition of sonic flow (relative to the wave) for the downstream state. He further noted that this sonic solution also corresponds to a minimum value of the entropy change across the detonation wave. Thus, the sonic-flow or the minimum-entropy requirement can provide a criterion for selecting the solution from the conservation laws. Crussard (1907) later demonstrated the equivalence of the three solutions of minimum velocity, sonic flow, and minimum entropy of the downstream state. Apart from the agreement between the CJ detonation velocity and experimentally measured values, no theoretical justification was given by either Chapman or Jouguet. Thus, subsequent investigations were concerned with providing a more rigorous justification for the CJ criterion, because it predicts detonation velocities in good agreement with experiments, even near the detonation limits.

For any detonation velocity greater than the minimum or CJ value, there correspond two solutions: the upper strong (or overdriven) and the lower weak detonation solutions. It is generally agreed by all investigators that the upper strong detonation solution is unstable for unsupported or freely propagating detonations. Thus, the strong detonation solution can be ruled out from stability considerations based on the fact that the flow is subsonic behind the strong detonation and expansion waves behind it can penetrate the reaction zone to attenuate the wave.

The lower weak detonation solution, however, is not so easily discarded. Early attempts were based on entropy arguments. Becker (1917, 1922a, 1922b) pointed out that the entropy increase across a strong detonation is higher than that across a weak detonation (for the same detonation velocity). Becker associated the entropy change with the probability of occurrence of a flow. He claimed that the strong detonation is the more probable solution. However, because strong detonations are unstable, Becker then arrived at the conclusion that the solution must be the sonic or the minimum-velocity solution as given by the CJ criterion.

Scorah (1935) also used an entropy argument to justify the CJ criterion. He pointed out that because the CJ velocity is the minimum, this would lead to a minimum time rate of degradation of available energy. In a later study, Duffey (1955) invoked the principle of minimum entropy production of steady state irreversible thermodynamics and argued that a steadily propagating detonation should also abide by this thermodynamic principle. Zeldovich (1940, 1950), however, refuted any thermodynamic argument by pointing out that the increase in entropy across a shock wave is not sufficient to guarantee that a shock wave will occur. The mechanism that produces the shock wave, such as a piston pushing the gas behind it, is also required. Jouguet also argued that a weak detonation is unstable because disturbances in its wake will be left behind, the flow being supersonic behind the weak detonation wave. Again, Zeldovich countered this argument by pointing out that instability requires that the disturbances grow in amplitude with time. The fact that disturbances are left behind a weak detonation does not prove that the weak detonation is unstable. In fact, dissipative processes will attenuate the disturbances.

Perhaps the most convincing argument for the possibility of the weak detonation solution was provided by von Neumann (1942). He first assumed that partially reacted Hugoniot curves exist. Defining n as a measure of the progress of the chemical reactions (where $0 \leq n \leq 1$ between initial reactants and final products), the partially reacted Hugoniot curve represents the locus of possible states corresponding to a particular value of n. Between the shocked state immediately behind the leading shock, where $n = 0$, and the final equilibrium product state, where $n = 1$, there exists a family of Hugoniot curves for different values of n as shown in Fig. 2.4.

The Rayleigh line from the initial state must necessarily intersect all these Hugoniot curves as the reaction progresses from reactants to products. Von Neumann then showed that if none of the partially reacted Hugoniot curves intersect one another, then weak detonations can be ruled out. This is the case because as

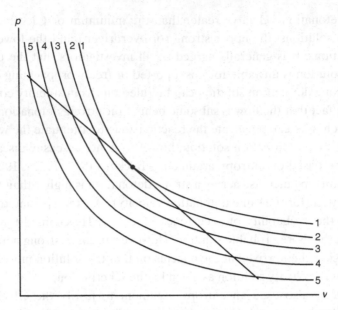

Figure 2.4. Partially reacted Hugoniot curves (von Neumann, 1942).

one progresses from the shocked state along the Rayleigh line, the upper strong
detonation solution will be encountered first. A subsequent transition following the
same Rayleigh line from the strong to the weak detonation solution would result in
an entropy decrease as, for example, in a rarefaction shock. This transition can be
ruled out in view of the second law of thermodynamics. However, if the reactions are
such that the partially reacted Hugoniot curves intersect one another, von Neumann
argued that there must exist an envelope of these partially reacted Hugoniot curves
(see Fig. 2.5). The solution must follow the Rayleigh line that is tangent to this envelope. However, this Rayleigh line would intersect the final equilibrium Hugoniot
curve ($n = 1$) on the weak detonation branch. Thus, weak detonations are possible
for those explosives that have partially reacted Hugoniot curves that intersect one
another.

Von Neumann then pointed out that the possibility of weak detonations can be
determined if a given explosive has intersecting Hugoniot curves. Weak detonations
obtained as a result of intersecting Hugoniot curves are referred to as pathological
detonations. It has been shown that systems that have an overshoot in temperature,
as a result of an initial rapid exothermic reaction followed by a slower endothermic reaction as the reaction approaches equilibrium, do result in pathological detonations (e.g., in a mixture of $H_2 + Cl_2$). The existence of pathological detonations
has been confirmed from analysis of the detailed chemical reactions in the reaction
zone as well as experimental measurements of the detonation velocity (Dionne $et\ al.$,
2000).

The above arguments used to exclude the weak detonation solution are based on
the assumption that the leading shock front initiates chemical reactions via adiabatic

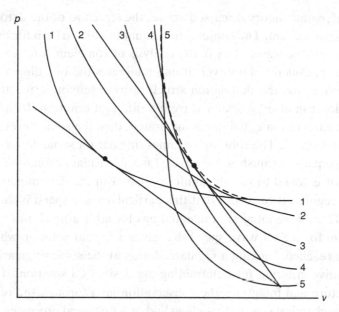

Figure 2.5. Partially reacted Hugoniot curves that intersect (von Neumann, 1942).

compression. There is the possibility that the weak detonation solution can also be reached directly from the initial state along the Rayleigh line to the equilibrium Hugoniot curve without first going to the von Neumann shocked state. In that case, the path would bypass the strong detonation solution. However, without shock heating, an alternate ignition mechanism must be provided to initiate the chemical reaction if the explosive is stable and does not undergo spontaneous autoignition at its initial state. Zeldovich had discussed various schemes, such as the sequential activation of a series of sparks along a tube to effect ignition. In that case, the wave speed would correspond to the artificial rate of the sequential spark firings and not to the physical properties of the explosive. It is clear that for detonations where shock ignition is the mechanism for initiating the chemical reactions, von Neumann's argument that the possibility of the weak detonation solution results from intersecting partially reacted Hugoniot curves is the most plausible one.

It is also of interest to point out that the existence of a steady detonation front should also depend on the possibility of being able to match the conditions behind a steady detonation wave to the nonsteady flow in the products. G.I. Taylor (1950) studied the dynamics of the combustion products behind a steadily propagating detonation wave and concluded that planar CJ detonations can be matched to the nonsteady Riemann expansion fan behind it. However, CJ spherical detonations result in an expansion singularity when matched to a progressive wave solution in its wake. Thus, diverging spherical CJ detonations are not possible. From mathematical considerations, it appears that only planar CJ detonations are compatible with the nonsteady flow of detonation products.

In the gasdynamic theory discussed above, the structure of the detonation wave is not taken into account. The conservation equations are written for the initial and final states across the wave. Only in the analysis of von Neumann are the partially reacted states considered. However, if one integrates the one-dimensional steady flow equations across the detonation structure by specifying a reaction rate law, the CJ solution can also be obtained from a different criterion. If the differential form of the conservation equations is integrated, then the sonic flow condition will eventually be reached. The solution becomes singular for sonic flow as the denominator of the equation vanishes. However, for a particular detonation wave speed, the numerator is found to vanish simultaneously with the denominator, rendering the solution regular. It turns out that this particular wave speed is the CJ detonation speed. Thus, if we integrate the equations for arbitrarily chosen wave speeds, we can iterate for the CJ wave speed that gives a regular solution when the sonic singularity is reached. Seeking a regular solution at the sonic singularity can serve as an alternative criterion for determining the desired CJ solution. If one considers the structure and integrates the conservation laws through the reaction zone, then different chemical reaction rate laws and other physical processes, such as heat and momentum losses, can also be included. This analysis of the equations for the detonation structure was first presented by Zeldovich (1940, 1950).

However, this method of determining the detonation solution requires an iterative procedure and numerical integration of the equations themselves. Thus, it is more complex than the algebraic gasdynamic theory, where only upstream and downstream states are considered and the detailed structure is bypassed when computing the detonation velocity.

2.8. RANKINE–HUGONIOT RELATIONS

Assuming a perfect gas and a constant chemical energy release (i.e., energy release that does not depend on the downstream state), it is possible to obtain algebraic expressions relating the downstream state (p_1, ρ_1, T_1, etc.) to the upstream state (p_0, ρ_0, T_0) and the detonation Mach number M_0. This was first carried out by Rankine and Hugoniot, and these relations are known as the Rankine–Hugoniot relations. From Section 2.3, the equation for the Rayleigh line can be written as

$$y = \left(1 + \gamma_0 M_0^2\right) - \left(\gamma_0 M_0^2\right) x, \tag{2.38}$$

and the Hugoniot curve is given by

$$y = \frac{\frac{\gamma_0+1}{\gamma_0-1} - x + 2q'}{\frac{\gamma_1+1}{\gamma_1-1}x - 1}, \tag{2.39}$$

where $q' = q/p_0 v_0$. Since a solution corresponds to the intersection of the Rayleigh line with the Hugoniot curve, we equate Eqs. 2.38 and 2.39 to obtain the following quadratic equation for the specific volume ratio, $x = v_1/v_0$:

$$x^2 - 2\frac{\gamma_1(\gamma_0 + \eta)}{\gamma_0(\gamma_1 + 1)}x + \frac{\gamma_1 - 1}{\gamma_1 + 1}\left[1 + 2\eta\left(\frac{1}{\gamma_0 - 1} + \bar{q}\right)\right] = 0, \qquad (2.40)$$

where $\eta = 1/M_0^2$ and $\bar{q} = q'/\gamma_0 = q/\gamma_0 p_0 v_0 = q/c_0^2$. Solving Eq. 2.40 for x, we obtain

$$x = \frac{v_1}{v_0} = \frac{\rho_0}{\rho_1} = \frac{\gamma_1(\gamma_0 + \eta \pm S)}{\gamma_0(\gamma_1 + 1)}, \qquad (2.41)$$

where

$$S = \sqrt{\left(\frac{\gamma_0}{\gamma_1} - \eta\right)^2 - K\eta}, \qquad (2.42)$$

in which

$$K = \frac{2\gamma_0(\gamma_1 + 1)}{\gamma_1^2}\left[\frac{\gamma_1 - \gamma_0}{\gamma_0 - 1} + \gamma_0(\gamma_1 - 1)\bar{q}\right]. \qquad (2.43)$$

The $\pm$ sign in Eq. 2.41 corresponds to the two roots of the quadratic equation. The positive sign refers to a weak detonation (or a strong deflagration), whereas the negative sign refers to a strong detonation (or a weak deflagration). When the two roots coincide, we obtain the CJ tangency solutions where $S = 0$, signified by $\eta = \eta^*$. From Eq. 2.42, we obtain for $S = 0$

$$\left(\frac{\gamma_0}{\gamma_1} - \eta^*\right)^2 - K\eta^* = 0. \qquad (2.44)$$

Solving for η^*, we obtain

$$\eta^* = \frac{1}{M_{CJ}^2} = \frac{\gamma_0}{\gamma_1}\left(1 - \frac{2}{1 \pm \sqrt{1 + \frac{4}{K}\frac{\gamma_0}{\gamma_1}}}\right). \qquad (2.45)$$

The $\pm$ sign in the above equation corresponds to the two tangency solutions of a CJ detonation ($+$ sign) and a CJ deflagration ($-$ sign).

For a real detonable mixture, $\bar{q}$ is generally of the order of 30 (for a fuel–air mixture) or higher (for a fuel–oxygen mixture). Thus, K is also a large number compared to unity, and $\frac{1}{K} \ll 1$. We may therefore expand the term under the square root sign in Eq. 2.45. Retaining the first term of the order of $\frac{1}{K}$, we get

$$\sqrt{1 + 4\left(\frac{\gamma_0}{\gamma_1}\right)\frac{1}{K}} = 1 + \frac{1}{2}4\left(\frac{\gamma_0}{\gamma_1}\right)\frac{1}{K} + O\left(\frac{1}{K^2}\right) = 1 + \frac{2}{K}\left(\frac{\gamma_0}{\gamma_1}\right).$$

Thus, Eq. 2.45 for a CJ detonation (taking the positive sign) can be written as

$$\eta^* = \frac{1}{M_{CJ}^2} \approx \frac{\gamma_0}{\gamma_1}\left(1 - \frac{2}{1 + \left[1 + \frac{2}{K}\frac{\gamma_0}{\gamma_1} + \cdots\right]}\right)$$

$$\approx \left(\frac{\gamma_0}{\gamma_1}\right)^2\frac{1}{K}.$$

From Eq. 2.43 and taking $\gamma_0 \approx \gamma_1$, we obtain

$$K \approx 2\left(\gamma_1^2 - 1\right)\bar{q}.$$

Thus,

$$\eta^* \approx \frac{1}{2\left(\gamma_1^2 - 1\right)\bar{q}}$$

or

$$(M_{CJ})_{\text{detonation}} \approx \sqrt{2\left(\gamma_1^2 - 1\right)\bar{q}}. \tag{2.46}$$

For a CJ deflagration, we consider the negative sign in Eq. 2.45. Expanding in powers of $\frac{1}{K}$, we obtain an analogous approximation of

$$(M_{CJ})_{\text{deflagration}} \approx \frac{1}{\sqrt{2\left(\gamma_1^2 - 1\right)\bar{q}}}, \tag{2.47}$$

where we have also used the approximation $\gamma_0 \approx \gamma_1$.

For the general case where $S \neq 0$, the density ratio ρ_1/ρ_0 can be obtained from Eq. 2.41 as

$$\frac{\rho_1}{\rho_0} = \frac{\gamma_0\left(\gamma_1 + 1\right)}{\gamma_1\left(\gamma_0 + \eta \pm S\right)}. \tag{2.48}$$

Using the above equation and the conservation of mass (i.e., $\rho_0 u_0 = \rho_1 u_1$), the particle velocity ratio can be obtained as

$$\frac{u_1}{u_0} = \frac{\gamma_1\left(\gamma_0 + \eta \pm S\right)}{\gamma_0\left(\gamma_1 + 1\right)}. \tag{2.49}$$

The pressure ratio can also be obtained using the momentum equation (i.e., $p_0 + \rho_0 u_0^2 = p_1 + \rho_1 u_1^2$) as

$$y = \frac{p_1}{p_0} = \frac{\gamma_0 + \eta \mp \gamma_1 S}{\left(\gamma_1 + 1\right)\eta}. \tag{2.50}$$

The temperature and sound speed ratios across the detonation can be obtained from Eqs. 2.48 and 2.50 using the equation of state (i.e., $p = \rho R T$).

Equations 2.48 to 2.50 are based on a coordinate system that is fixed to the propagating detonation. Consider a detonation propagating at a velocity D relative to a fixed laboratory coordinate system as shown below:

The equations across the wave can readily be obtained from Eqs. 2.48 to 2.50 via a coordinate transformation where we write $u_0 = D$ and $u_1 = D - u_1'$ (u_0 and u_1 are the velocities with respect to the coordinate system fixed relative to the wave):

$$u_1 = D - u_1 \qquad\qquad u_0 = D$$

$$p_1, \rho_1 \qquad\qquad\qquad p_0, \rho_0$$

The density and pressure ratios across the wave in the laboratory fixed coordinate system remain the same as given by Eqs. 2.48 and 2.50. However, the particle velocity ratio as obtained from the conservation of mass [i.e., $\rho_0 D = \rho_1(D - u_1')$] is

$$\frac{u_1'}{D} = \frac{\gamma_0 - \gamma_1\,(\eta \pm S)}{\gamma_0\,(\gamma_1 + 1)}. \tag{2.51}$$

The pressure p_1 is generally normalized with respect to $\rho_0 D^2$ instead of p_0. From Eq. 2.50, we obtain

$$\frac{p_1}{\rho_0 D^2} = \frac{\gamma_0 + \eta \mp \gamma_1 S}{\gamma_0\,(\gamma_1 + 1)}. \tag{2.52}$$

Equations 2.48 to 2.52 are referred to as the Rankine–Hugoniot relationships across a detonation wave and are analogous to the relationships across a normal shock wave in a non-reacting medium, which can also be obtained from the same equations by taking $\bar{q} = 0$ in Eq. 2.43. For the case of a non-reacting medium, the Rankine–Hugoniot relationships for a normal shock in a perfect gas can be obtained as

$$\frac{\rho_1}{\rho_0} = \frac{\gamma + 1}{(\gamma - 1) + 2\eta}, \tag{2.53}$$

$$\frac{u_1}{u_0} = \frac{(\gamma - 1) + 2\eta}{\gamma + 1}, \tag{2.54}$$

$$\frac{u_1'}{D} = \frac{2}{\gamma + 1}(1 - \eta), \tag{2.55}$$

and

$$\frac{p_1}{p_0} = 1 + \frac{2\gamma}{\gamma + 1}\left(\frac{1}{\eta} - 1\right), \tag{2.56}$$

or

$$\frac{p_1}{\rho_0 D^2} = \frac{\gamma + 1}{(\gamma - 1) + 2\eta}, \tag{2.57}$$

where $\eta = 1/M_0^2 = c_0^2/D^2$.

For CJ detonations (i.e., $S = 0$) where $M_{CJ} >> 1$ (so that $\eta_{CJ} << 1$), Eqs. 2.48, 2.51, and 2.52 reduce to the following limiting forms:

$$\frac{\rho_1}{\rho_0} = \frac{\gamma + 1}{\gamma}, \tag{2.58}$$

$$\frac{u_1'}{D} = \frac{1}{\gamma + 1}, \tag{2.59}$$

$$\frac{p_1}{\rho_0 D^2} = \frac{1}{\gamma + 1}, \tag{2.60}$$

which differ from the strong shock relationships for a non-reacting gas, where

$$\frac{\rho_1}{\rho_0} = \frac{\gamma + 1}{\gamma - 1}, \tag{2.61}$$

$$\frac{u_1'}{D} = \frac{p_1}{\rho_0 D^2} = \frac{2}{\gamma + 1}. \tag{2.62}$$

2.9. DEFLAGRATIONS

Solutions on the lower deflagration branch of the Hugoniot curve have not received as much attention as those on the upper detonation branch, due to the lack of a practical reason to consider them. Unlike detonations, where the CJ velocity is practically always observed experimentally, there is generally no unique steady deflagration velocity for a given explosive mixture. The deflagration speed also depends strongly on boundary conditions. The determination of the deflagration speed requires a consideration of the propagation mechanism of the combustion wave, rather than just the energetics of the mixture as for detonations. Furthermore, steadily propagating deflagrations are seldom realized experimentally, because deflagrations are unstable: they tend to self-accelerate and then transit to detonations under appropriate boundary conditions. Nevertheless, it is of interest to discuss the possibility of steady deflagration solutions on the lower deflagration branch of the Hugoniot curve.

In Sections 2.4 and 2.5, we have already demonstrated that entropy is at a maximum at the tangency CJ state. Behind a weak deflagration the velocity is subsonic, whereas behind a strong detonation it is supersonic. A Rayleigh line from the initial state will first intersect the weak deflagration branch. A transition from the weak to the strong deflagration branch along the Rayleigh line will constitute a rarefaction shock wave, which is impossible to realize physically. Therefore, in general, only solutions on the weak deflagration branch of the Hugoniot curve are possible.

For weak deflagrations, the flows both ahead and behind the wave are subsonic (relative to the wave). Disturbances from the rear of the wave can propagate upstream and influence the condition ahead of the wave. Thus, the back boundary condition must be taken into consideration when determining the propagation speed of weak deflagrations. For example, if the deflagration propagates from the closed end of a tube where the particle velocity must vanish, then the flow of the reactants ahead of the deflagration must be such that the velocity of the product gases is zero downstream of the wave. On the other hand, if the deflagration propagates from an open-ended tube, then the pressure in the products downstream of the deflagration must correspond to the pressure of the environment at the open end.

To illustrate the effect of boundary conditions on the propagation of deflagrations, consider a weak deflagration propagating from the closed end of a tube. The particle velocity will be zero behind the wave. Due to the density decrease (or increase in specific volume) across the reaction front, the reactants ahead of the wave will be displaced in the same direction by the product expansion as the deflagration wave propagates. If we consider the initial velocity of the reactants to be zero, then the displacement effect of the deflagration is equivalent to that of a piston generating a shock wave ahead of the reaction front. The particle velocity behind the shock is such that the expansion across the deflagration gives zero particle velocity behind it. This is illustrated in the following sketch:

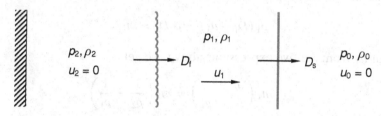

The displacement particle velocity can be determined from the conservation of mass across the deflagration, that is

$$\rho_1 (D_f - u_1) = \rho_2 D_f. \tag{2.63}$$

Solving for u_1, we get

$$\frac{u_1}{D_f} = 1 - \frac{\rho_2}{\rho_1} = 1 - \frac{v_1}{v_2}. \tag{2.64}$$

Since the density ratio across a deflagration is typically 6 or 7 for stoichiometric fuel–air mixtures, the displacement particle velocity is typically about 85% of the deflagration speed.

For different flame speeds D_f, the displacement velocity u_1 ahead of the deflagration is different and the strength of the precursor shock will vary accordingly. Since the Hugoniot curve depends on the initial state, we see that for each deflagration speed, the equilibrium Hugoniot curve will also be different. From Eq. 2.13, the Hugoniot curve is given by

$$y = \frac{\frac{\gamma_1+1}{\gamma_1-1} - x + 2\frac{q}{p_1 v_1}}{\frac{\gamma_2+1}{\gamma_2-1}x - 1}, \tag{2.65}$$

where we have adjusted the subscripts to take account of the state behind the precursor shock. In general, the change of γ across the precursor shock is small, because the precursor shock Mach number is not large. For convenience, we will consider

$\gamma_0 = \gamma_1 = \gamma_2 = \gamma$. The effective chemical energy release on the Hugoniot curve is dependent on the shock strength, because q is normalized with respect to the shocked state ahead of the deflagration. Thus, for a stronger precursor shock, the value of $p_1 v_1$ increases and the nondimensional chemical energy release $\bar{q}$ is reduced.

To solve a weak deflagration problem, both the back boundary condition and the flame speed must be specified. The precursor shock strength is then determined to satisfy the conservation equations across the shock and the reaction front. If we consider a CJ deflagration, then the sonic condition behind the deflagration provides an extra equation to close the set of conservation equations across the deflagration. One needs only to find the precursor shock strength to satisfy the condition across the CJ deflagration front, downstream of which it is sonic. The conservation of mass across the shock is given by

$$\rho_0 D_s = \rho_1 (D_s - u_1) = \dot{m}_s, \tag{2.66}$$

and for that across the reaction front we write

$$\rho_1 (D_f - u_1) = \rho_2 D_f = \dot{m}_f. \tag{2.67}$$

Solving for u_1 from the two expressions above, we get

$$u_1 = \dot{m}_s \left(\frac{1}{\rho_0} - \frac{1}{\rho_1} \right) = \dot{m}_f \left(\frac{1}{\rho_2} - \frac{1}{\rho_1} \right). \tag{2.68}$$

Defining the flame and flow Mach numbers, respectively, as

$$M_f = \frac{D_f - u_1}{c_1} \quad \text{and} \quad M_1 = \frac{D_s - u_1}{c_1},$$

Eq. 2.68 becomes

$$M_1 \left(\frac{\rho_1}{\rho_0} - 1 \right) = M_f \left(\frac{\rho_1}{\rho_2} - 1 \right). \tag{2.69}$$

The density ratio across the shock front, as given by the Rankine–Hugoniot equation, is

$$\frac{\rho_1}{\rho_0} = \frac{(\gamma + 1) M_s^2}{2 + (\gamma - 1) M_s^2}, \tag{2.70}$$

where $M_s = D_s / c_0$. The flow Mach number behind a normal shock can also be written as

$$M_1^2 = \frac{2 + (\gamma - 1) M_s^2}{2\gamma M_s^2 - (\gamma - 1)},$$

or alternatively

$$M_s^2 = \frac{2 + (\gamma - 1) M_1^2}{2\gamma M_1^2 - (\gamma - 1)}. \tag{2.71}$$

Solving for $(\rho_1/\rho_0 - 1)$ using Eqs. 2.70 and 2.71, we obtain

$$\frac{\rho_1}{\rho_0} - 1 = \frac{2}{\gamma + 1} \left(\frac{1}{M_1^2} - 1 \right). \tag{2.72}$$

As shown previously in Eq. 2.41, the density ratio across a CJ wave is given by

$$x_{CJ} = \frac{\rho_1}{\rho_2} = \frac{\gamma + \eta_{CJ}}{\gamma + 1} = \frac{\gamma M_{CJ}^2 + 1}{(\gamma + 1) M_{CJ}^2}, \tag{2.73}$$

where $M_{CJ} = M_f = (D_f - u_1)/c_1$. Thus, considering these last two expressions, Eq. 2.69 can be written as

$$\frac{2}{\gamma + 1} \left(\frac{1}{M_1} - M_1 \right) = \frac{1}{\gamma + 1} \left(\frac{1}{M_{CJ}} - M_{CJ} \right),$$

or

$$\left(\frac{1}{M_1} - M_1 \right) = \frac{1}{2} \left(\frac{1}{M_{CJ}} - M_{CJ} \right), \tag{2.74}$$

which relates the displacement flow velocity ahead of the CJ deflagration to the CJ Mach number.

Using Eqs. 2.43 and 2.44, we obtain

$$(1 - \eta_{CJ})^2 - K\eta_{CJ} = 0,$$

or

$$\left(M_{CJ} - \frac{1}{M_{CJ}} \right)^2 = K = 2 \left(\gamma^2 - 1 \right) \bar{q}, \tag{2.75}$$

and Eq. 2.74 then becomes

$$\left(\frac{1}{M_1} - M_1 \right)^2 = \frac{(\gamma^2 - 1)\bar{q}}{2} = \frac{\gamma^2 - 1}{2} \frac{q}{c_1^2}, \tag{2.76}$$

where c_1 is the sound speed of the gas between the deflagration and the precursor shock. It is more interesting to express the precursor shock Mach number, M_s, in terms of the heat release instead of the flow Mach number, M_1. Using Eqs. 2.70 and 2.71, we obtain

$$\left(\frac{1 - M_1^2}{M_1} \right)^2 = \left(\frac{\rho_1}{\rho_0} \right)^2 \left(\frac{M_1}{M_s} \right)^2 \left(\frac{M_s^2 - 1}{M_s} \right)^2. \tag{2.77}$$

From the conservation of mass across the shock (Eq. 2.66), we get

$$M_s = \frac{D_s}{c_0} = \left(\frac{\rho_1}{\rho_0} \right) \left(\frac{D_s - u_1}{c_1} \right) \left(\frac{c_1}{c_0} \right) = \left(\frac{\rho_1}{\rho_0} \right) M_1 \left(\frac{c_1}{c_0} \right),$$

or

$$\left(\frac{c_0}{c_1} \right)^2 = \left(\frac{\rho_1}{\rho_0} \right)^2 \left(\frac{M_1}{M_s} \right)^2.$$

Using this result in Eq. 2.77, we get

$$\left(\frac{1 - M_1^2}{M_1}\right)^2 = \left(\frac{c_0}{c_1}\right)^2 \left(\frac{M_s^2 - 1}{M_s}\right)^2.$$

Combining this expression with Eq. 2.76, we obtain

$$\left(\frac{c_0}{c_1}\right)^2 \left(\frac{M_s^2 - 1}{M_s}\right)^2 = \frac{\gamma^2 - 1}{2} \frac{q}{c_1^2},$$

or

$$\left(\frac{M_s^2 - 1}{M_s}\right)^2 = \frac{\gamma^2 - 1}{2} \frac{q}{c_0^2}. \tag{2.78}$$

For a CJ detonation in the same mixture, Eq. 2.46 shows us that

$$M_{CJ}^2 = 2\left(\gamma^2 - 1\right) \frac{q}{c_0^2},$$

which means that

$$\left(\frac{M_s^2 - 1}{M_s}\right)^2 = \frac{M_{CJ}^2}{4}.$$

Finally, neglecting unity compared to M_s^2, we get

$$\frac{M_s}{M_{CJ}} \approx \frac{1}{2}, \tag{2.79}$$

which means that the Mach number of the precursor shock for a CJ deflagration is approximately one-half the CJ detonation Mach number in the same explosive mixture.

The velocity of a CJ deflagration corresponds to the maximum possible deflagration speed. There are indications that prior to the transition from deflagration to detonation, the deflagration has accelerated to its maximum possible speed. Since there is no solution to permit a continuous transition from the lower branch of the Hugoniot curve to the upper detonation branch, the DDT process is abrupt (i.e., discontinuous).

2.10. CLOSING REMARKS

The gasdynamic theory of detonation is based solely on the conservation laws (the Rankine–Hugoniot relation) that link the upstream and downstream equilibrium states across the wave. The physics of the transition process does not have to be considered in the gasdynamic theory. Although the Rankine–Hugoniot relation are obtained by integrating the differential forms of the conservation equations, only the limits of integration (upstream and downstream states) are involved if the variations within the transition zone are not required. The gasdynamic theory does give the possible solutions even though the particular solution of interest is not indicated. The CJ criterion, however, provides the extra information for choosing the desired solution, namely, the minimum-velocity solution for tangency of the Rayleigh line to the equilibrium Hugoniot curve.

The CJ criterion is a postulate and cannot be derived from the conservation laws. However, by considering the properties of the possible solutions, that is, strong and week detonations, physical arguments can be put forward to justify the choice of the CJ solution. The good agreement between the CJ velocity and experimental values is fortuitous.

Although the strong detonation solution can be eliminated by means of stability considerations for self-propagating detonations (not supported at the back by piston motion), the weak detonation solution cannot be ruled out by entropy arguments. Von Neumann devised an example of intersecting intermediate Hugoniot curves of partially reacted states that permit weak detonations to be realized without violating the entropy argument. By considering intermediate Hugoniot curves, von Neumann essentially involved the detonation structure in the selection of the solution of the Rankine–Hugoniot equations across the front. However, if the structure were to be considered, one can simply iterate for the correct detonation velocity that gives a regular solution when the sonic singularity is encountered . In this case, the CJ criterion is no longer required. It should also be noted that the boundary condition at the front must also be compatible with the solution for the nonsteady flow of the detonation products behind the front. Thus, the rear boundary condition influences the selection of the detonation velocity (e.g., for overdriven detonations). The Rankine–Hugoniot equations within the CJ criterion that focuses only on the detonation front itself are incomplete. The flow within the structure as well as the flow in the products must be considered for a complete theory. Furthermore, the unstable transient three-dimensional cellular structure of detonations in almost all practical explosive mixtures renders the steady one-dimensional CJ theory invalid in principle. However, the CJ theory remains an excellent approximate theory for the prediction of the detonation velocity, even for highly unstable near-limit detonations.

Bibliography

Becker, R. 1917. *Z. Electrochem.* 23:40–49, 93–95, 304–309. See also 1922. *Z. Tech. Phys.* 152–159, 249–256; 1922. *Z. Phys.* 8:321–362.

Chapman, D.L. 1889. *Phil. Mag.* 47:90–104.

Crussard, J.C. 1907. *C. R. Acad. Sci. Paris* 144:417–420.

Dionne, J.P., R. Duquette, A. Yoshinaka, and J.H.S. Lee. 2000. *Combust. Sci. Technol.* 158:5.

Döring, W. 1943. *Ann. Phys. 5e Folge* 43:421–436.

Duffey, G.H. 1955. *J. Chem. Phys.* 23:401.

Hugoniot, H. 1887–1889. *J. Ecole Polytech.* Cahiers 57, 58.

Jouguet, E. 1904. *Co. R. Acad. Sci. Paris* 140:1211.

Kistiakowsky, G., and E.B. Wilson. 1941. The hydrodynamic theory of detonation and shock waves. O.S.R.D. Rept. 114.

McBride, B.J., and S. Gordon. 1996. Computer program for calculation of complex chemical equilibrium compositions and applications II. User's manual and program description. NASA Rept. NASA RP-1311-P2.

Mikelson, V.A. 1890. On the normal ignition velocity of explosive gaseous mixtures. Ph.D. dissertation, Moscow University.

Rankine, W.J. 1870. *Phil. Trans.* 160:277–288.

Reynolds, W.C. 1986. *The element potential method for chemical equilibrium analysis: Implementation on the interactive program STANJAN*, 3rd ed. Mech. Eng. Dept., Stanford University.

Scorah, R.L. 1935. *J. Chem. Phys.* 3:425.

Taylor, G.I. 1950. *Proc. R. Soc. Lond.* A 200:235.

von Neumann, J. 1942. Theory of detonation waves. O.S.R.D. Rept. 549.

Zeldovich, Ya. B. 1940. *Zh. Exp. Teor. Fiz.* 10(5):542–568. English translation, NACA TN No. 1261 (1950).

3 Dynamics of Detonation Products

3.1. INTRODUCTION

A solution to the steady conservation equations across a detonation wave alone does not guarantee that a steady CJ wave can be realized experimentally. In the previous chapter, physical arguments (e.g., stability and entropy considerations) were presented in an attempt to rule out certain solutions of the conservation equations. However, solutions to the conservation equations for detonations must also be compatible with the rear boundary conditions in the combustion products, as in the case of deflagrations. For strong detonations where the flow downstream is subsonic, disturbances from the rear can propagate upstream to influence the wave, and thus it is clear that the solution must satisfy the rear boundary condition. Even for a Chapman–Jouguet (CJ) detonation, where the CJ criterion permits the solution of the conservation equations to be determined independently of the downstream flow of the combustion products, the solution for the nonsteady expansion of the products must nevertheless satisfy the sonic condition at the CJ plane. However, this is not always the case. For example, for diverging cylindrical and spherical detonations, the sonic condition behind a CJ detonation results in a singularity, which leads to the question of whether or not diverging CJ detonations can exist. Thus, the existence of a solution to the steady conservation laws also depends on the compatibility of the solution with the dynamics of the combustion products behind the wave. In this chapter, we shall consider the dynamics of the combustion products behind a steadily propagating detonation, for the existence of a detonation wave requires that both the conservation equations across the detonation front and the equations that govern the nonsteady flow of the detonation products be satisfied simultaneously. Solutions for the dynamics of the combustion products behind detonation waves were first obtained by G.I. Taylor (1950) and Zeldovich (1942).

3.2. BASIC EQUATIONS

In the following analysis we shall assume a perfect gas with a constant specific heat ratio, γ. The one-dimensional gasdynamic equations for planar, cylindrical, and spherical geometry are given as

$$\frac{\partial \rho}{\partial t} + \rho \frac{\partial u}{\partial r} + u \frac{\partial \rho}{\partial r} + \frac{j\rho u}{r} = 0, \tag{3.1}$$

$$\frac{\partial u}{\partial t} + u \frac{\partial u}{\partial r} + \frac{1}{\rho} \frac{\partial p}{\partial r} = 0, \tag{3.2}$$

$$\left(\frac{\partial}{\partial t} + u \frac{\partial}{\partial r}\right) \frac{p}{\rho^\gamma} = 0, \tag{3.3}$$

where $j = 0, 1$, or 2 for planar, cylindrical, and spherical symmetry, respectively. We have also assumed particle isentropic flow. Restricting ourselves to the propagation of a steady CJ detonation wave at constant velocity D, the entropy increase across the steady detonation will be the same for every particle. Thus, we have isentropic flow throughout in the products (i.e., $p/\rho^\gamma = $ constant). For a perfect gas ($p = \rho RT$), we obtain the following isentropic relationship:

$$\frac{T}{p^{\frac{\gamma-1}{\gamma}}} = \text{constant}, \qquad \frac{T}{\rho^{\gamma-1}} = \text{constant}. \tag{3.4}$$

Since $c^2 = \gamma RT$, we can replace T with the sound speed and write

$$\frac{c}{p^{\frac{\gamma-1}{2\gamma}}} = \text{constant}, \qquad \frac{c}{\rho^{\frac{\gamma-1}{2}}} = \text{constant}, \tag{3.5}$$

and from Eq. 3.5 we obtain

$$\frac{dc}{c} = \frac{\gamma-1}{2\gamma} \frac{dp}{p} = \frac{\gamma-1}{2} \frac{d\rho}{\rho}.$$

For isentropic flow, we can replace p and ρ by the sound speed c, and the particle velocity u. There are now two dependent variables. Equations 3.1 and 3.2 can be written as

$$\frac{\partial c}{\partial t} + c \frac{\gamma-1}{2} \frac{\partial u}{\partial r} + u \frac{\partial c}{\partial r} + \frac{\gamma-1}{2} \frac{jcu}{r} = 0, \tag{3.6}$$

$$\frac{\partial u}{\partial t} + u \frac{\partial u}{\partial r} + c \frac{2}{\gamma-1} \frac{\partial c}{\partial r} = 0. \tag{3.7}$$

For a freely propagating CJ detonation wave, we are seeking a simple progressive wave solution for the product flow behind the detonation. In other words, if c and u are known at one point at one instant in time, then they will have the same value at a later time t at a distance $(u + c)t$ from the initial point. To seek a simple wave solution, we make the following nondimensional transformation of dependent and independent variables:

$$\xi = \frac{r}{Dt}, \qquad \phi(\xi) = \frac{u}{D}, \quad \text{and} \quad \eta(\xi) = \frac{c}{D}.$$

Equations 3.6 and 3.7 become

$$\frac{2}{\gamma - 1} (\phi - \xi) \eta' + \eta \phi' - \frac{j\eta\phi}{\xi} = 0, \tag{3.8}$$

$$(\phi - \xi) \phi' + \frac{2}{\gamma - 1} \eta\eta' = 0, \tag{3.9}$$

where the primes denote differentiation with respect to ξ. Solving for the derivatives ϕ' and η', we obtain

$$\phi' = \left(\frac{j\phi}{\xi}\right) \frac{-\eta^2}{(\phi - \xi)^2 - \eta^2}, \tag{3.10}$$

$$\eta' = -\left(\frac{\gamma - 1}{2}\right) \left(\frac{j\eta\phi}{\xi}\right) \frac{\phi - \xi}{(\phi - \xi)^2 - \eta^2}. \tag{3.11}$$

For the planar case where $j = 0$, either we have the trivial solutions where $\phi' = \eta' = 0$, or else ϕ' and η' are both nonzero but $(\phi - \eta)^2 - \eta^2 = 0$. For the trivial solution $\phi' = \eta' = 0$, we have $\phi(\xi) = $ constant and $\eta(\xi) = $ constant. This solution corresponds to uniform flow behind the planar CJ detonation, which can only be realized if a piston is moving behind the detonation at the same particle velocity as that behind the CJ detonation. For the second solution, where $(\phi - \eta)^2 - \eta^2 = 0$, we obtain $\phi - \eta = \pm\eta$. Dividing Eq. 3.10 by Eq. 3.11, we get

$$\frac{\phi'}{\eta'} = -\left(\frac{2}{\gamma - 1}\right) \frac{\eta}{\phi - \xi},$$

and substituting $\phi - \eta = \pm\eta$, we get

$$\frac{\phi'}{\eta'} = \mp\frac{2}{\gamma - 1},$$

which integrates to yield

$$\phi = \mp\frac{2}{\gamma - 1}\eta + \text{constant}.$$

Since $\eta = \pm(\phi - \eta)$, this equation can also be written as

$$\phi = \frac{2}{\gamma + 1}\xi + \text{constant},$$

and the solution for $\eta(\xi)$ can be obtained as

$$\eta(\xi) = \mp\frac{\gamma - 1}{\gamma + 1}\xi + \text{constant}.$$

The constants can be evaluated from the boundary conditions at the CJ front $\xi = 1$ where $\phi = \phi_1$ and $\eta = \eta_1$. Thus,

$$\phi(\xi) = \left(\frac{2}{\gamma + 1}\right)(\xi - 1) + \phi_1, \tag{3.12}$$

$$\eta(\xi) = \left(\frac{\gamma - 1}{\gamma + 1}\right)(\xi - 1) + \eta_1. \tag{3.13}$$

The appropriate signs for the solution are chosen for an expansion flow where, for $\xi \leq 1$, both the particle velocity and the sound speed (i.e., temperature) decrease behind the CJ detonation front. With a known sound speed (or temperature) distribution, the pressure and the density profiles can be readily obtained from the isentropic relationships (i.e., Eq. 3.5).

Since a detonation is a compression front, the particle velocity behind it moves in the same direction of propagation as the detonation wave. Thus, the expansion waves will accelerate the particles in the opposite direction. From Eq. 3.12, we can determine the value of $\xi(\phi = 0)$ when $\phi(\xi) = 0$:

$$\xi(\phi = 0) = 1 - \frac{\gamma + 1}{2}\phi_1.$$

For a strong CJ detonation (i.e., $M_{CJ} >> 1$) where the particle velocity can be approximated by

$$\phi_1 \approx \frac{1}{\gamma + 1},$$

we obtain $\xi(\phi = 0) = 0.5 = x_0/Dt$. In other words, the particle velocity decreases to zero at about half the distance that the detonation has propagated from the closed end of the tube. For $0 \leq \xi \leq \xi(\phi = 0)$ where $\phi = 0$, the products remain stationary and the other thermodynamic properties take on constant values. The CJ condition at the detonation front gives $u_1 + c_1 = D$ or $\phi_1 + \eta_1 = 1$. Hence, $\eta = \frac{\gamma}{\gamma + 1}$, and at $\xi = \xi(\phi = 0) = \frac{1}{2}$, the value of the sound speed is obtained as $\eta(\phi = 0) = 0.5$. Thus, the pressure in the stagnant region of the detonation products can be obtained as

$$\frac{p(\phi = 0)}{p_1} = \left(\frac{c(\phi = 0)}{c_1}\right)^{\frac{2\gamma}{\gamma - 1}} = \left(\frac{\eta(\phi = 0)}{\eta_1}\right)^{\frac{2\gamma}{\gamma - 1}}.$$

For $\eta(\phi = 0) = \frac{1}{2}$ and $\eta_1 = \frac{\gamma}{\gamma + 1}$, the value of the pressure in the stagnant region is $p(\phi = 0) \simeq 0.34\, p_1$ (assuming $\gamma = 1.4$) or about one-third the CJ detonation pressure.

If the detonation propagates from an open end of a tube at $x = 0$, the particle velocity will continue to decrease past zero for $\xi < \xi(\phi = 0)$ (i.e., the flow reverses and propagates away from the detonation toward the tube exit). At the open end of the tube, the value of the particle velocity can be obtained as

$$\phi(0) = \phi_1 - \frac{2}{\gamma + 1} = \frac{1}{\gamma + 1} - \frac{2}{\gamma + 1} = -\frac{1}{\gamma + 1},$$

which is the same value as that behind the CJ detonation front, but in the opposite direction. The sound speed at the open end of the tube, $\xi = 0$, can also be obtained as

$$\eta(0) = \eta_1 - \frac{\gamma - 1}{\gamma + 1} = \frac{\gamma}{\gamma + 1} - \frac{\gamma - 1}{\gamma + 1} = \frac{1}{\gamma + 1}.$$

The pressure at the open end is thus

$$\frac{p(0)}{p_1} = \left(\frac{\eta(0)}{\eta_1}\right)^{\frac{2\gamma}{\gamma-1}} = \left(\frac{1}{\gamma}\right)^{\frac{2\gamma}{\gamma-1}},$$

and for $\gamma = 1.4$, the pressure at the open end $x = 0$ is $p(0) \simeq 0.1p_1$, which is about 10% of the CJ detonation pressure at the front.

As mentioned previously, the trivial solution of $\phi' = 0$ and $\eta' = 0$ gives $\phi(\xi) =$ constant, $\eta(\xi) =$ constant, and evaluating the constants at the front, we have

$$\phi(\xi) = \phi_1 = \text{constant} \quad \text{and} \quad \eta(\xi) = \eta_1 = \text{constant}.$$

This solution requires a piston to follow the detonation, moving at the necessary speed to maintain a uniform state in the products. Thus, the piston velocity will be $\phi_p = \phi_1$, and the location of the piston will be at $\xi_p = \phi_p = \phi_1$. If we have a piston moving at a velocity slower than the value of ϕ_1 that is necessary to maintain a constant state in the detonation products, then there will be an expansion fan behind the detonation front to reduce ϕ_1 to a value compatible with the piston velocity. If the piston is stationary, there will be a full expansion, bringing $\phi(\xi)$ from its value at the detonation of ϕ_1 to zero, as in a closed-end tube. However, for a piston velocity greater than the CJ value of ϕ_1 at the detonation front, the flow is uniform from the piston to the detonation front, but the detonation will be overdriven in order to match the piston velocity.

3.3. DIVERGING CYLINDRICAL AND SPHERICAL CJ DETONATIONS

For the cylindrical and spherical geometries, the value of j does not equal zero. Equations 3.10 and 3.11 must now be solved for the isentropic expansion of the detonation products behind a cylindrical (or spherical) CJ detonation wave.

We shall assume that at time $t = 0$, the cylindrical (or spherical) detonation is initiated instantaneously at $r = 0$ and propagates thereafter at a constant velocity that corresponds to the CJ velocity of the explosive mixture. At the detonation front, $\xi = 1$, where $\phi = \phi_1$ and $\eta = \eta_1$, the CJ condition $\phi_1 + \eta_1 = 1$ must be satisfied. Thus, at the CJ detonation front, the denominators in Eqs. 3.10 and 3.11 are both $(\phi_1 - 1)^2 - \eta_1^2 = 0$, and both equations become singular. The derivatives ϕ' and η' are finite for the planar case, because the numerators vanish for $j = 0$. However, for diverging detonations the numerators do not vanish, and ϕ' and η' tend to infinity as $\xi \to 1$. Although the values of ϕ_1 and η_1 are finite at $\xi = 1$, their gradients are infinite. Therefore, Eqs. 3.10 and 3.11 cannot be integrated numerically starting at $\xi = 1$. To proceed with the integration, we must seek an analytical solution in the neighborhood of the singularity so that we can start the numerical integration at a small distance away from the front to avoid the singularity. We assume that the solution in the neighborhood of the front is given by the following series:

$$\phi(\xi) = a_0 + a_1 (1 - \xi)^\alpha + \cdots, \tag{3.14}$$

$$\eta(\xi) = b_0 + b_1 (1 - \xi)^\beta + \cdots, \tag{3.15}$$

and from the boundary condition at $\xi = 1$ we obtain $a_0 = \phi_1$ and $b_0 = \eta_1$. The coefficients a_1 and b_1, as well as the exponents α and β, need to be determined next. The exponents α and β must be positive for ϕ and η to be finite at the front where $\xi = 1$. From Eqs. 3.14 and 3.15, we obtain

$$\phi' = -a_1\alpha(1-\xi)^{\alpha-1} + \cdots, \tag{3.16}$$

$$\eta' = -b_1\beta(1-\xi)^{\beta-1} + \cdots. \tag{3.17}$$

Because $\phi' \to \infty$ and $\eta' \to \infty$ as $\xi \to 1$, we see that α and β must be less than unity.

Substituting the expressions for ϕ, η, ϕ', and η' as given by Eqs. 3.14 to 3.17 into the basic equations (i.e., Eqs. 3.10 and 3.11), we obtain

$$\left\{ 2\alpha a_1^2 b_0 (1-\xi)^{2\alpha-1} + 2a_1\alpha b_0 b_1 (1-\xi)^{\alpha+\beta-1} + \cdots \right\}$$
$$- j\left\{ a_0^2 b_0^2 + a_1 b_0^2 (1-\xi)^2 + \cdots \right\} = 0 \tag{3.18}$$

and

$$\left\{ 2\beta b_0 b_1^2 (1-\xi)^{2\beta-1} + 2\beta a_1 b_0 b_1 (1-\xi)^{\alpha+\beta-1} + \cdots \right\} - j\frac{\gamma-1}{2}a_0 b_0^2 + \cdots = 0. \tag{3.19}$$

To satisfy Eq. 3.18, we see that the exponents must be such that $2\alpha - 1 = 0$ or $\alpha + \beta - 1 = 0$. To satisfy Eq. 3.19, $2\beta - 1 = 0$ or $\alpha + \beta - 1 = 0$. Therefore, we can conclude that $\alpha = \beta = \frac{1}{2}$ will simultaneously satisfy all the conditions. Setting $\alpha = \beta = \frac{1}{2}$ in Eqs. 3.18 and 3.19, we can obtain the first-order coefficients of a_1 and b_1 of the expansion of Eqs. 3.14 and 3.15 as

$$a_1 = \pm\sqrt{\frac{2j\phi_1\eta_1}{\gamma+1}}, \tag{3.20}$$

$$b_1 = \pm\frac{\gamma-1}{2}\sqrt{\frac{2j\phi_1\eta_1}{\gamma+1}}. \tag{3.21}$$

The solutions of $\phi(\xi)$ and $\eta(\xi)$ in the neighborhood of the front can now be written as

$$\phi(\xi) = \phi_1 \pm \sqrt{\frac{2j\phi_1\eta_1}{\gamma+1}}\sqrt{1-\xi} + \cdots, \tag{3.22}$$

$$\eta(\xi) = \eta_1 \pm \left(\frac{\gamma-1}{2}\right)\sqrt{\frac{2j\phi_1\eta_1}{\gamma+1}}\sqrt{1-\xi} + \cdots. \tag{3.23}$$

The two $\pm$ signs in the expressions above indicate that there are two possible solutions for the isentropic flow behind a diverging cylindrical or spherical detonation. If we take the negative sign, then this will correspond to an isentropic expansion solution behind the CJ detonation. If, on the other hand, we take the positive sign, then that gives an isentropic compression solution where both ϕ and η increase further behind the CJ detonation front. The compression solution will require a constant-velocity piston moving behind the detonation, and the piston location is given by

$\phi_p = \xi_p$. Thus, we proceed with the numerical integration until $\phi = \xi$, which corresponds to the boundary condition at the piston face where the particle velocity equals the piston velocity. Note that for planar CJ detonation, the flow is uniform behind the front up to the piston face. For diverging detonation, the flow is nonuniform. An additional compression is required to compensate for the area divergence associated with the cylindrical or spherical geometry.

The singular nature of the solution behind diverging CJ detonations raises the question of the existence of steady diverging CJ waves. Physically speaking, the existence of an infinite expansion gradient immediately behind the detonation throws some doubt on the validity of the conservation laws themselves. The global conservation laws are derived from the integration of the differential equations of motion across the wave, which requires that the gradients of the fluid quantities be negligible upstream and downstream of the wave. However, the presence of an infinite expansion gradient behind a diverging CJ detonation certainly violates this negligible-gradient requirement for the steady conservation equations to be valid. This is the view advanced by both Jouguet (1917) and Courant and Friedrichs (1950). Acknowledging the difficulties associated with the presence of the infinite gradient behind a CJ detonation front, G.I. Taylor (1950) argued that the singularity does not prevent the motion described by the solution from being a good approximation of the real product flow. Taylor mentioned that the error is likely to be of the order of the ratio of the thickness of the reaction zone to the radius of the detonation front. The latter diminishes as the detonation expands.

Note that if we take curvature and a finite thickness of the detonation into consideration, the detonation may no longer be a CJ detonation. If the boundary condition behind the detonation is not governed by the CJ condition, then the singularity problem at the front does not occur. An alternative argument proposed by Lee (1965) is based on the fact that a rather large ignition energy must be used to initiate a spherical detonation wave in general. Thus, the detonation wave is bound to be strongly overdriven initially and decays as it expands. Because overdriven detonations do not result in a singularity at the front, the question of the existence of spherical detonations does not arise if the initiation energy is taken into account. However, we no longer have a constant-velocity CJ detonation throughout, and a CJ detonation can only be obtained asymptotically as $R \to \infty$.

3.4. PISTON MOTION BEHIND DIVERGING DETONATIONS

From Eqs. 3.22 and 3.23, we note that there are two possible solutions behind a diverging CJ detonation. The compression solution requires a piston to follow behind the diverging CJ detonation. The piston velocity must correspond to the value of the particle velocity at the surface of the piston, where $\phi = \xi$. However, if the piston velocity is less than this critical value (which gives a continuous isentropic compression from behind the detonation to the piston face), we have an expansion solution followed by a secondary shock wave located between the piston and the CJ detonation front. For the planar case, no such shock wave is necessary, and the expansion wave simply reduces the particle velocity to match the piston velocity. For the diverging geometry, a secondary shock is required to match the compression flow generated by the piston to the expansion flow behind the detonation due to the area divergence. It is of interest to investigate this problem and to determine the strength of the secondary shock required to match the flow for various piston velocities.

For a constant piston velocity behind a CJ detonation, we still have self-similar flow from behind the detonation to the secondary shock and from the secondary shock to the piston surface. The strength of the secondary shock wave is such that these two regions of self-similar flow can be matched:

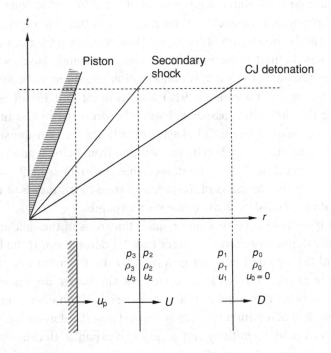

The self-similar equations of motion are as given previously by Eqs. 3.10 and 3.11, which are valid for the isentropic expansion flow bounded by the detonation front and the secondary shock wave, $\xi_s \leq \xi \leq 1$. Equations 3.10 and 3.11 are also valid for the isentropic compression flow between the secondary shock and the piston surface, $\xi_p \leq \xi_s \leq 1$. The conservation equations across the secondary shock can be written as

$$\rho_2(U - u_2) = \rho_3(U - u_3),$$

$$p_2 + \rho_2(U - u_2)^2 = p_3 + \rho_3(U - u_3)^2,$$

$$h_2 + \frac{(U - u_2)^2}{2} = h_3 + \frac{(U - u_3)^2}{2},$$

where $h = e + \frac{p}{\rho}$ is the enthalpy.

The conservation of mass gives

$$\frac{\rho_2}{\rho_3} = \frac{U - u_3}{U - u_2},$$ (3.24)

and the density ratio across a normal shock, as given by the Rankine–Hugoniot relations, is

$$\frac{\rho_2}{\rho_3} = \frac{(\gamma - 1)M_s^2 + 2}{(\gamma + 1)M_s^2},$$ (3.25)

where

$$M_s = \frac{U - u_2}{c_2}.$$ (3.26)

In terms of the dimensionless variables used in the similarity equations, we can write

$$M_s = \frac{\xi_s - \phi_2}{\eta_2},$$ (3.27)

where $\xi_s = r_s/Dt = Ut/Dt = U/D$ is the shock location, $\phi_2 = u_2/D$ is the particle velocity, and $\eta_2 = c_2/D$ is the sound speed just in front of the secondary shock wave. Equations 3.24 and 3.25 give

$$\frac{\rho_2}{\rho_3} = \frac{U - u_3}{U - u_2} = \frac{\xi_s - \phi_3}{\xi_s - \phi_2} = \frac{(\gamma - 1)M_s^2 + 2}{(\gamma + 1)M_s}.$$ (3.28)

At the detonation front, $\xi = 1$, the particle velocity and sound speed are given by the Rankine–Hugoniot equations across the CJ detonation front as

$$\phi_1 = \frac{1 - \frac{1}{M_{CJ}^2}}{\gamma + 1} \quad \text{and} \quad \eta = \frac{\gamma + \frac{1}{M_{CJ}^2}}{\gamma + 1}.$$ (3.29)

To obtain a solution with a piston behind a diverging CJ detonation front, we integrate Eqs. 3.10 and 3.11 numerically for the isentropic expansion flow behind the detonation (using the CJ boundary conditions given by Eq. 3.29 and the perturbation solutions given by Eqs. 3.22 and 3.23 near the front, to avoid the singularity). Assuming a certain shock location, ξ_s, we find the particle velocity ϕ_2 and sound speed in front of the shock, η_2, from the numerical integration. Equation 3.27 gives the shock Mach number M_s, and Eq. 3.28 gives the density ratio. The particle velocity and sound speed as a function of the Mach number for a normal shock in an ideal gas are given by the following Rankine–Hugoniot equations:

$$\phi_3 = \frac{2}{\gamma + 1}\left(1 - \frac{1}{M_s^2}\right),$$ (3.30)

$$\eta_3 = \sqrt{\frac{2}{(\gamma + 1)^2}\left(\gamma - 1 + \frac{2}{M_s^2}\right)\left(1 - \frac{\gamma - 1}{2\gamma M_s^2}\right)}.$$ (3.31)

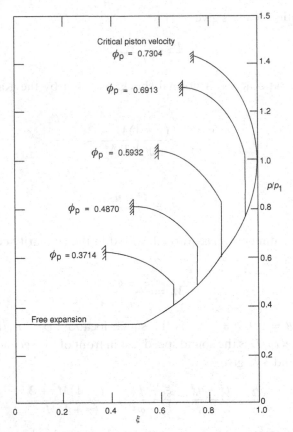

Figure 3.1. Pressure distribution behind a cylindrical detonation for various piston velocities.

With ϕ_3 and η_3 determined, we continue the integration until the piston surface, where the boundary condition $\phi_p = \xi_p$ has to be satisfied. The pressure distribution is illustrated in Fig. 3.1.

We see that for a given piston velocity, the products first expand isentropically from behind the CJ detonation front (at $\xi = 1$) to the secondary shock (at $\xi = \xi_s$). Behind the secondary shock, the flow then switches to an isentropic expansion until the piston surface, where $\phi = \xi$. Note that as the piston velocity increases, the secondary shock moves toward the detonation front, while its strength decreases. At a critical piston velocity, ϕ_p^*, the secondary shock merges with the CJ front, and its strength decreases to that of an isentropic compression wave. The flow behind the detonation is now isentropic throughout from the front to the piston surface. In the other limit as the piston velocity decreases to zero (i.e., $\phi_p \to 0$), the location of the secondary shock tends toward $\xi_s = \xi_0$, where $\phi = 0$. Again, its strength decreases to that of an isentropic wave as the secondary shock approaches $\xi_s \to \xi_0$. Thus, the strength of the secondary shock vanishes at the two limits ($0 \le \phi \le \phi_1$) and reaches a maximum for $\xi_0 \le \xi_s \le 1$.

For the planar case, we can demonstrate readily that the secondary shock wave is a sound wave (i.e., there is no secondary shock) since the isentropic expansion

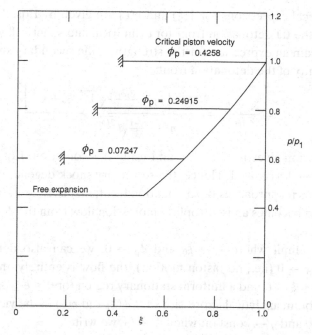

Figure 3.2. Pressure distribution behind a planar detonation for various piston velocities.

behind a planar detonation is given by analytical expressions

$$\phi(\xi) = \frac{2}{\gamma + 1} (\xi - 1) + \phi_1,$$

$$\eta(\xi) = \frac{\gamma - 1}{\gamma + 1} (\xi - 1) + \eta_1.$$

The strength of the secondary shock, given by

$$M_s = \frac{U - u_2}{c_2} = \frac{\xi_s - \phi_2}{\eta_2},$$

can be written as

$$M_s = \frac{\xi_s - \left(\frac{2}{\gamma+1}(\xi_s - 1) + \phi_1\right)}{\frac{\gamma-1}{\gamma+1}(\xi_s - 1) + \eta_1} \tag{3.32}$$

by using the analytical expansion solution behind the planar detonation. As a result of the CJ condition of $u_1 + c_1 = D$ or $\phi_1 + \eta_1 = 1$, Eq. 3.32 gives $M_s = 1$. Thus, for the planar geometry, the expansion flow behind the detonation simply terminates at a value of ξ, which gives a particle velocity of ϕ that matches the piston velocity (see Fig. 3.2). Without the area divergence between the piston and secondary shock wave, there is simply a uniform region from the secondary shock (which is now an isentropic compression wave, $M_s = 1$) to the piston surface. When $\phi_p \to 0$, the location of the secondary shock $\xi_s \to \xi_0 \approx \frac{1}{2}$.

Since analytical expressions for $\phi(\xi)$ and $\eta(\xi)$ are given by Eqs. 3.22 and 3.23 in the vicinity of the CJ detonation front for cylindrical and spherical geometries, it is possible to obtain an expression for the strength of the secondary shock, M_s, when it is in the vicinity of the detonation front:

$$M_s = \frac{\xi_s - \phi_2}{\eta_2} = \frac{\xi_s - \left\{\phi_1 - \frac{2j\phi_1\eta_1}{\gamma+1}\sqrt{1-\xi_s} + \cdots\right\}}{\eta_1 - \frac{\gamma-1}{2}\sqrt{\frac{2j\phi_1\eta_1}{\gamma+1}}\sqrt{1-\xi_s}}. \tag{3.33}$$

The CJ criterion gives $\eta_1 = 1 - \phi_1$, and using this condition, Eq. 3.33 yields the result that $M_s \to 1$ as $\xi_s \to 1$. Hence, the secondary shock degenerates into a compression wave as it approaches the detonation front; that is, $\xi_s \to 1$. In that case, the entire flow field becomes an isentropic compression flow from the detonation to the piston surface, ϕ_p^*.

In the other limit where $\xi_s \to \xi_0$ and $\phi_p \to 0$, we can also demonstrate that $M_s \to 1$. As $\phi_p \to 0$ (i.e., no piston motion), the flow is entirely an isentropic expansion for $\xi_0 \le \xi \le 1$, and a uniform stationary region for $0 \le \xi \le \xi_0$, where $\phi = 0$. We can also obtain analytical expressions for $\phi(\xi)$ and $\eta(\xi)$ in the vicinity of $\eta = \eta_0$. Because $\phi \to 0$ and $\eta \to$ constant when $\xi \to \xi_0$, we write

$$\phi(\xi) = \phi^{(1)}(\xi - \xi_0) + \cdots,$$

$$\eta(\xi) = \eta^{(0)} + \eta^{(1)}(\xi - \xi_0) + \cdots \tag{3.34}$$

in the vicinity of $\xi = \xi_0$. The coefficients of the expansion in $\xi - \xi_0$ of Eq. 3.34 can be obtained by substituting Eq. 3.34 into the similarity equations 3.10 and 3.11. For our purpose, it suffices to determine the coefficient η_0 only, and we obtain $\eta(0) = \xi_0$. Substituting Eq. 3.34 into Eq. 3.27 for M_s, we obtain

$$M_s = \frac{\xi_s - \phi^{(1)}(\xi_s - \xi_0) + \cdots}{\xi_0 + \eta^{(1)}(\xi_s - \xi_0) + \cdots}.$$

As $\xi_s \to \xi_0$, we see again that $M_s \to 1$. Thus, the strength of the secondary shock vanishes at the two limits of compression or expansion throughout the nonsteady flow region behind the CJ detonation front. The variation in the strength of the secondary shock with its location is illustrated in Fig. 3.3 for a cylindrical detonation.

The flow behind a steady CJ cylindrical or spherical detonation can be isentropic compression or an expansion within the limits of piston velocities $0 \le \phi_p \le \phi_p^*$. For piston velocities between the two limits, a secondary shock wave must be present, dividing the flow into expansion and compression regions. The CJ boundary condition at the detonation front generates a singularity, which throws doubt on the existence of steady diverging CJ detonations. However, the assumptions of (1) an instantaneous CJ detonation originating from the center of symmetry and (2) the detonation front being of vanishing thickness, so that curvature has no influence on the detonation state behind it, are not realistic. Acceptance of these assumptions is required to obtain the self-similar solution for the dynamics of the combustion products behind diverging waves. The consequence of having a singularity behind

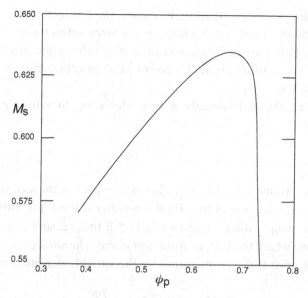

Figure 3.3. Variation of the secondary shock velocity with piston velocity for a cylindrical detonation.

the front does not warrant serious consideration of the physical implications of this singularity. As G.I. Taylor pointed out, the self-similar solution could serve as a good approximation for the actual flow field, thus alleviating the need to debate the physical problems arising from this singularity.

3.5. DIVERGING DETONATIONS IN A NONUNIFORM MEDIUM

In Section 3.3, we investigated the dynamics of the combustion products behind diverging (cylindrical or spherical) CJ detonation waves and found that the solution is singular at the CJ front. This singularity, in the form of an infinite gradient behind the detonation, raises the question of whether steady diverging CJ detonations exist or not. From a mathematical point of view, the singularity is not acceptable, because it renders the integrated form of the conservation laws across the detonation wave invalid. However, if the detonation front is sufficiently thin so that it can be treated as a discontinuity, then one may argue (as G.I. Taylor did) that the expansion singularity may be of the same order of magnitude as the compression discontinuity that one assumes for the detonation front. Hence, the solution might still be a good representation of the motion in reality. It is to be noted that the singularity is a consequence of the vanishing denominator in the equations of motion (Eqs. 3.10 and 3.11) at the CJ detonation front, whereas the numerator does not vanish for diverging detonations, because the value of the geometrical factor j is nonzero ($j = 1, 2$ for cylindrical and spherical geometries, respectively). However, if we consider an initially nonuniform density medium, an extra term appears in the numerator of the similarity equation due to the initial density gradient field. This provides an extra degree of freedom to enable the numerator to vanish at the CJ front, and it may be possible to seek singularity-free solutions for the propagation of diverging CJ detonations. These solutions may offer some physical insight into the singular behavior of diverging CJ waves in a uniform density medium, as analyzed in Section 3.3.

We shall assume the initial density of the explosive medium to be given by a power law of the form

$$\rho(r) = A r^\omega. \tag{3.35}$$

Note that if we assume a perfect gas, that is $p = \rho R T$, a density variation implies either that the pressure varies in a similar manner if the temperature is kept uniform or that the temperature varies as $T \propto 1/r^\omega$ if the pressure is uniform. We shall consider the case where the temperature is uniform. The similarity equations for the conservation of mass and of momentum for the variable density medium become

$$\frac{2}{\gamma - 1} (\phi - \xi) \eta' + \eta \phi' - \frac{j \eta \phi}{\xi} = 0, \tag{3.36}$$

$$\frac{2}{\gamma - 1} \left(1 + \frac{\omega}{\gamma(j+1)}\right) \eta \eta' + \left[(\phi - \xi) + \frac{\eta^2 \omega}{\gamma(j+1)(\phi - \xi)}\right] \phi'$$

$$+ \frac{\eta^2 \omega}{\gamma \xi} \left(1 - \frac{\phi}{(j+1)(\phi - \xi)}\right) = 0. \tag{3.37}$$

Solving for the derivatives ϕ' and η', we obtain

$$\phi' = \frac{\eta^2 \left(j \frac{\phi}{\xi} + \frac{\omega}{\gamma}\right)}{(\phi - \xi)^2 - \eta^2}, \tag{3.38}$$

$$\eta' = \frac{-\frac{\gamma - 1}{2} \eta \left(j \frac{\phi(\phi - \xi)}{\xi} + \frac{\omega}{\gamma} \frac{\eta^2}{\phi - \xi}\right)}{(\phi - \xi)^2 - \eta^2}. \tag{3.39}$$

When $\omega = 0$ and $\rho_0 = A = $ constant, Eqs. 3.38 and 3.39 reduce to Eqs. 3.10 and 3.11 for the uniform-density case analyzed previously in Section 3.3.

The value of ω cannot be arbitrary. If we consider the mass integral [i.e., conserving the total mass enclosed by the detonation at any instant when its radius is $R_s(t)$], we write

$$\int_0^{R_s} k_j \rho(r) \, r^j \, dr = \int_0^{R_s} k_j \rho_0(r) \, r^j \, dr,$$

where $k_j = 2\pi$ and 4π for $j = 1$ and 2, respectively. With $\rho_0(r) = A r^\omega$, the above equation becomes

$$\int_0^{R_s} \rho(r) \, r^j \, dr = \frac{A R_s^{(j + \omega + 1)}}{j + \omega + 1}. \tag{3.40}$$

From the above equation, we see that if the mass enclosed by the detonation has to be finite, then $j + \omega + 1 > 0$ and thus

$$\omega > -(j+1), \tag{3.41}$$

that is, $\omega \geq -2$ for cylindrical and $\omega \geq -3$ for spherical geometries.

For CJ detonations, the boundary conditions for ϕ and η at the front are given by

$$\phi_1 = \frac{M_{CJ}^2 - 1}{M_{CJ}^2(\gamma + 1)},$$

$$\eta_1 = \frac{\gamma M_{CJ}^2 + 1}{M_{CJ}^2(\gamma + 1)},$$

where $M_{CJ}^2 = 2\left(\gamma^2 - 1\right) q/c_0^2$.

The CJ criterion also requires that $\phi_1 = \eta_1 = 1$, and thus the denominator of Eqs. 3.38 and 3.39 vanishes at the front ($\xi = 1$), resulting in a singularity. For arbitrary values of ω (that satisfy Eq. 3.41), the numerator may not vanish at the CJ front, and we again have a singularity as in the uniform-density case. To proceed with the integration of Eqs. 3.38 and 3.39, we again have to seek a perturbation solution near the CJ front ($\xi = 1$) to obtain the starting condition. Carrying out a similar analysis as in Section 3.3, we obtain the follows results:

$$\phi(\xi) = \phi_1 \pm \sqrt{\frac{2\eta_1\left(j\phi_1 + \frac{\omega}{\gamma}\right)}{\gamma + 1}} \sqrt{1 - \xi} \pm \cdots, \tag{3.42}$$

$$\eta(\xi) = \eta_1 \pm \frac{\gamma - 1}{2}\sqrt{\frac{2\eta_1\left(j\phi_1 + \frac{\omega}{\gamma}\right)}{\gamma + 1}} \sqrt{1 - \xi} \pm \cdots. \tag{3.43}$$

These equations reduce to Eqs. 3.22 and 3.23 when $\omega = 0$.

Using Eqs. 3.42 and 3.43, starting values for ϕ and η in the immediate vicinity of the CJ detonation front $\xi = 1$ can be obtained, and we can then integrate the equations numerically to obtain the solution for $\phi(\xi)$ and $\eta(\xi)$ for given values of ω. The two signs in Eqs. 3.42 and 3.43 correspond to the expansion and the compression solutions behind diverging CJ detonation waves. Note that in Eqs. 3.42 and 3.43, we must have $\omega \geq -\gamma j \phi_1$; otherwise, the coefficients become imaginary.

For a nonuniform-density medium, it is also possible to seek a singularity-free solution by requiring the numerators of Eqs. 3.38 and 3.39 to vanish at the CJ front. Examining Eqs. 3.38 and 3.39, we note that for the particular value

$$\omega = -j\gamma\phi_1, \tag{3.44}$$

the numerator also vanishes at the CJ front as the denominator goes to zero. To obtain the solution for this case, we must again seek a perturbation solution in the neighborhood of the front, because ϕ' and η' are now both indeterminate. Writing the solution near $\xi = 1$ as a series expansion

$$\phi(\xi) = \phi_1 + a_1 (1 - \xi) + \cdots, \tag{3.45}$$

$$\eta(\xi) = \eta_1 + b_1 (1 - \xi) + \cdots \tag{3.46}$$

and substituting these expressions into Eqs. 3.38 and 3.39, we obtain the following equations for the coefficients:

$$a_1 = \frac{j\eta_1 (a_1 + \phi_1)}{2(a_1 + b_1 + 1)}, \tag{3.47}$$

$$b_1 = -\frac{j(\gamma - 1)}{4} \left[\frac{2\phi_1(a_1 + b_1 + 1) - \eta_1(a_1 + \phi_1)}{a_1 + b_1 + 1} \right]. \tag{3.48}$$

In obtaining these expressions, use has been made of the CJ condition $\phi_1 + \eta_1 = 1$ and also of Eq. 3.44 for the particular value of ω that renders the numerators of Eqs. 3.38 and 3.39 equal to zero. Using Eq. 3.47, Eq. 3.48 can be written as

$$b_1 = -\frac{\gamma - 1}{2} (j\phi_1 - a_1),$$

and substituting that expression for b_1 into Eq. 3.48 yields a quadratic equation for the coefficient a_1:

$$(\gamma + 1) a_1^2 + [2 - j (\eta_1 - \phi_1 + \gamma\phi_1)] a_1 - j\phi_1\eta_1 = 0. \tag{3.49}$$

This equation gives two roots for a_1 and hence two corresponding values for b_1. The two roots again denote expansion and compression solutions behind the CJ front. With the coefficients of the perturbation expressions determined, values for $\phi(\xi)$ and $\eta(\xi)$ in the vicinity of the front can now be obtained. Using these as starting values, Eqs. 3.38 and 3.39 can now be integrated numerically to provide a singularity-free solution for diverging CJ waves in a nonuniform-density medium.

It is interesting to note that there exists a very simple solution to Eqs. 3.38 and 3.39 of the form

$$\phi(\xi) = \phi_1\xi, \tag{3.50}$$

$$\eta(\xi) = \eta_1\xi. \tag{3.51}$$

To find the particular value of ω that corresponds to this solution, we substitute Eqs. 3.50 and 3.51 into Eq. 3.38 to obtain

$$\phi_1 = \frac{\eta_1^2 \left(j\phi_1 + \frac{\omega}{\gamma}\right)}{(\phi_1 - 1)^2 - \eta_1^2}. \tag{3.52}$$

Solving for η_1^2 in this expression yields

$$\eta_1^2 = \frac{\phi_1 (\phi_1 - 1)^2}{(j + 1)\phi_1 + \frac{\omega}{\gamma}}. \tag{3.53}$$

A second expression for η_1^2 can be obtained by substituting Eqs. 3.50 and 3.51 into Eq. 3.39 to yield

$$1 = \frac{-\left(\frac{\gamma-1}{2}\right)\left(j\phi_1(\phi_1 - 1) + \frac{\eta_1^2 \omega}{\gamma(\phi_1-1)}\right)}{(\phi_1 - 1)^2 - \eta_1^2}. \tag{3.54}$$

Dividing Eq. 3.52 by Eq. 3.54 gives

$$\phi_1 = \frac{\eta_1^2\left(j\phi_1 + \frac{\omega}{\gamma}\right)}{-\left(\frac{\gamma-1}{2}\right)\left(j\phi_1(\phi_1 - 1) + \frac{\eta_1^2 \omega}{\gamma(\phi_1-1)}\right)},$$

which can be solved for η_1^2:

$$\eta_1^2 = \frac{-j\phi_1^2\left(\frac{\gamma-1}{2}\right)(\phi_1 - 1)^2}{j\phi_1(\phi_1 - 1) - \frac{\omega}{\gamma}}. \tag{3.55}$$

Equation 3.55 can then be equated to Eq. 3.53 to eliminate η_1^2:

$$\phi_1 = \frac{2}{2 + (\gamma - 1)(j + 1)}. \tag{3.56}$$

The values for ϕ and η at the front are given by Eqs. 3.56 and 3.53, respectively, when the solutions for $\phi(\xi)$ and $\eta(\xi)$ are of the form of Eqs. 3.50 and 3.51. However, these values for ϕ_1 and η_1 at the detonation front must also satisfy the Rankine–Hugoniot relationships for a detonation wave:

$$\phi_1 = \frac{1 + S}{\gamma + 1}, \tag{3.57}$$

$$\eta_1^2 = \frac{\gamma(1 + S)(\gamma - S)}{(\gamma + 1)^2}, \tag{3.58}$$

where

$$S = \sqrt{1 - 2(\gamma^2 - 1)\eta\frac{q}{c_0^2}} \tag{3.59}$$

and $\eta = 1/M_s^2$. In the above equations, we have assumed that $\eta << 1$, and hence we neglect η as compared to unity. However, $\eta q/c_0^2$ is of the same order, and we cannot neglect $\eta q/c_0^2$ as compared to unity. For CJ detonations where $S = 0$, we get

$$M_{CJ}^2 = 2(\gamma^2 - 1)\frac{q}{c_0^2}$$

by solving Eq. 3.59.

Note that we have also taken the appropriate sign in the Rankine–Hugoniot equations to correspond to an overdriven detonation. We ignore the sign for the weak detonation solution. Furthermore, in the simple solutions given by Eqs. 3.50 and 3.51, the overdriven detonation front propagates at a constant velocity. This is a consequence of the initial density gradient ahead of the detonation. A decreasing density ahead of the detonation front tends to amplify the detonation and compete with the attenuating effect of area divergence. The constant-velocity wave corresponds to the balance between these two competing effects.

To solve for the particular value of ω, we now equate the values for ϕ_1 and η_1 from the conservation equations to those at the front given by the Rankine–Hugoniot equations. Using Eq. 3.57 to eliminate S from Eq. 3.58, we obtain

$$\eta_1^2 = \gamma\phi_1(1 - \phi_1). \tag{3.60}$$

Replacing η_1^2 in Eq. 3.53 with Eq. 3.60 and solving for ω, we get

$$\omega = (1 - \phi_1) - \gamma\phi_1(j + 1).$$

Substituting Eq. 3.56 for ϕ_1 into this equation gives

$$\omega = -\frac{(\gamma + 1)(j + 1)}{2 + (\gamma - 1)(j + 1)}. \tag{3.61}$$

Hence, for diverging spherical waves,

$$\omega = -\frac{3(\gamma + 1)}{3\gamma - 1}, \tag{3.62}$$

and for cylindrical waves,

$$\omega = -\frac{\gamma + 1}{\gamma}. \tag{3.63}$$

To find the detonation velocity, we substitute Eq. 3.56 for ϕ_1 into Eq. 3.57 and obtain the value for S as

$$S = \frac{2 - \gamma(j - 1)}{2 + (\gamma - 1)(j + 1)}, \tag{3.64}$$

which is particularly simple for the cylindrical geometry where $j = 1$:

$$S = \frac{1}{\gamma}.$$

The detonation Mach number can be found by equating the expressions for S in Eqs. 3.64 and 3.59.

Thus, we see that if the initial density of the explosive medium is nonuniform, we can obtain solutions that are not singular at the CJ detonation front by adjusting the value of the density exponent, ω. We can even have a constant-velocity overdriven detonation when the amplification effect of the density gradient just balances the attenuation effect due to area divergence.

It is important to note that although the classical CJ theory does not require consideration of the solution for the nonsteady flow of the products in its wake, a more rigorous theory requires the solution for the products to be considered as well. Only when the detonation velocity gives a singularity-free solution in the flow behind the detonation can we accept the solution as physically possible.

From mathematical considerations, we shall accept that constant-velocity cylindrical or spherical CJ detonations do not exist, due to the presence of the singularity. From a physical point of view, the assumption of zero initiation energy is perhaps a more serious issue. Instantaneous direct initiation of cylindrical or spherical detonation requires a substantial amount of ignition energy to be deposited at the center of symmetry. Taking the initiation energy into consideration, the initial propagation of the detonation is, in fact, a decaying strong blast wave. It is therefore of importance to examine the asymptotic decay of this reacting blast wave and to observe its approach to a CJ detonation far from the center of symmetry. The study of this asymptotic decay should elucidate whether the self-similar solution presented here does indeed provide a good approximate description of diverging detonations, as suggested by G.I. Taylor.

3.6. CLOSING REMARKS

The CJ theory permits the detonation velocity to be determined without a consideration of the rear boundary conditions (except for the strong detonation solution, where the flow is subsonic). For CJ detonations, however, their existence requires that a solution for the nonsteady flow of the products match the steady boundary conditions behind the CJ detonation. For the planar case, the Riemann solution satisfies the sonic condition at the rear boundary of the detonation, and the solution is continuous. For diverging (cylindrical or spherical) detonations, it is found that matching the solution of the nonsteady expansion of the products to the sonic condition behind the CJ detonation results in a singularity (i.e., there is an infinite expansion gradient). This is, in principle, an unacceptable situation, and thus steady diverging CJ detonations are considered not to exist. Diverging strong and weak detonations, on the other hand, do not give rise to the singularity, but are ruled out due to other considerations. The requirement of a regular solution at the sonic singularity becomes in fact a more rigorous mathematical criterion for the choice of the correct solution from the conservation laws. For the pathological detonation of von Neumann and for nonideal detonations where friction, curvature, and heat transfer are considered, it is this criterion of a regular solution when the sonic singularity is reached that leads to the determination of the desired solution. This is referred to as the *generalized CJ criterion*.

For normal planar CJ detonations, the use of the generalized CJ criterion leads to the same result, for when the sonic condition is reached, the numerator of the integral equation also vanishes (i.e., becomes zero), indicating that chemical equilibrium has been obtained as required by CJ theory. Thus, the study of the dynamics of the detonation products, and the compatibility of the solution for the products with the boundary conditions at the detonation front, led to the generalized CJ criterion. This more rigorous criterion from a mathematical standpoint extends to cases where the planar CJ theory breaks down (i.e., for pathological detonations and nonideal detonations).

Bibliography

Courant, R., and K.O. Friedrichs. 1948. *Supersonic flow and shock waves*, Interscience Publ.

Jouguet, E. 1917. *La mécanique des explosifs*, 359–366. Paris: Ed. Doin.

Lee, J.H.S. 1965. *Proc. Combust. Inst.* 10:805.

Taylor, G.I. 1940. Report to the Ministry of Home Security. Also 1950 *Proc. R. Soc. Lond. A* 200:235.

Zeldovich, Ya. B., 1942. *Zh. Eksp. Teor. Fiz.* 12:389–406.

4 Laminar Structure of Detonations

4.1. INTRODUCTION

In the gasdynamic analysis of detonation waves discussed in Chapter 2, the conservation equations are based on the upstream and downstream equilibrium states only. The detailed transition through the structure of the detonation was not considered. For a given initial upstream state, the locus of possible downstream states is given by the Hugoniot curve. Although the transition from the initial to the final state follows the Rayleigh line, the intermediate states along the Rayleigh line need not be considered if only the relationships between upstream and downstream are desired. It suffices to have the final state correspond to the intersection of the Rayleigh line with the Hugoniot curve.

Because the gasdynamic theory of detonation is concerned with the relationships between the upstream and downstream equilibrium states only, shocks, detonations, and deflagrations can all be analyzed using the conservation equations across the front. The conservation equations do not require the mechanism for the transition across the wave to be specified. However, to describe this transition zone, a model for the structure of the detonation wave must be defined. The model specifies the physical and chemical processes that are responsible for transforming the initial state to the final state. Most of the early pioneers of detonation research proposed the mechanism responsible for the propagation of the detonation implicitly in their discussion of the phenomenon. Mallard and Le Châtelier (1881) stated that a detonation propagates as a sudden adiabatic compression wave that initiates the chemical reactions and that its propagation velocity should be comparable to the sound speed of the products. Berthelot (1882) compared the detonation velocity with the mean molecular velocity of the product gases (and thus the sound speed) based on the temperature of an adiabatic isobaric combustion process. Dixon (1893) pointed out that the temperature of the detonation products is likely to be closer to an adiabatic isochoric combustion process than to an isobaric process. Vieille (1899) described the detonation as a discontinuity supported by chemical reactions induced by this discontinuity and computed a pressure of 40 bar (for a mixture of $2H_2 + O_2$), which corresponds closely to the von Neumann pressure spike behind the leading shock. Becker (1917) also recognized the role of shock heating in the initiation of the chemical reactions in a detonation wave. However, he rejected the view of Le Châtelier that chemical reactions were brought to the ignition temperature by adiabatic compression alone in the shock front. He postulated that heat conduction from the reaction zone should also be responsible for initiating the rapid chemical reaction in the shock-heated gases. However, in contrast with a shock wave, heat conduction and viscous effects play a minor role (if any) in the detonation structure, as the reaction zone is located much farther downstream and is separated from the shock by a relatively long induction zone. The chemical reactions must therefore be initiated by the adiabatic heating of the leading shock front.

The model for the detonation structure is formally credited to Zeldovich (1940), von Neumann (1942), and Döring (1943) (and is generally referred to as the ZND model), but their ideas essentially evolved from those of earlier investigators. The ZND model is shown in the following sketch:

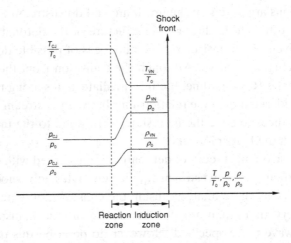

It formally states that the detonation structure consists of a leading shock front that adiabatically compresses and heats the reactants to the ignition temperature. This is followed by an induction zone where active radical species are generated by thermal dissociation of the shock-heated molecules. When sufficient concentrations of the active radical species are produced, rapid chain-branching reactions convert the reactants to products. In general, the induction zone is almost thermally neutral, and thus the thermodynamic state of the shock-heated mixture remains relatively constant through the induction zone. The rapid chemical energy release in the reaction zone results in a further rise in temperature and a corresponding drop in pressure and density. For an unsupported detonation wave, the pressure decreases further in the expansion fan that trails behind the reaction zone. It is this expansion that accelerates the gases backwards away from the front, which produces the forward thrust that supports the propagation of the leading shock. Thus, in the ZND model, both the ignition and the driving mechanisms for the detonation wave are specified.

In the original work of Zeldovich (1940), he was interested in the effect of heat and momentum losses (i.e., friction) in the reaction zone on the propagation of the detonation. He studied the influence of the tube wall on the detonation velocity, and also the existence of a detonability limit when the tube diameter is decreased beyond a certain minimum size. In von Neumann's investigation (1942), he investigated the possibility of weak detonations, and this led him to consider the intermediate states corresponding to the different degrees of completion of the chemical reactions within the structure. He assumed the existence of partially reacted Hugoniot curves that represent the locus of states corresponding to a particular degree of the completeness of the reaction. A consideration of the path through these partially reacted states then led von Neumann to establish the condition that permits the weak detonation solution to be realized. Von Neumann, however, did not pursue the analysis further by integrating the conservation equations that describe the variation of the thermodynamic states within the structure. Döring, a student of Becker, continued Becker's earlier work on detonations and carried out a more detailed analysis of the detonation structure. He integrated the conservation equations and obtained the profiles for the thermodynamic state across the detonation zone.

It should be noted that the steady one-dimensional planar structure of a detonation is seldom realized experimentally, because detonations are intrinsically unstable for almost all practical explosive mixtures. However, the ZND structure still serves as an important model where the detailed chemical kinetics of the explosive reactions can be studied under the gasdynamic conditions that correspond to detonation processes. The analysis of the laminar structure of a ZND detonation also provides a characteristic chemical length scale for the detonation process, which can be correlated with real detonation parameters.

4.2. THE ZND STRUCTURE FOR AN IDEAL GAS

We shall first analyze the ZND structure for the simplest case of a perfect gas with constant γ. The chemical reaction will be described by a single-step Arrhenius rate law. Referring to the sketch for the flow across a detonation with respect

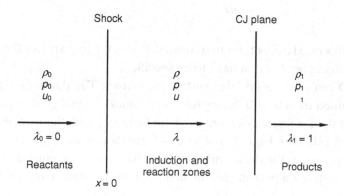

to a coordinate fixed to the detonation wave, the conservation equations for steady one-dimensional flow can be written as

$$\frac{d}{dx}(\rho u) = 0, \tag{4.1}$$

$$\frac{d}{dx}(p + \rho u^2) = 0, \tag{4.2}$$

$$\frac{d}{dx}\left(h + \frac{u^2}{2}\right) = 0, \tag{4.3}$$

where

$$h = \frac{\gamma}{\gamma - 1}\frac{p}{\rho} - \lambda Q, \tag{4.4}$$

in which Q is the chemical energy per unit mass, and $0 \leq \lambda \leq 1$ is the reaction progress variable. Differentiating Eqs. 4.3 and 4.4 and combining them gives

$$\frac{dh}{dx} + u\frac{du}{dx} = \frac{\gamma}{\gamma - 1}\left(\frac{1}{\rho}\frac{dp}{dx} - \frac{p}{\rho^2}\frac{d\rho}{dx}\right) - \frac{d\lambda}{dx}Q + u\frac{du}{dx} = 0. \tag{4.5}$$

From Eqs. 4.1 and 4.2, we obtain

$$\frac{d\rho}{dx} = -\frac{\rho}{u}\frac{du}{dx} \quad \text{and} \quad \frac{dp}{dx} = -\rho u\frac{du}{dx}.$$

Substituting the above expressions into Eq. 4.5 to eliminate $\frac{dp}{dx}$ and $\frac{d\rho}{dx}$, we obtain

$$\frac{du}{dx} = \frac{(\gamma - 1)\,uQ\frac{d\lambda}{dx}}{c^2(1 - M^2)}, \tag{4.6}$$

where we have used the expressions $c^2 = \gamma p/\rho$ and $M = u/c$. Since $dx = u\,dt$, we can also write Eq. 4.6 as

$$\frac{du}{dx} = \frac{(\gamma - 1)\,Q\frac{d\lambda}{dx}}{c^2(1 - M^2)}. \tag{4.7}$$

If we now specify the reaction rate law, then the above equation can be integrated numerically. A simple rate law that is widely used in theoretical and numerical detonation studies is the single-step Arrhenius expression,

$$\frac{d\lambda}{dt} = k(1 - \lambda)\,e^{-E_a/RT}. \tag{4.8}$$

The Arrhenius rate law contains two constants (viz., the preexponential factor k and the activation energy E_a) that have to be specified.

The ZND structure can be determined as follows: The detonation velocity can first be obtained from the CJ theory for given values of γ and Q. The von Neumann state behind the leading shock front can then be found from the Rankine–Hugoniot normal shock relations. Equations 4.7 and 4.8 can then be integrated simultaneously using the expressions for the sound speed as a function of pressure and density, the equation of state for a perfect gas, the definition of the Mach number that relates u

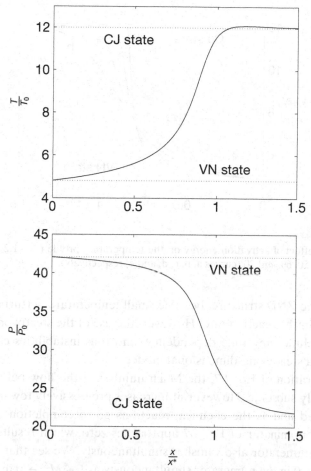

Figure 4.1. Typical temperature and pressure profiles for a ZND detonation ($\gamma = 1.2$, $E_a = 50RT_0$, and $Q = 50RT_0$), where x^* is the half-ZND-reaction zone length.

and c, and the conservation of mass and momentum, which links the density and the pressure with the particle velocity. The integration proceeds from the leading shock to the CJ plane, where the CJ state has already been determined from CJ theory. Typical temperature and pressure profiles are shown in Fig. 4.1.

For the case of a single-step Arrhenius rate law, the most important parameter that controls the ZND structure is the activation energy, which is a measure of the temperature sensitivity of the chemical reaction. For low activation energies, the reaction proceeds in a gradual manner behind the leading shock. For high activation energies, on the other hand, the reaction rate is very slow initially but accelerates rapidly when the temperature exceeds a value of the order of E_a/R. This gives rise to a long induction period followed by a relatively short reaction time where the reaction goes rapidly to completion. A comparison of the temperature profiles corresponding to different values of activation energy is shown in Fig. 4.2. High temperature sensitivity (i.e., large activation energy) also tends to give rise to

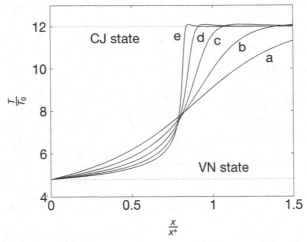

Figure 4.2. The effect of activation energy on the temperature profile ($\gamma = 1.2$ and $Q = 50RT_0$); $E_a/RT_0 = 30, 40, 50, 60$, and 70 (curves a, b, c, d, and e, respectively).

instability of the ZND structure, because small temperature perturbations result in large changes in the reaction rate. However, the use of the steady one-dimensional equations precludes any time dependency, and thus instabilities cannot manifest themselves in a steady one-dimensional model.

In the integration of Eq. 4.7, the Mach number of the flow behind the leading shock is initially subsonic. However, it increases progressively towards unity as energy is released and as the chemical reactions go to completion. As $\lambda \to 1$ and $M \to 1$, the denominator of Eq. 4.7 approaches zero, which results in a singularity unless the numerator also vanishes simultaneously. We see that this will be the case if $d\lambda/dt \to 0$ in the numerator simultaneously as $1 - M^2 \to 0$ in the denominator if the CJ velocity is used. In CJ theory, the sonic plane (i.e., the tangency point between the Rayleigh line and the equilibrium Hugoniot curve) corresponds to the chemical equilibrium plane (i.e., $\lambda = 1$, $d\lambda/dt = \phi$) where all chemical reactions are completed. Thus, the use of the detonation velocity as determined from CJ theory to compute the von Neumann state to start the numerical integration of Eq. 4.7 guarantees the approach to chemical equilibrium when the sonic plane $M = 1$ is reached.

Alternatively, we can also use the *regularity condition* (i.e., that the numerator vanishes as the denominator approaches zero) as a criterion to obtain the desired solution for the detonation velocity. In this case, we do not need to have prior knowledge of the detonation velocity. Instead, we first assume an arbitrary leading shock velocity to compute the von Neumann state and then proceed to integrate Eq. 4.7 for the ZND structure. As the integration proceeds and $M \to 1$, the denominator vanishes, but the numerator may not necessarily vanish simultaneously for the arbitrarily chosen value of the detonation velocity. Thus, we have to iterate for the correct detonation velocity that satisfies the regularity condition at the sonic singularity. In this manner, we can also determine the detonation velocity, which should also coincide with the value from the CJ theory if the explosive mixture does not have intersecting intermediate Hugoniot curves. The detonation state obtained in this manner is often referred to as an *eigenvalue detonation*.

In general, the detonation state determined from imposing the regularity condition should be identical to that from CJ theory if equilibrium is reached when $1 - M^2 \to 0$. If this is the case, eigenvalue detonations are identical to CJ detonations. However, there are kinetic mechanisms that can give values for the detonation velocity that differ from the equilibrium CJ velocity. Also, for cases where there are source terms in the conservation equations that take curvature, friction, and heat losses into account, the regularity condition is the only way to obtain a steady detonation state, because CJ theory breaks down when additional source terms are present in the one-dimensional steady conservation equations across the detonation front.

Equation 4.6 can also be transformed into the more familiar equation for one-dimensional steady compressible flow with heat addition where the variation of the flow Mach number is given as a function of the heat release (or increase in the stagnation enthalpy). We may write the heat release as

$$d\left(\lambda Q\right) = dq = dh_0 = c_p\, dT_0 = \frac{dc_0^2}{\gamma - 1},$$

where h_0, T_0, and c_0 denote the stagnation enthalpy, temperature, and sound speed, respectively. To transform Eq. 4.6 into an equation for the flow Mach number variation, we use the following expressions. From the definition of the Mach number, $M = u/c$, we get

$$\frac{dM}{M} = \frac{du}{u} - \frac{dc}{c}.$$

Similarly, for the sound speed and the perfect-gas equation of state ($c^2 = \gamma RT$ and $p = \rho RT$, respectively), we obtain

$$2\frac{dc}{c} = \frac{dT}{T} \quad \text{and} \quad \frac{dp}{p} = \frac{d\rho}{\rho} + \frac{dT}{T}.$$

From the continuity and momentum equations, we get

$$d(\rho u) = 0, \qquad \frac{du}{u} = -\frac{d\rho}{\rho}$$

and

$$d(p + \rho u^2) = 0, \qquad \frac{dp}{p} = -\gamma M^2 \frac{du}{u}.$$

Using the preceding expressions, Eq. 4.6 can then be written as

$$\frac{dM}{M} = \frac{\left(1 + \gamma M^2\right)\left(1 + \frac{\gamma - 1}{2}M^2\right)}{2\left(1 - M^2\right)} \frac{dT_0}{T_0}, \tag{4.9}$$

which gives the Mach number variation with the stagnation temperature as heat is added to the flow. This equation can be integrated to yield

$$\frac{T_0}{T_0^*} = \frac{2(\gamma+1)M^2\left(1+\frac{\gamma-1}{2}M^2\right)}{(1+\gamma M^2)^2}, \tag{4.10}$$

where T_0^* corresponds to the stagnation temperature when $M = 1$. For a given initial Mach number, the maximum amount of heat that can be added to a flow occurs when $M \to 1$. Thus, the maximum heat addition can be written as

$$q_{max} = h_0^* - h_0 = c_p(T_0^* - T_0) = c_p T_0 \left(\frac{T_0^*}{T_0} - 1\right),$$

so that

$$\frac{q_{max}}{c_p T_0} = (\gamma-1)\frac{q_{max}}{c_0^2} = \frac{T_0^*}{T_0} - 1.$$

Using Eq. 4.10 for T_0^*/T_0, this expression becomes

$$2(\gamma^2-1)\frac{q_{max}}{c_0^2} = \frac{(M^2-1)^2}{M^2\left(1+\frac{\gamma-1}{2}M^2\right)}, \tag{4.11}$$

which gives the initial Mach number M for which a maximum amount of heat, q_{max}, can be added to just choke the flow (i.e., for $M \to 1$). In other words, if we specify q_{max}, then M corresponds to the initial flow Mach number for which the flow becomes sonic when q_{max} has been added to the flow. The flow Mach number is given by the slope of the Rayleigh line, and q_{max} determines the equilibrium Hugoniot curve. Equation 4.11 gives the CJ Mach number and hence the slope of the Rayleigh line. However, there are two possible ways to get from the initial state to the final state on the equilibrium Hugoniot curve. Either we can go directly from the initial state to the final state along the Rayleigh line, or we can first go along the Rayleigh line to the shock Hugoniot curve, and then proceed from the shocked state (i.e., the von Neumann state) back down the same Rayleigh line to the equilibrium Hugoniot curve as heat is released. Because thermodynamic states are path-independent, it would be of interest to demonstrate the equivalence of the two paths from the initial state to the final state on the equilibrium Hugoniot curve.

If we first go from the initial state to the von Neumann state, then the flow Mach number behind the normal shock, M_1, is given by

$$M_1^2 = \frac{2+(\gamma-1)M_s^2}{2\gamma M_s^2 - (\gamma-1)},$$

where M_s is the shock Mach number. If we substitute that expression into Eq. 4.11 by replacing M with M_1, we get

$$2(\gamma^2-1)\frac{q_{max}}{c_0^2} = \frac{(M_s^2-1)^2}{M_s^2\left(1+\frac{\gamma-1}{2}M_s^2\right)}, \tag{4.12}$$

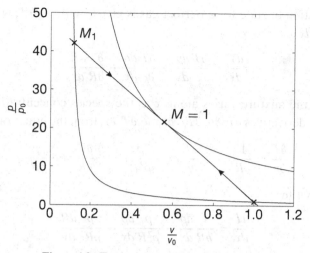

Figure 4.3. Equivalence of Eqs. 4.11 and 4.12.

which is identical to Eq. 4.11. In Eq. 4.11, M denotes the initial Mach number of the supersonic flow for which heat addition by the amount q_{max} will just choke the flow. However, for Eq. 4.12, M_s denotes the initial shock Mach number that yields a subsonic flow of M_1 behind it. The heat release q_{max} will then bring the subsonic flow to sonic conditions. The two results should be identical, for the paths for the heat addition are on the same Rayleigh line, as seen in Fig. 4.3.

Although the discussion in this section is based on a perfect gas with constant γ, generalization to a multicomponent perfect-gas mixture with detailed multistep kinetic reactions can be readily carried out. The conservation equations given by Eqs. 4.1 to 4.3 still apply, but the expression for the specific enthalpy per unit mass (Eq. 4.4) has to be modified. We can define h_i as the specific enthalpy of the ith species per unit mass, that is,

$$h_i = h_{f_i}^\circ + \int_{298\,\mathrm{K}}^{T} c_{p_i}\, dT,$$

where $h_{f_i}^\circ$ is the enthalpy of formation and c_{p_i} is the heat capacity of the ith species. The enthalpy per unit mass of the mixture is given by

$$h = \sum_i X_i h_i(T), \tag{4.13}$$

where X_i is the mass fraction of the ith species. Using Eq. 4.13, the energy equation can be written as

$$\frac{dh}{dx} + u\frac{du}{dx} = \sum_i \left(h_i \frac{dX_i}{dx} + X_i c_{p_i} \frac{dT}{dx} \right) + u\frac{du}{dx} = 0. \tag{4.14}$$

Since the equation of state for a perfect gas is given by $p = \rho RT$, we can write the derivative dT/dx as

$$\frac{dT}{dx} = \frac{\partial T}{\partial p}\frac{dp}{dx} + \frac{\partial T}{\partial \rho}\frac{d\rho}{dx} + \frac{\partial T}{\partial R}\frac{dR}{dx},$$

because R for the mixture varies along x as the species concentration X_i changes. Evaluating the derivatives $\partial T/\partial p$, $\partial T/\partial \rho$, and $\partial T/\partial R$ from the equation of state yields

$$\frac{\partial T}{\partial p} = \frac{1}{\rho R}, \qquad \frac{\partial T}{\partial \rho} = -\frac{p}{\rho^2 R}, \qquad \frac{\partial T}{\partial R} = -\frac{p}{\rho R^2},$$

and hence we write

$$\frac{dT}{dx} = \frac{1}{\rho R}\frac{dp}{dx} - \frac{p}{\rho^2 R}\frac{d\rho}{dx} - \frac{p}{\rho R^2}\frac{dR}{dx}.$$

Using the conservation of mass and momentum (Eqs. 4.1 and 4.2), the derivatives $d\rho/dx$, dp/dx can be obtained as

$$\frac{d\rho}{dx} = -\frac{\rho}{u}\frac{du}{dx} \quad \text{and} \quad \frac{dp}{dx} = -\rho u\frac{du}{dx}.$$

Thus, the derivative dT/dx becomes

$$\frac{dT}{dx} = -\frac{u}{R}\frac{du}{dx} + \frac{T}{u}\frac{du}{dx} - \frac{T}{R}\frac{dR}{dx}, \tag{4.15}$$

where we have replaced $p/\rho R$ by T using the equation of state.

The gas constant for the mixture can be written as $R = \sum_i X_i R_i$, where $R_i = \bar{R}/W_i$ is the gas constant of the ith species, $\bar{R}$ is the universal gas constant, and W_i is the molecular weight of the ith species. Thus,

$$\frac{dR(X_i)}{dx} = \sum_i \frac{\partial R}{\partial X_i}\frac{dX_i}{dt}\frac{1}{u}, \tag{4.16}$$

where we have replaced $dx = u\, dt$. Substituting Eqs. 4.15 and 4.16 into Eq. 4.14 yields

$$\frac{dh}{dx} = \sum_j X_j \frac{dh_j}{dT}\left(-\frac{u}{R}\frac{du}{dx} + \frac{T}{u}\frac{du}{dx} - \frac{T}{R}\sum_i \frac{\partial R}{\partial X_i}\frac{1}{u}\frac{dX_i}{dt}\right) + \sum_i \frac{h_i}{u}\frac{dX_i}{dt} + u\frac{du}{dx} = 0.$$

Solving for du/dx yields

$$\frac{du}{dx} = \frac{\frac{1}{u}\left\{\sum_i T\frac{\partial R}{\partial X_i} - \frac{\sum_i R h_i}{\sum_j X_j\frac{dh_j}{dt}}\right\}\frac{dX_i}{dt}}{\frac{R u}{\sum_j X_j\frac{dh_j}{dt}} + \frac{RT}{u} - u}. \tag{4.17}$$

The above can be simplified by noting that

$$\frac{dh_i}{dT} = c_{p_i}, \quad \sum_i X_i c_{p_i} = c_p = \frac{\gamma R}{\gamma - 1}, \quad c_f^2 = \frac{\gamma p}{\rho} = \gamma RT, \quad \text{and} \quad \gamma = \frac{c_p}{c_v}.$$

Eq. 4.17 then becomes

$$\frac{du}{dx} = \frac{(\gamma - 1)\left\{\sum_i \frac{\gamma T}{\gamma - 1}\frac{\partial R}{\partial X_i} - \sum_i h_i\right\}\frac{dX_i}{dt}}{c_f^2 - u^2},$$ (4.18)

where $c_f^2 = \gamma RT$ is defined as the *frozen* sound speed of the mixture. Equation 4.18 can be integrated simultaneously with kinetic rate equations for the various chemical species. We assume that at each position x inside the reaction zone, the variables are related by the integrated form of the conservation law, and the equation of state also applies locally.

Thus,

$$\rho u = \rho_0 u_0, \qquad \rho = \frac{\rho_0 u_0}{u},$$

$$p = p_0 + \rho_0 u_0 (u_0 - u),$$

$$T = \frac{p}{\rho R},$$

and $R = \sum X_i R_i$ with the local values of X_i determined from the kinetic rate equations. Note that the local sound speed is based on the local values of the species concentration and refers to the frozen sound speed in Eq. 4.18.

To start the integration, we may choose an arbitrary detonation (or leading shock) velocity and compute the von Neumann state behind the shock. We then proceed with the integration of Eq. 4.18 together with the kinetic rate equations. The correct solution is determined by iteration using the regularity requirement that when $M = u/c_f \to 1$, the numerator vanishes simultaneously with the denominator. The so-called eigenvalue detonation velocity obtained in this manner may or may not be identical to the CJ value based on chemical equilibrium conditions at the downstream plane. For example, if the reaction mechanism contains both exothermic and endothermic reactions with the faster exothermic reactions occurring first, followed by slower endothermic reactions in the approach to chemical equilibrium, the eigenvalue detonation velocity obtained will be higher than the equilibrium CJ velocity. Such detonations are referred to as pathological detonations and will be elaborated on further in the next section.

4.3. PATHOLOGICAL DETONATIONS

The possibility of steady detonations with velocities different than those predicted by the CJ theory was first pointed out by von Neumann (1942). He noted that the considerations leading to the CJ criterion were only based on the equilibrium Hugoniot curve (where chemical reactions have been completed) and that it is not possible to achieve a proper understanding of the CJ hypothesis from a consideration of that curve alone. The entire family of intermediate Hugoniot curves corresponding to the different degrees of reaction completeness must be considered. In so doing,

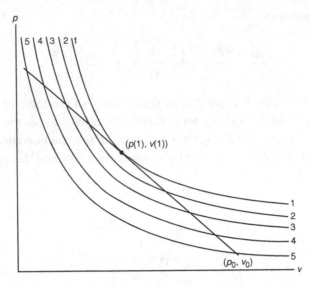

Figure 4.4. Family of intermediate Hugoniot curves (von Neumann, 1942).

von Neumann demonstrated the existence of pathological detonations for a certain morphology of partially reacted Hugoniot curves.

In von Neumann's analysis, the entire family of Hugoniot curves for the different values of λ (i.e., $0 \le \lambda \le 1$) is considered. Von Neumann first assumed that for each point within the reaction zone, there corresponds a local equilibrium state $(p(\lambda), v(\lambda))$ that lies on a Hugoniot curve for that particular value of λ. The Rayleigh line from the initial state must therefore intersect all partially reacted Hugoniot curves in sequence. The initial state is first brought to an elevated state via the leading normal shock, so that rapid explosive reactions can be initiated. Thus, we first have a normal shock transition from the initial state (p_0, v_0) to the von Neumann state $(p(0), v(0))$ on the same $\lambda = 0$ (or the shock) Hugoniot curve along the Rayleigh line. Then, as reactions proceed (i.e., for $\lambda > 0$), the state $(p(\lambda), v(\lambda))$ will follow the Rayleigh line, intersecting the family of intermediate Hugoniot curves. For a family of Hugoniot curves where no curve intersects another, the intermediate $(p(\lambda), v(\lambda))$ values go progressively along the Rayleigh line from $(p(0), v(0))$ on the unreacted shock Hugoniot curve to $(p(1), v(1))$ on the fully reacted equilibrium Hugoniot curve where $\lambda = 1$, as shown in Fig. 4.4. From previous arguments, $(p(1), v(1))$ must necessarily be the tangency point of the Rayleigh line with the equilibrium Hugoniot curve. The slope of the Rayleigh line, ϕ, cannot be less than $\phi^*(1)$ (the slope of the line tangent to the $\lambda = 1$ Hugoniot curve); otherwise the Rayleigh line would not intersect the equilibrium Hugoniot curve. That cannot be the case, because, once started, the chemical reactions must go to completion.

If the Hugoniot curves are not as shown in Fig. 4.4 but intersect one another, then it is possible for the Rayleigh line to be tangent at some intermediate value $\lambda = \lambda_0$ yet still intersect all the Hugoniot curves for $0 \le \lambda \le 1$. In this case, there must be an envelope that touches all of the Hugoniot curves and is tangent to them at the point of contact. Thus, the Rayleigh line that is tangent to the envelope at the

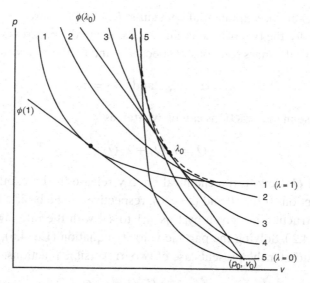

Figure 4.5. Hugoniot curves corresponding to different reaction completenesses in the case of a pathological detonation; the envelope of the Hugoniot curves is shown as a dotted line (von Neumann, 1942).

contact point $\lambda = \lambda_0$ must also be tangent to the Hugoniot curve corresponding to the value λ_0, as illustrated in Fig. 4.5. In the case shown, it is clear that $\phi(\lambda_0) > \phi(1)$. Hence, the detonation velocity is greater than the CJ velocity based on the completely reacted ($\lambda = 1$) Hugoniot curve. Because the Rayleigh line is tangent to the λ_0 Hugoniot curve, we have sonic conditions at the plane $\lambda = \lambda_0$ inside the reaction zone rather than at the end of the reaction zone $\lambda = 1$ on which the CJ hypothesis is based. Thus, von Neumann was able to prove that it is not sufficient to just consider the equilibrium Hugoniot curve alone and that there exist steady detonation velocities greater than the CJ velocity when the family of Hugoniot curves indexed by λ intersect. In this case, the steady detonation velocity is determined by the tangency of the Rayleigh line to the intermediate ($\lambda = \lambda_0$) Hugoniot curve, giving a detonation velocity larger than the CJ value based on tangency to the equilibrium ($\lambda = 1$) Hugoniot curve. In order to determine the detonation velocity for these pathological detonations, we must consider the details of the reaction zone structure and the reaction mechanisms that give rise to the family of intermediate Hugoniot curves.

To demonstrate in more detail the existence of pathological detonations, we consider a simple two-step chemical rate law first suggested by Fickett and Davis (1979). We assume an irreversible exothermic reaction A→B, followed by another irreversible but endothermic reaction B→C, so the overall reaction is from reactants A to products C. We also assume each reaction is governed by a simple Arrhenius reaction rate law of the form

$$\frac{d\lambda_1}{dt} = k_1 (1 - \lambda_1) e^{-E_{a,1}/RT}, \tag{4.19}$$

$$\frac{d\lambda_2}{dt} = k_2 (\lambda_1 - \lambda_2) e^{-E_{a,2}/RT}, \tag{4.20}$$

where k_1 and k_2 are preexponential constants, $E_{a,1}$ and $E_{a,2}$ are the respective activation energies for the two reactions, and λ_1 and λ_2 are the progress variables, which can be related to the mass fractions of species A and C. We can write

$$\lambda_1 = 1 - x_A \quad \text{and} \quad \lambda_2 = x_C.$$

The heat release in the reactions can be written as

$$Q = \lambda_1 Q_1 + \lambda_2 Q_2, \tag{4.21}$$

where Q_1 and Q_2 denote the chemical energy release in the exothermic and endothermic reactions ($Q_1 > 0$ and $Q_2 < 0$), respectively. The basic equations for the steady ZND structure are given by Eqs. 4.1 to 4.4 with the rate law now given by Eqs. 4.19 and 4.20. Solving the particle velocity equation (Eq. 4.6), we get the following expression for the present case of two irreversible reactions:

$$\frac{du}{dx} = \frac{(\gamma - 1)\left(\dot{\lambda}_1 Q_1 + \dot{\lambda}_2 Q_2\right)}{c^2 - u^2}. \tag{4.22}$$

To integrate this equation, we assume a shock velocity and compute the von Neumann state behind the normal shock using the Rankine–Hugoniot relationships. We then proceed to integrate Eq. 4.22 simultaneously with rate equations given by Eqs. 4.19 and 4.20 (with the various parameters, e.g., k_1, k_2, $E_{a,1}$, $E_{a,2}$, Q_1, Q_2, specified). As the integration progresses, we shall eventually arrive at the singularity $c^2 - u^2 = 0$, and to achieve a continuous (regular) solution we must require the numerator $\dot{\lambda}_1 Q_1 + \dot{\lambda}_2 Q_2 \to 0$ simultaneously as $u \to c$. This condition cannot be met for arbitrary values of the detonation velocity, and we must therefore iterate for the particular detonation velocity (i.e., the eigenvalue) that gives a continuous solution at the sonic plane, $u = c$, inside the reaction zone. Once we find this eigenvalue, the

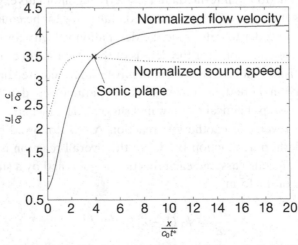

Figure 4.6. The particle velocity and sound speed for a pathological detonation.

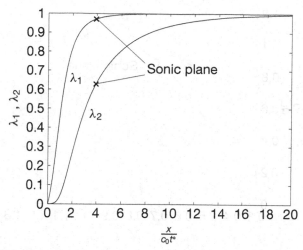

Figure 4.7. Variation in λ_1 and λ_2 for a pathological detonation.

integration can proceed past the sonic plane until complete equilibrium, when both $\lambda_1 = \lambda_2 = 1$ have been attained.

For the particular case of $k_1 = k_2 = 100$, $\gamma = 1.2$, $E_{a,1} = 22$, $E_{a,2} = 32$, $Q_1/c_0^2 = 50$, and $Q_2/c_0^2 = -10$, the variation of the particle velocity and sound speed behind the shock is shown in Fig. 4.6. The sonic plane when $u = c$ is indicated in the figure and occurs before equilibrium is reached. The corresponding variations of λ_1 and λ_2 behind the shock are illustrated in Fig. 4.7.

It can be seen that neither λ_1 nor λ_2 is at its equilibrium value (i.e., unity) at the sonic plane. An overshoot in the chemical energy release can be observed in Fig. 4.8. The overshoot results from the more rapid exothermic reaction, and thus sonic conditions occur at the peak heat release. The heat release decreases past the

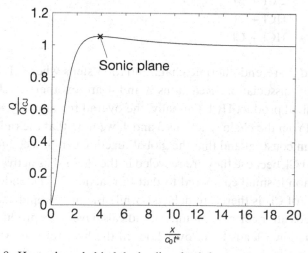

Figure 4.8. Heat release behind the leading shock for a pathological detonation.

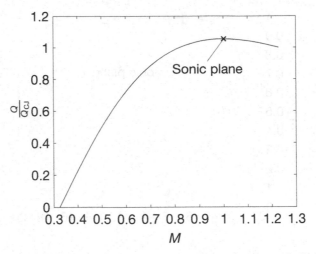

Figure 4.9. Heat release as a function of the local Mach number for a pathological detonation.

sonic plane as the slower endothermic reactions eventually bring the system to equilibrium, where $\lambda_1 = \lambda_2 = 1$. The variation of the local Mach number with the heat release is shown in Fig. 4.9. At the maximum energy release, the local Mach number is unity ($u = c$ at the sonic plane), and it continues to increase to a supersonic value of $M = 1.2302$, which corresponds to a weak detonation solution on the equilibrium Hugoniot curve.

The existence of pathological detonations in a H_2–Cl_2 system was first pointed out by Zeldovich and Ratner (1941), who noted that, according to the Nernst chain, two molecules of HCl can be produced from H_2 and Cl_2 without a change in the radical concentration of H or Cl under the following reaction scheme:

1. $H_2 + M \rightarrow 2\,H + M$
2. $Cl_2 + M \rightarrow 2\,Cl + M$
3. $H_2 + Cl \rightarrow HCl + H$
4. $H + Cl_2 \rightarrow HCl + Cl$

Reactions 1 and 2 are endothermic chain initiation steps where H and Cl radicals are produced by dissociation. Reactions 3 and 4 are exothermic chain reactions, producing the final product HCl. Globally, the overall reaction $H_2 + Cl_2 \rightarrow 2\,HCl$ is exothermic. From the chain reactions 3 and 4, we note that the radical pools of H and of Cl remain constant and that the global reaction can proceed for any concentration of H and Cl, because they are recycled in the chain. The activation energy for the chain reaction is small compared to that of reaction 2. The endothermic dissociation reaction of Cl_2 is therefore delayed, while the more rapid exothermic chain reaction proceeds after certain amounts of radicals have been produced. The highly exothermic reaction 3 leads to an overshoot in the heat release, which eventually equilibrates as the energy is used in the slower endothermic dissociation of Cl_2. With an overshoot in the heat release, the detonation velocity then corresponds to the value at the peak heat release, rather than to the equilibrium CJ value.

Using the detailed kinetic mechanisms of the H_2–Cl_2 reaction and the well-known rate constants for these reactions, pathological detonation velocities have been computed by Guénoche et al. (1981) and Dionne et al. (2000). Experimental verification of these results has also been obtained by Dionne et al. Therefore, von Neumann's argument that the entire family of Hugoniot curves must be considered (instead of just the equilibrium curve) has been verified by the demonstration of the existence of pathological detonations in H_2–Cl_2 systems. This demonstration also has fundamental significance in that it is necessary to consider the detailed detonation structure rather than just the global conservation laws that involve the two equilibrium states of the reactants and products as in the CJ theory. This is especially true when determining steady detonation velocities that may not be based on chemical equilibrium.

4.4. NONIDEAL DETONATIONS

It is clear that the CJ theory, which only considers solutions that lie on the equilibrium Hugoniot curve, is inadequate to predict even the steady detonation state in general (e.g., there are pathological detonations). Hence, as von Neumann had suggested, the structure must be considered even for determining the steady detonation speed. In analyzing the structure to determine the steady detonation state, we have to iterate for the particular detonation velocity that gives a continuous solution across the sonic singularity. If we analyze the structure, then we are no longer restricted by the assumption of no momentum or heat loss in the detonation zone as in CJ theory. We can also include the effect of curvature by including an area divergence term in the continuity equation. In contrast to *ideal* detonations, where we consider a strictly planar detonation without momentum and heat losses in the reaction zone, we shall define *nonideal* detonations as those with curvature, momentum, and heat losses. It is clear that classical CJ theory can no longer be used to predict the detonation state for nonideal detonations. We now have to consider the differential form of the conservation equations for the detonation structure and include the appropriate source terms to take account of curvature, momentum, and heat losses.

Although real detonations have a three-dimensional, transient, cellular structure with velocity and pressure fluctuations, the mean flow in the direction of propagation can still be considered as one-dimensional and steady. The three-dimensional turbulent fluctuations can then be modeled by source terms in the momentum and energy equations, as in turbulence modeling. Thus, the analysis of one-dimensional, steady, nonideal detonation is a generalized theory for describing the propagation of a steady detonation wave.

With respect to a fixed coordinate system, the conservation equations can be written as

$$\frac{\partial \rho}{\partial t'} + \frac{\partial \rho u'}{\partial x'} = m,$$

$$\frac{\partial (\rho u')}{\partial t'} + \frac{\partial (\rho u'^2 + p)}{\partial x'} = f,$$

$$\frac{\partial \rho e'}{\partial t'} + \frac{\partial u' (\rho e' + p)}{\partial x'} = q,$$

where $e' = \frac{p}{(\gamma-1)\rho} + \frac{1}{2}u' - \lambda Q$ and the primed quantities refer to the fixed coordinate system. The above equations can be transformed to a coordinate system attached to the detonation wave propagating at a velocity D, using

$$t = t', \quad x = x_D(t) - x',$$

$$u = D - u', \quad \frac{\partial}{\partial t'} = \frac{\partial}{\partial t} + D\frac{\partial}{\partial x},$$

$$\frac{\partial}{\partial x'} = -\frac{\partial}{\partial x},$$

where the unprimed quantities refer to the shock-fixed coordinates.

Using the above transformations, the set of conservation equations becomes

$$\frac{\partial \rho}{\partial t} + \frac{\partial (\rho u)}{\partial x} = m,$$

$$\frac{\partial (\rho u)}{\partial t} + \frac{\partial (\rho u^2 + p)}{\partial x} = Dm - f + \rho\frac{dD}{dt},$$

$$\frac{\partial (\rho e)}{\partial t} + \frac{\partial u (\rho e + p)}{\partial x} = \frac{1}{2}D^2 m - Df + q + \rho u\frac{dD}{dt}.$$

Setting all the time derivatives equal to zero for the steady flow across this steadily propagating wave at velocity D, we have

$$\frac{d (\rho u)}{dx} = m,$$

$$\frac{d (\rho u^2 + p)}{dx} = Dm - f,$$

$$\frac{du (\rho e + p)}{dx} = \frac{1}{2}D^2 m - Df + q,$$

and the source terms m, f, and q represent area divergence, momentum, and heat losses, respectively. In a quasi-one-dimensional flow, the source term m in the continuity equation represents the area divergence $A(x)$, and we can show that $m = \frac{-\rho u}{A}\frac{dA}{dx}$.

The area divergence term can also be used to allow for the curvature of the detonation front. In a curved wave front, the particle crossing the shock front is subjected to a lateral expansion. The lateral expansion can be modeled by an area change $A(x)$. For example, in a cylindrical front, $A(r) = 2\pi r$, and the source term becomes

$$m = \frac{-\rho u}{A}\frac{dA}{dr} = \frac{-\rho u}{r}.$$

For the spherical case,

$$m = \frac{-\rho u}{A}\frac{dA}{dr} = \frac{-2\rho u}{r},$$

in accord with the continuity equation for cylindrical and spherical geometries. Note that the area divergence source term also manifests itself in the momentum and energy equations as well, for the area divergence leads to momentum and energy losses from the x-direction of propagation.

The source term, f, in the momentum equation can be used to represent frictional losses as in viscous pipe flow. However, it should be noted that friction arises from the shear stress at the wall and is a two-dimensional mechanism. When it is represented by a force term f in the one-dimensional momentum equation, this force term becomes a body force term. As a result, there will be a term in the energy equation representing the rate of work done by f. However, in reality the friction force due to viscous shear does not do any work, since the particle velocity vanishes at the wall, where the shear force acts. Thus, if we model wall friction within the context of a one-dimensional flow, there should not be a term in the energy equation representing the rate of work done by it. We shall, however, include a rate-of-work term in the energy equation to preserve a general treatment of the detonation structure when additional source terms in the continuity, momentum, and energy equations are present. Heat loss to the wall can be modeled by some convective heat transfer expression, or, using the Reynolds analogy, the heat loss can be linked to the momentum losses.

We shall first leave the source terms unspecified and proceed with the analysis of the detonation structure. Following the previous analysis in Section 4.2, we may eliminate $d\rho/dx$ and dp/dx. With the help of the equation of state, we can derive an equation for the variation of the particle velocity u inside the reaction zone:

$$\frac{du}{dx} = \frac{(\gamma - 1)\left[\rho\lambda Q + q\right] + m\left[-\gamma u(D - u) + c^2\right] + f\left[\gamma u - D(\gamma - 1)\right]}{\rho(c^2 - u^2)},$$

where $\dot\lambda = d\lambda/dt$ and $dt = dx/u$. The speed of sound is given by $c^2 = \gamma p/\rho = \gamma RT$, where p and ρ are obtained from the integrated form of the conservation law. If we assume a shock velocity, the von Neumann state can be determined from the Rankine–Hugoniot equations, and the above ZND equation can then be integrated simultaneously with the specified reaction rate law [e.g., $\dot\lambda = k(1 - \lambda)\exp(-E/RT)$].

For an arbitrarily chosen shock velocity, the sonic singularity $c^2 - u^2 \to 0$ is reached without the numerator of the ZND equation vanishing simultaneously with the denominator. The criterion for a regular solution is to require the numerator to vanish as the sonic singularity is reached, that is,

$$(\gamma - 1)\left[\rho\lambda Q + q\right] + m\left[-\gamma u(D - u) + c^2\right] + f\left[\gamma u - D(\gamma - 1)\right] = 0.$$

This is referred to as the *generalized* CJ criterion. The usual CJ criterion is that when $c^2 - u^2 \to 0$, we have equilibrium conditions, that is, $\dot{\lambda} = 0$; thus, the numerator vanishes simultaneously with the denominator. When source terms m, f, and q are absent, we note that the ZND equation reduces to

$$\frac{du}{dx} = \frac{(\gamma - 1)\dot{\lambda}Q}{c^2 - u^2}.$$

The ZND equation can also be rewritten using the local Mach number, $M = u/c$, as the dependent variable, as in previous sections. The resulting expression is identical to the equation for steady quasi-one-dimensional flow with area change, heat transfer, and friction. All the source terms tend to drive the flow toward sonic conditions, and, with the various source terms operating simultaneously, it becomes a question of which source term dominates in bringing the flow to sonic conditions. Thus, complete chemical reaction is no longer assured, and the resulting eigenvalue detonation velocity can differ significantly from the CJ velocity, which is based on complete chemical energy release.

To understand the consequences of the competition between the various source terms, it is best to consider some examples so that some general features can be deduced. In essence, since we are considering the propagation of a detonation wave, a term for the heat release from chemical reactions must be present. The source term due to heat loss can be absorbed into the term for chemical energy release, and we can use a net energy release term to take account of heat transfer. Thus, we essentially need to consider only the competition between heat release and area divergence (or curvature), or between heat release and friction (i.e., momentum losses).

Consider first the effect of curvature on the propagation of detonation (without friction). In the presence of area divergence within the reaction zone, a competition exists between the chemical heat release rate, which accelerates the flow to sonic conditions at the CJ plane, and the diverging nozzle flow due to the lateral area divergence, which leads to a slowdown in the subsonic flow acceleration and exothermicity. The solution of the steady ZND equations for the structure with curvature yields a Z-shaped curve for the detonation velocity, as shown in Fig. 4.10. Note that the lowest branch of the Z-shaped curve is physically irrelevant and therefore not shown in the figure. The curve reveals the nonlinear dependence between the wave curvature and the detonation velocity. For increasing values of the wave curvature κ, the detonation has a higher velocity deficit. The curve also displays a turning point (κ^*, D^*), corresponding to the maximum permissible curvature beyond which a sonic plane does not exist. Beyond this limit, the physical interpretation is that any disturbance behind the detonation wave can catch up to it and quench it, hence not permitting the existence of a self-sustained detonation. For $\kappa < \kappa_{\text{cr}}$, however, the solution for the detonation velocity is multivalued, giving the Z-shape behavior. The general trend of decreasing detonation velocity with increasing curvature is in accord with experimental observations and also makes physical sense. However,

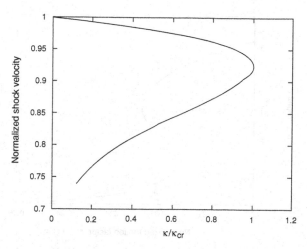

Figure 4.10. Detonation velocity versus curvature for steady curved detonation with $\gamma = 1.2$, $E_a/RT = 50$, $Q = 50$.

the existence of multiple solutions for one value of curvature may not be physically meaningful. Typically, the lower branch of the curve is unstable, so only the top branch has a physical meaning.

Similarly, we can consider only friction, with no curvature. Nonideal detonations arising from friction were first considered by Zeldovich in his pioneering study of detonation structure. The source term for friction suggested by Zeldovich is given by

$$f = k_f \rho u \left| u \right|,$$

where u is the particle velocity in the fixed reference frame and k_f is the friction factor. An example of k_f for a rough tube is given by Schlichting as

$$k_f = 2\left[2\log\left(\frac{R}{k_s}\right) + 1.74\right],$$

where k_s is the equivalent sand roughness and R is the tube radius. The absolute value of the velocity used above ensures that friction is always opposing the flow. For simplicity, the friction factor is often assumed to be a constant. The ZND equation with friction alone, that is, for $m = q = 0$, reduces to

$$\frac{du}{dx} = \frac{(\gamma - 1)\,\rho\dot{\lambda}Q + f\left[\gamma u - D(\gamma - 1)\right]}{\rho\left(c^2 - u^2\right)}.$$

Because the pressure now is no longer an algebraic function of the particle velocity when a source term is present in the momentum equations, it is necessary to integrate simultaneously the differential equation for the conservation of momentum together with the ZND equation and reaction rate law. As in the nonideal case of area divergence, the eigenvalue solution is again determined by the mathematical requirement of regularity at the sonic singularity.

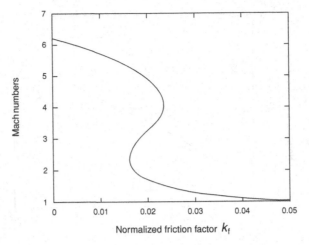

Figure 4.11. Detonation velocity versus friction factor for steady detonation with $E_a/RT = 32$.

An example of possible steady state solutions as a function of the friction factor k_f is shown in Fig. 4.11. The variation of the detonation velocity with the friction factor shows a similar Z-shaped curve to the case of front curvature. The decrease in detonation velocity with increasing friction makes physical sense, and it is in accord with experimental observation. A critical value of the friction factor exists beyond which no steady solution for the detonation structure is possible. As in the case of curvature, there also exist multiple solutions when the friction factor is less than the critical value.

With friction, Brailovsky and Sivashinsky (1997, 2000) and Dionne (2000) pointed out that there exists a critical value of the shock velocity, D_{cr}, at the lower branch, below which no solution satisfying the generalized CJ criterion can be found. For this minimum case, it is interesting to note that the state at the sonic plane corresponds to a relative particle velocity equal to the shock velocity, $u = D$. This implies that the absolute particle velocity becomes zero there, and that the shock velocity is equal to the sound speed of the products. When the flow is at rest, the friction term f and the numerator can go to zero only if the sonic plane coincides with the end of the reaction zone and the shock velocity is equal to the sound speed in the burned products. Note that an analytical solution can readily be obtained for D_{cr}, namely,

$$D_{cr} = \sqrt{(\gamma - 1)Q + 1}.$$

It is also of interest to note that the solution $D = D_{cr}$ represents a propagating constant volume explosion; that is, the sonic plane of the detonation front corresponds to a moving boundary of an expanding constant volume explosion. This can be seen from the fact that because the flow is at rest behind the detonation, the normalized specific volume is unity from the conservation of mass. As the mixture is stationary in front of and behind the detonation front, the detonation then becomes a propagating front where the explosive mixture undergoes a constant volume explosion as the detonation propagates.

For $D > D_{cr}$, the generalized CJ criterion can be used to determine the steady detonation solution, because the sonic plane exists within the reaction zone. The flow can smoothly undergo a transition from subsonic to supersonic velocity relative to the shock front and eventually satisfies the boundary conditions at the end wall. For the case $D = D_{cr}$, the regularity condition is automatically met, because the shock velocity coincides with the sonic velocity of the combustion products. However, if $D < D_{cr}$, the shock velocity is now below the sound speed of the products. Thus, the solution no longer encounters a sonic singularity because the flow is subsonic throughout (relative to the shock). Therefore, for $D < D_{cr}$, the generalized CJ criterion can no longer be used to determine an eigenvalue velocity for the detonation. To continue the solution beyond $D < D_{cr}$, an alternate criterion was proposed by Brailovsky and Sivashinsky and was also used by Dionne. Noting that for the case where $D = D_{cr}$ the flow is at rest when chemical equilibrium is reached, this condition can thus be used as a criterion to seek an eigenvalue detonation solution. The continuation of the Z-shape curve below $D < D_{cr}$ in Fig. 4.11 is based on this criterion. It is important to note that all these criteria to determine the steady solution for nonideal detonations are based on mathematical considerations, and their validity has to be established via comparisons with experiments.

For both steady state analyses of area divergence and friction, multivalued behavior appears to be possible for certain cases of nonideal detonations where source terms are present in the conservation laws for the detonation structure. However, there is no doubt that only one steady detonation solution (or none) is generally observed in an experiment. To resolve this dilemma, the nonsteady equations must be used. Starting with specified initial conditions that can initiate an overdriven detonation, the asymptotic decay of the detonation is then obtained by solving the nonsteady reactive Euler equations. If steady solutions exist, then the nonsteady solution should approach this steady state asymptotically. The use of the steady equations for the ZND structure presupposes the existence of steady state solutions. Thus, the multiple solutions indicated by the Z-shaped curve are a consequence of the assumption that steady solutions exist without taking into consideration the stability of these solutions. In other words, it is possible that none of the steady state solutions from the ZND equation can be realized when starting from a transient initial state.

4.5. CLOSING REMARKS

The CJ theory permits the detonation velocity to be determined without considering the propagation mechanism, which is a consequence of the CJ criterion that relates the minimum-velocity solution to the conservation laws. There is really no justification for this particular choice, although one could argue from stability considerations that the strong detonation solution can be eliminated because expansion waves can penetrate the reaction zone to attenuate the wave. However, the weak detonation solution is not as easily discarded on the basis of entropy arguments or of the inability to find an ignition mechanism if the path follows the Rayleigh line directly from the initial to the final state on the weak detonation branch of the Hugoniot curve. If the path went to a shocked state first to effect ignition and then returned along the Rayleigh line to the weak solution, a rarefaction shock would be encountered, and a transition involving an entropy decrease would result. Thus, on the basis of just the conservation laws and equilibrium thermodynamics, weak detonations cannot be excluded and thus the CJ theory is incomplete.

However, von Neumann presented an argument for the existence of steady weak detonations when the partially reacted Hugoniot curves intersect one another. Tangency of the Rayleigh line is now with the envelope of partially reacted Hugoniot curves instead of with the equilibrium Hugoniot curve of completed reactions. The Rayleigh line then intersects the equilibrium Hugoniot curve at the weak detonation branch, and thus a weak detonation is obtained. Tangency of the Rayleigh line to a partially reacted Hugoniot curve means that the sonic condition is reached before chemical equilibrium is reached. A singularity results when the sonic condition is reached, for the denominator of the ZND equations vanishes. As there is no reason to expect a discontinuous jump in the fluid properties, a regular solution is sought when the sonic condition is reached (i.e., the numerator and the denominator vanish simultaneously as $M \to 1$), which then provides a criterion for determining the detonation velocity. This is referred to as the "generalized CJ criterion" even though the mathematical requirement has nothing to do with the original postulate of Chapman and Jouguet. For mixtures where the intermediate Hugoniot curves do not intersect and the Rayleigh line is tangent to the equilibrium Hugoniot curve, the generalized CJ criterion can still be employed to obtain the detonation velocity. From the ZND equation (Eq. 4.7), we note that when $M \to 1$ and the denominator vanishes, the numerator also vanishes when chemical equilibrium is obtained and $d\lambda/dt \to 0$. Thus, the generalized CJ criterion can be used for all explosive mixtures, but the detonation velocity has to be determined by integrating the ZND equation for the structure and by iteration to satisfy the generalized CJ criterion.

Von Neumann's work implies that, in general, the structure must be considered to find the steady detonation solution, and the criterion for obtaining the desired solution is now a mathematical one. A complete theory for detonation must include a consideration of the detonation structure, which contains the chemical mechanism.

Bibliography

Becker, R. 1917. *Z. Electrochem.* 23:40–49, 93–95, 304–309. See also 1922. *Z. Tech. Phys.* 3:152–159, 249–256.

Berthelot, M. 1882. *C. R. Acad. Sci. Paris* 94:149–152.

Brailovsky, I., and G.I. Sivashinsky. 1997. *Combust. Sci. Technol.* 130:201–231.

Brailovsky, I., and G.I. Sivashinsky. 2000. *Combust. Flame* 122:130–138.

Dionne, J.P. 2000. Ph.D. thesis, McGill University, Montreal, Quebec.

Dionne, J.P., R. Duquette, A. Yoshinaka, and J.H.S. Lee. 2000. *Combust. Sci. Technol.* 158:5–14.

Dixon, H. 1893. *Phil. Trans. A* 184:97–188.

Döring, W. 1943. *Ann. Phys. 5e Folge* 43:421–436.

Fickett, W., and W.C. Davis. 1979. *Detonation.* University of California Press.

Guénoche, H., P. Le Diuzet, and C. Sedes. 1981. In *Dynamics of Explosions*, 387–407. New York: AIAA.

Mallard, E., and H. Le Châtelier. 1881. *C. R. Acad. Sci. Paris* 93:145–148.

Vieille, P. 1899. *C. R. Acad. Sci. Paris* 130:413–416.

von Neumann, J. 1942. Theory of detonation waves. O.S.R.D. Rept. 549.

Zeldovich, Ya. B. 1940. *Zh. Exp. Teor. Fiz.* 10(5):542–568. English translation, NACA TN No. 1261 (1950).

Zeldovich, Ya. B., and S.B. Ratner. 1941. *Zh. Exp. Teor. Fiz.* 11:170.

5   Unstable Detonations: Numerical Description

5.1. INTRODUCTION

Self-propagating one-dimensional ZND detonations are unstable and hence not observed experimentally in general. However, a solution for the laminar structure of a ZND detonation can always be obtained from the steady one-dimensional conservation equations, irrespective of the activation energy, which controls the temperature sensitivity and thus the stability of the detonation. The use of the steady one-dimensional conservation equations excludes any time-dependent multidimensional solution that describes the instability of the detonation wave. The classical method of investigating the stability of a steady solution is to impose small multidimensional perturbations on the solution and see if the amplitude of the perturbations grows. The assumption of small perturbations permits the perturbed equations to be linearized and integrated, and thus the unstable modes to be determined. As with most hydrodynamic stability analyses, the dispersion relation is rather involved and cannot be expressed analytically, which obscures the physical basis of the stability mechanism.

An alternative method is to start with the time-dependent, nonlinear, reactive Euler equations and then integrate them numerically for given initial conditions. Stability is indicated when a steady ZND solution is achieved asymptotically at large times. Unlike linear stability analyses, direct numerical simulations have the advantage that the full nonlinearity of the problem is retained. Furthermore, stability in one, two, and three dimensions can be separately investigated, which facilitates the interpretation of the numerical results. Current numerical techniques and modern computers can readily handle the integration of the multidimensional, time-dependent, reactive Euler equations.

However, direct numerical simulations have their own limitations in that near the stability limits, where the growth and decay rate of the perturbations are very slow, the approach to steady state can take an extremely long time. Problems that require very fine temporal and spatial resolutions are still outside the scope of direct numerical simulations. Nevertheless, a numerical description of unstable detonations can provide more detailed information on the flow field that cannot be obtained from experimental measurements.

In the description of the real unstable structure of detonations, it is best to start with observations from numerical experiments, with which we can progress from one-dimensional pulsating instabilities to two- and three-dimensional cellular instabilities. Although three-dimensional unstable detonations such as spinning detonations can also be described numerically, the difficulty in displaying and analyzing the numerical results renders such simulations of limited value at present. Until theoretical models for cellular detonations can be developed that allow the detailed numerical results from three-dimensional simulations to be analyzed effectively, one- and two-dimensional simulations suffice in providing good qualitative insight into the unstable structure of detonations.

In this chapter, a brief overview of linear stability analysis of the steady ZND structure and the different analyses under various asymptotic limits is first given, to provide a complete perspective of the unstable structure of detonations. We shall then present results of time-dependent numerical simulations of one-dimensional pulsating detonations, as well as two-dimensional cellular detonations.

5.2. LINEAR STABILITY ANALYSIS

The linear stability problem consists of superimposing three-dimensional time-dependent perturbations on the steady ZND solution to see if the perturbations grow or decay with time. Assuming that the perturbations are small, the basic equations can be linearized about the steady one-dimensional ZND solution. Integration of the linearized equations for the perturbations then permits us to determine the stability boundary and growth rate as a function of the wavenumber of the perturbations. The dependence of the stability limit on the relevant parameters that affect stability can also be obtained from such a linearized stability analysis. A hydrodynamic stability analysis of detonations was first formulated by Fay (1962), and an extensive study of this problem was undertaken by Erpenbeck (1962, 1964, 1967) using an initial value Laplace transform approach. An excellent summary of Erpenbeck's results is given in Fickett and Davis (1979). A formal normal-mode approach to the problem of linear detonation stability was given later by Lee and Stewart (1990), and an extensive body of literature on similar studies includes those of Namah *et al.* (1991), Bourlioux and Madja (1992), He and Lee (1995), Sharpe (1997), Short (1997a, b), Short and Stewart (1998), and Liang *et al.* (2004).

5.3. NORMAL-MODE LINEAR ANALYSIS

In the general normal-mode formulation, the equations governing small (linear) perturbations to the steady ZND wave are constructed through a transformation to a new spatial coordinate system $x = x^l - t - \Psi(y, t)$, $y^l = y$, where $x^l = t + \Psi(y^l, t)$

is the shock locus in laboratory frame, which now becomes $x = 0$. We seek a normal-mode decomposition,

$$\Psi = \Psi_0 \exp(\alpha t + iky), \qquad z = z^* + \Psi_0 z'(x) \exp(\alpha t + iky),$$

for the growth rate and frequency eigenvalue α and the wavenumber k. The parameter $z = (v, u_1, u_2, p, \lambda)$ represents the vector of dependent variables, the asterisk $*$ refers to the underlying steady wave solution, the primed quantities indicate the spatially (x) dependent eigenfunctions, and $\Psi_0 \ll 1$. The resulting linear stability is characterized by a dispersion equation, which can be solved to evaluate the spectrum for α and $z'(x)$. The result is a boundary value problem that is bounded by the linearized shock relations at $x = 0$ and an additional closure condition specified on the perturbation behavior. In most cases, this closure condition must come from a spatial boundedness requirement on the eigenfunctions. Such a condition is derived by suppressing any disturbance on the forward characteristics far downstream of the steady wave. In general, this analysis is somewhat technical, especially for Chapman–Jouguet (CJ) waves. A good detailed review of the generalized normal-mode stability formulation can be found in Gorchkov *et al.* (2007).

To determine the neutrally stable and unstable eigenvalues α, the dispersion equation is integrated from $x = 0$ (using the shock relations) into the region $x < 0$. The numerical strategy is based on a shooting method to determine an eigensolution that is satisfied at the equilibrium or sonic point for the detonation wave given by the closure condition just mentioned. An alternative reverse integration strategy has also been developed recently by Sharpe (1999). This approach has proven to be more efficient to determine the spectrum of α, especially for CJ detonation waves, by first seeking analytically a linearized perturbation solution at the CJ point as initial conditions for the integration, and then shooting back numerically to satisfy the condition at the shock front.

In the past, much of the analysis of the linear stability of detonation has been conducted for one-step, irreversible Arrhenius models. There are five main bifurcation parameters in the problem: the degree of the detonation overdrive, f; the activation energy, E_a; the heat release, Q; the ratio of specific heats, γ; and the reaction order. Variations in any of these quantities affect the domains between instability and stability of planar ZND detonations. These changes may be understood in the context of the variation in the underlying ZND profile that each change in a bifurcation parameter invokes. This was first discussed in detail by Short and Stewart (1998) and can be related to the change in the rate of heat release in the ZND profile; the more rapid the heat release, the greater the tendency to instability. Although there are exceptions to these general trends, increasing overdrive, decreasing activation energy, decreasing heat release, or increasing reaction order basically tends to stabilize the detonation.

Within the context of the one-step reaction, other studies have also considered the normal-mode linear stability of quasi-steady, weakly curved detonations (Watt & Sharpe, 2004) and spinning detonations (Kasimov & Stewart, 2002). In summary, the study of the linear stability of steady detonation waves via a normal mode analysis is well established. It has also been extended to calculate the neutral stability boundaries for more complex reaction mechanisms (Short & Dold, 1996; Sharpe, 1999; Liang *et al.*, 2004) and for an arbitrary equation of state (Gorchkov *et al.*, 2007).

With exact numerical solutions for the linear stability problem, we can find basic information like the frequency of the unstable modes of the system and the ranges of the parameters that render the system stable. It should be noted, however, that linear stability analyses are valid only for the initial growth of the perturbations and cannot describe results far from the stability limits or the final nonlinear unstable structure of the detonation. In general, the exact linearized equations for the perturbed quantities are quite complex and require numerical integration. Thus, it is rather difficult to obtain physical insight into the stability mechanisms. Furthermore, linear stability analyses do not reveal the gasdynamic mechanisms behind the generation of the instability.

5.4. ASYMPTOTIC MODELING OF UNSTABLE DETONATION

In the past two decades, progress has been made in detonation stability theory through an analytical asymptotic approach. Analytical theories, even based on limiting values of the characteristic parameters, can provide insight into the physical mechanisms that drive detonation instabilities. The early study of Zaidel (1961) was based on a square-wave model, which is essentially the asymptotic limit of the activation energy going to infinity. In later studies, Abouseif and Toong (1982, 1986) used a model with a large-activation-energy induction zone coupled with a finite-rate heat-release zone, which is similar to Zaidel's square-wave model. Numerous studies were later carried out using more formal asymptotic approaches to the detonation stability problem. Buckmaster and Ludford (1987) and Buckmaster and Neves (1988) investigated the linear stability of detonations in the limit of a large activation energy. Bourlioux *et al.* (1991) carried out a weakly nonlinear analysis about neutral stability points. Yao and Stewart (1996), Short (1996, 1997a, b) and Short and Stewart (1997) analyzed the dynamics of the instability mechanism for low- to moderate-frequency perturbations and small wave curvature. Clavin *et al.* (1997), and Short and Stewart (1999) studied the stability problem in the limit of high degree of overdrive and weak heat release. Many of these studies tend to be somewhat mathematical in nature: they involve complicated algebraic manipulation of the reactive Euler equations. The mathematical details tend to obscure the physics and defeat the objective of this book. Thus we will only briefly review these asymptotic analyses and their main findings.

As mentioned previously, the key characteristic parameters that govern the stability of the steady detonation structure include the activation energy, E_a; the ratio of specific heats, γ; the degree of overdrive, $f = (D/D_{CJ})^2$; and the chemical heat release, Q. An analytical asymptotic investigation of detonation stability involves taking a combination of these four parameters to an asymptotic limit. The analysis also requires an assumption concerning the characteristic disturbance frequency and wavelength. The result of the asymptotic expansion is a decomposition of the steady detonation structure.

5.5. HIGH ACTIVATION ENERGY AND THE NEWTONIAN LIMIT

The early analytical study by Buckmaster and Neves (1988) presented a formal asymptotic formulation of the full detonation stability problem in the limit of high activation energy. This limit reduces the detonation model to Zaidel's square-wave model for the detonation structure where conditions are nearly uniform in the induction zone. Their study was concerned with the evolution of one-dimensional disturbances with wavelengths of the order of the induction-zone length. They found that the stability spectrum can be obtained by an analysis confined to the behavior of perturbations within the induction zone only. The evolution equations for the disturbances have the form of the classical linearized acoustic equations with a singular forcing term to account for the presence of weak chemical activity within the induction zone. To obtain the instability spectrum in the limit of high activation energy, a compatibility relation is required to avoid the presence of a strong singularity in the perturbation variables. As a result, the temperature perturbation at the flame front is zero.

An interesting finding by Buckmaster and Neves (1988) is that the one-dimensional spectrum displays the same pathological behavior as Zaidel's nonasymptotic analysis of the square-wave detonation model (i.e., the spectrum consists of an infinite number of unstable oscillatory modes). Using a different asymptotic strategy, Short (1996) derived an analytical dispersion relation for the classical limit of a square-wave detonation, resulting from the approximations of both a large activation energy and a ratio of specific heats that approaches unity. The latter limit is generally referred to as the Newtonian limit. It was first introduced in the context of an analytical solution of Clarke's induction equation in a shock-induced ignition problem by Blythe and Crighton (1989). The asymptotic dispersion relation derived by Short reproduces the numerical results of Buckmaster and Neves. Despite the fact that stability studies using large activation energies have been criticized in the past for possessing an infinite number of unstable oscillatory modes, Short (1997a, b) confirmed that this result is indeed a feature of the standard detonation model in the limit of infinite activation energy.

It is interesting to point out that the analysis of Buckmaster and Neves reveals a nonuniformity in the high-frequency expansions. This occurs when the disturbance frequency is of the order of the inverse particle transit time through the main reaction layer. It was later shown by Short (1997a, b) that this disturbance (with a frequency higher than the scales defined by the heat-release thickness) is stable. Hence, this analysis cannot predict the behavior of modes with high-frequency disturbances, because the reaction zone cannot be treated as a discontinuity. The spatial structure must be taken into account (Clavin *et al.* 1997). Therefore, the nonuniformity in the high-frequency expansion must be resolved for a correct asymptotic description. A rescaling of the characteristic disturbance parameters would be required to describe the high-frequency portions of the stability spectrum, taking into account the large frequencies that affect the quasi-steady reaction-zone structure. Such an asymptotic study could verify that sufficiently high-frequency disturbances are stable.

In fact, it is not apparent that one needs to be concerned with high-frequency disturbances. Recent studies using numerical simulations suggest that high-frequency disturbances play only a minor role in the dynamics of detonation. Many of the salient features of the nonlinear pulsation of the unstable detonation appear to be determined by the low-frequency spectra (Bourlioux & Majda, 1995; Short & Wang, 2001; Short, 2005).

5.6. ASYMPTOTIC ANALYSIS OF MULTIDIMENSIONAL INSTABILITIES AND CELL SPACING PREDICTION

Detonations generally manifest multidimensional instabilities. A significant number of asymptotic studies have attempted to relate the instability mechanisms to the origin of the cellular detonation structure. These asymptotic analyses generally focus on two-dimensional, slowly varying, small-wavenumber disturbances. For example, Buckmaster and Ludford (1987) considered long-wavelength perturbations of the order of the activation energy multiplied by the induction-zone length. In addition, the time scale of the evolution of the instability was also taken to be of the order of the activation energy multiplied by the particle transit time through the induction zone. Under this limit, they found a single real unstable mode, which has a growth rate that increases monotonically with increasing wavenumber. Therefore, the result does not lead to a prediction of a characteristic cell spacing.

Buckmaster (1989) later included curvature of the detonation front in the evolution of a slowly evolving instability. He carried out an analysis in which the scale of the disturbance wavelength is of the order of the square root of the activation energy multiplied by the induction-zone length. The curvature of the front then plays a crucial role in determining the induction-zone structure, which is a significant effect that is missing in the longer wavelength analysis of Buckmaster and Ludford. The results of Buckmaster (1989) also show a single unstable real mode. However, this mode now demonstrates an explicit maximum growth rate at a finite wavenumber. For sufficiently short wavelengths, the slowly evolving instability is found to be stabilized by the weak curvature effect. At an intermediate wavelength, a maximum growth rate is obtained. Above this critical wavelength, linear disturbances decay and stability prevails. Buckmaster suggested that the maximum growth rate can be associated with the prediction of a unique detonation cell spacing. However, although the unstable mode is real, it does not propagate transversely to the front. Hence, the theory does not explain the formation of the transverse waves, because the imaginary part of the growth rate is zero. Nevertheless, the study of Buckmaster demonstrated a possible physical mechanism for cell formation. The later

normal-mode analysis by Short (1997a, b) indicated that the asymptotic results of
Buckmaster and Ludford (1987) and Buckmaster (1989) can only be recovered numerically in the limit of very large activation energies.

Using the normal-mode approach and again considering a combination of high activation energy and the Newtonian limit $\gamma \to 1$, Short (1996) and Short and Stewart
(1997) derived an analytical dispersion relationship for the linear stability of a CJ
detonation that describes the evolution of two-dimensional disturbances. Note that
in the study of Short and Stewart, the dispersion relationship is valid for instability
evolution time scales that are long compared to the particle passage time through
the main reaction layer. It is also valid for transverse disturbances with wavelengths
that are long compared to the main reaction-layer thickness. For long-wavelength
disturbances, the reaction layer can be considered as a discontinuity with heat release. For sufficiently large activation energies, the disturbance frequencies described by the analysis can be much higher than the inverse time scale of particle
passage through the induction zone. The wavelengths can also be much smaller than
the thickness of the induction zone.

Since the dispersion relationship is a function of the four fundamental detonation
parameters (γ, Q, E, f), the analytical relationship derived by Short and Stewart
possesses all the desirable characteristics of the exact numerical solution of the linear instability spectra. This includes the prediction of critical finite wavenumbers
at which maximum modal growth rates are attained and above which disturbances
decay and stability prevails. Taking the maximum growth of the lowest-frequency
mode to correspond to the cell size, the asymptotic theory developed by Short provides an estimate of the cell size in the limit of large activation energy. Simplified
polynomial forms of the dispersion relation were also derived later by Short (1997a,
b). He obtained a parabolic linear evolution equation that is of third order in time
and sixth order in space, which governs the initial development of cellular instabilities. It also highlights the important role played both by acoustic wave propagation
in the induction zone and by curvature of the detonation front.

Yao and Stewart (1996) also derived a weakly nonlinear equation that describes
the evolution of a near-CJ detonation wave in the limit of large activation energy.
This equation is only valid for a weakly curved wave front and for instabilities that
are slowly evolving relative to the particle passage time scale through the steady
wave. Their analysis is based on the asymptotic theory of *detonation shock dynamics* (Bdzil & Stewart, 1989; Stewart & Bdzil, 1988), which showed that the evolution
of a detonation shock can be predicted by the solution to a wholly intrinsic partial
differential equation. The result is an evolution equation that is of third order in
time and second order in space. It is a relation between the normal shock velocity,
the shock curvature, and the first normal time derivative of the shock curvature.
Numerical solutions of the one-dimensional version of this equation demonstrate
the presence of nonlinear limit cycle oscillations. In certain domains of the detonation parameters, numerical solutions of the two-dimensional evolution equation

also generate cellular front patterns that are qualitatively similar to those observed in experiments (Stewart *et al.*, 1996). In the same spirit, one-dimensional evolution equations that govern the dynamics of CJ detonation waves for a two-step, chain-branching-like reaction model with a spatially distributed main heat-release layer have also been derived by Short (2001) and Short and Sharpe (2003), assuming a low-frequency evolution.

5.7. ASYMPTOTIC LIMIT OF LARGE OVERDRIVE

Besides the combination of high activation energy and the Newtonian limit $\gamma \to 1$, an alternative asymptotic approach is to consider detonation waves with high degrees of overdrive (Clavin & He, 1996a, b). The Mach number of the overdriven detonation wave is now assumed to be large, and the specific heat ratio is again assumed to be close to unity. If the heat release is taken at these limits to be of the order of the thermal enthalpy of sufficiently overdriven waves, the flow Mach number throughout the detonation will then be small. As a result, the flow behind the shock can be taken to be quasi-isobaric to the leading order of the solution. Together with the assumption of high-temperature sensitivity at the shock, a nonlinear integral equation was derived in the quasi-isobaric limit for small velocity fluctuations from the steady detonation speed. The resulting equation can describe pulsating types of instabilities. Clavin and He (1996a, b) have also included compressibility effects (acoustics) in this approach as a higher-order perturbation. They showed that these effects have a stabilizing influence. This is so because in the high-overdrive limit, the instability mechanism is dominated by fluctuations at the shock that perturb the heat release through their passage along entropy waves. The instability is not due to the amplification of acoustic disturbances by chemical energy release. Based on these results, Clavin suggests that the same mechanism will hold even for CJ detonations. This generalization requires further investigation, and it is suggested that the presence of thermoacoustic instabilities should not be eliminated in view of results of a nonuniform limit as the CJ wave speed is approached (Short, 2005). More analysis is therefore needed to identify the instability mechanisms for CJ waves. Clavin *et al.* (1997) have also carried out similar linear stability analyses of sufficiently overdriven waves subject to multidimensional disturbances.

5.8. ASYMPTOTIC LIMIT OF WEAK HEAT RELEASE

The asymptotic limit of weak heat release (i.e., the heat release is small compared to the post-shock thermal enthalpy) has also been considered by several authors in their studies of overdriven detonation wave stability (Short & Stewart, 1999; He, 2000; Short & Blythe, 2002; Daou & Clavin, 2003). This limit can lead to several cases, ranging from finite overdrive to the limit of infinite overdrive for finite heat release with various temperature sensitivities of the reaction. Overall, the resulting simplification from the weak heat-release limit is that the leading-order solution to the detonation structure is given by the uniform inert shocked state. Notable results were obtained by Short and Blythe (2002), who identified several limits [between small heat release ($\gamma - 1 \ll 1$) and the inverse activation energy of the thermally sensitive reaction] that lead to instability. Clavin and Denet (2002) showed that numerical simulations of a weakly nonlinear integral differential equation, derived in the limit of large overdrive and weak heat release, can qualitatively reproduce cellular front patterns. Clavin and Williams (2002) also considered the limit of weak heat release, using an integral-equation formulation for detonations with small overdrive near the CJ regime.

In addition to these formal asymptotic studies, it is interesting to mention that more traditional one-dimensional low-frequency, as well as two-dimensional (Ginzburg–Landau), weakly nonlinear evolution equations have been derived to investigate analytically the problem of detonation instability (Erpenbeck, 1970; Bourlioux *et al.*, 1991; Majda & Roytburd, 1992). Other analytical theories for estimating the detonation cell spacing are the early studies of Strehlow and Fernandes (1965) and of Barthel (1974) and a more formal asymptotic analysis by Majda (1987) based on the propagation dynamics of high-frequency disturbances using acoustic ray tracing.

Although one- and two-dimensional asymptotic analyses and theoretical modeling yield some important information regarding the nature of detonation instability, the mathematical complexities are such that the studies are limited to very simple chemistry and to limiting values of the characteristic stability parameters. With recent advances in numerical techniques, direct numerical simulations appear to be a more profitable approach in the description of the propagation of unstable detonations.

5.9. DIRECT NUMERICAL SIMULATION OF UNSTABLE DETONATION

Advances in numerical techniques and computer technology in recent years have permitted the integration of the nonlinear reactive Euler equations to a high degree of accuracy with sufficient spatial and temporal resolutions to resolve the fine-scale structures in the detonation reaction zone. Thus, the stability phenomenon can perhaps be investigated more readily by direct numerical simulation, which retains the full nonlinear behavior, in contrast to stability analysis. Unlike experiments, numerical simulations also have the advantage that the number of spatial dimensions can easily be controlled. Thus we can study one-dimensional, longitudinal instabilities in isolation and then progress to more complex instabilities in higher dimensions. This permits a better interpretation of the numerical results obtained. In a real experiment, it is difficult to suppress the instability in other dimension(s) to simplify the interpretation of the results. Numerical studies can also provide more details of the entire flow field that are difficult, if at all possible, to measure experimentally.

Restricting ourselves to two space dimensions with a simplified one-step chemistry model, the governing equations are given by the reactive Euler equations as

$$\frac{\partial \rho}{\partial t} + \frac{\partial (\rho u)}{\partial x} + \frac{\partial (\rho v)}{\partial y} = 0, \tag{5.1}$$

$$\frac{\partial (\rho u)}{\partial t} + \frac{\partial (p + \rho u^2)}{\partial x} + \frac{\partial (\rho uv)}{\partial y} = 0, \tag{5.2}$$

$$\frac{\partial (\rho v)}{\partial t} + \frac{\partial (\rho uv)}{\partial x} + \frac{\partial (p + \rho v^2)}{\partial y} = 0, \tag{5.3}$$

$$\frac{\partial (\rho e)}{\partial t} + \frac{\partial (\rho ue + pu)}{\partial x} + \frac{\partial (\rho ve + pv)}{\partial y} = 0, \tag{5.4}$$

$$\frac{\partial (\rho \lambda)}{\partial t} + \frac{\partial (\rho u\lambda)}{\partial x} + \frac{\partial (\rho v\lambda)}{\partial y} = \dot{\omega} \tag{5.5}$$

with

$$\dot{\omega} = -k\rho\lambda \exp\left(-\frac{E_a}{RT}\right), \tag{5.6}$$

where

$$e = \frac{p}{(\gamma - 1)\rho} + \frac{u^2 + v^2}{2} + \lambda Q \quad \text{and} \quad p = \frac{\rho RT}{M}.$$

In these equations, p, u, ρ, T, and e denote pressure, density, particle velocity, temperature, and energy per unit mass, respectively. λ is a reaction progress variable that varies between $0 \le \lambda \le 1$ as reactants ($\lambda = 1$) proceed to products ($\lambda = 0$). In Eqs. 5.1 to 5.5, we have assumed a perfect gas and a constant specific heat ratio γ. We have also assumed a simple, single-step Arrhenius rate law (Eq. 5.6). However, if a more realistic rate model is used, we need only to replace Eq. 5.6 by the appropriate rate equation.

Prior to numerical integration, the basic equations are nondimensionalized, with the flow variables of the uniform unreacted gas ahead of the detonation front. The preexponential factor k in Eq. 5.6 is an arbitrary parameter that defines the spatial and temporal scales. It is usually chosen such that the half-reaction zone length $L_{1/2}$ of the corresponding ZND detonation wave is scaled to unit length. The activation energy E_a, the heat release Q, and the specific heat ratio γ are parameters that need to be specified.

A number of hierarchical adaptive codes have been developed for the integration of the reactive Euler equations. Details of these numerical methods, particularly for detonation simulations, are discussed in the dissertations of Bourlioux (1991) and Quirk (1991). Different initial conditions can be used to start the numerical integration. For example, the steady ZND solution can be superimposed onto the numerical grid. The smearing of the shock front over a subgrid results in truncation errors that are sufficient to trigger the instability. This then grows with time if the system is unstable. The transient development of the perturbations is highly grid-dependent, and hence these initial transients are not of physical significance.

When the numerical integration has been carried out to a sufficiently long time, the initial transient solution will converge asymptotically to some steady pulsating solution. Near the stability limit, however, the convergence may be very slow, and it becomes necessary to carry out the numerical integration for extremely long times to obtain the asymptotic solution. Also, in the numerical integration, the back boundary must be taken sufficiently far away from the front so that disturbances cannot return from it to catch up with the front and thus give spurious results by influencing the developing instability of the front (Hwang *et al.*, 2000).

The method of studying the growth of perturbations on the steady ZND solutions to get the long-time asymptotic unstable solution is equivalent to the normal-mode stability analysis by superimposing perturbations on the ZND solution and observing their growth or decay. If the detonation is highly unstable, there may not be a steady ZND solution at all if one starts from arbitrary initial conditions. Thus, a proper method to study unstable detonations is to start from quiescent initial conditions and initiate the detonation (e.g., blast or piston initiation). The time evolution of the developing detonation is then determined, and again the long-time asymptotic solution gives the unstable behavior of the detonation front. This method may be advantageous in improving the convergence of the solution by not relying on the grid-dependent initial perturbations such as shock smearing and truncation error. In this manner, we are not perturbing a steady ZND wave initially, and the question whether or not a steady ZND wave exists does not arise.

On the other hand, the slow convergence of the final steady asymptotic unstable solution remains, especially near the stability limits. Hence, the numerical integration must again be carried out for extremely long times (e.g., typically tens of thousands of the characteristic time of the half-ZND-reaction time) to ensure convergence. For highly unstable detonations (e.g., very high activation energies for the single-step Arrhenius reaction model), the phenomenon may be chaotic. In the case of extreme sensitivity to small errors in the initial conditions, the convergence to an asymptotic solution becomes unattainable. Thus, in general, the case of extremely unstable detonations cannot be solved numerically.

5.10. ONE-DIMENSIONAL INSTABILITY (ONE-STEP REACTION RATE MODEL)

If the basic equations have only one spatial dimension, the detonation front instability can only manifest itself as a longitudinal pulsation in the direction of propagation. The parameters that control the onset of instability are the activation energy, E_a (for the case of a single-step Arrhenius rate model); the degree of overdrive, $f = (D/D_{CJ})^2$, where D (and D_{CJ}) denote the velocity of the detonation front; the heat of reaction, Q (which controls the CJ detonation velocity, $D_{CJ} \propto \sqrt{Q}$); and the specific heat ratio, γ.

The temperature sensitivity of the reaction, as determined by the value of the activation energy, is perhaps the most important stability parameter. The detonation is unstable for high values of E_a (hence, highly temperature-sensitive), because small temperature perturbations result in large fluctuations in the reaction rate. The degree of overdrive, f, also influences the stability of the detonation, because a high degree of overdrive increases the shock temperature. An increase in the shock temperature T_s has the effect of lowering the temperature sensitivity of the reaction, because the exponential temperature dependence of the reaction rate law depends on the ratio of the activation energy to the shock temperature, that is, E_a/RT_s. An increase in the heat of reaction, Q, also renders the detonation more unstable, because the physical effects of the perturbations are enhanced for a higher value of the heat of reaction.

The role of γ in detonation stability is more mathematical than physical in nature. In stability analysis, the quantity $\gamma - 1$ appears in combination with the heat of reaction, Q, and as $\gamma \rightarrow 1$, we have $\gamma - 1 \ll 1$, which in effect reduces the heat release. Stability analyses using limiting values of small Q, small $\gamma - 1$, high degree of overdrive, and the like permit considerable simplification of the stability analysis, enabling an asymptotic analytical solution to be obtained, and are of value in understanding the mathematical nature of the instability phenomenon. We are primarily interested in more realistic CJ detonations, and in this case temperature sensitivity of the chemical reactions is generally the governing stability parameter. In the numerical study of Ng (2005), values $Q = 50$ and $\gamma = 1.2$ were chosen to investigate the stability of CJ detonations with the variation of the activation energy E_a.

Figure 5.1 shows the variation of the leading shock pressure of the detonation wave as a function of time for progressively increasing values of the activation energy. The shock pressure is normalized with respect to its value corresponding to the steady CJ detonation (i.e., the von Neumann pressure). For a low value of the activation energy, a steady shock front pressure is obtained after the initial transients have decayed (Fig. 5.1a). As the stability limit is approached, the decay of the initial perturbations takes a much longer time, but the detonation is still stable and the harmonic oscillations will eventually be damped out for sufficiently long times (Fig. 5.1b). Past the stability limit, the perturbations grow (Fig. 5.1c). However, the growth rate of the harmonic oscillation near the stability boundary is again very slow, and the computation must be carried out over a sufficiently long time. Thus, the numerical determination of the stability limit is a demanding task. Nevertheless, the stability limit determined numerically is found to agree extremely well with that determined via linear stability analysis when careful attention is paid to the resolution of the numerical computations to ensure that the small growth and decay rates near the stability boundary are properly described (Sharpe & Falle, 2000b). As the activation energy is increased further past the stability boundary, the perturbation grows and eventually converges to a steady, nonlinear oscillation corresponding to a limit cycle.

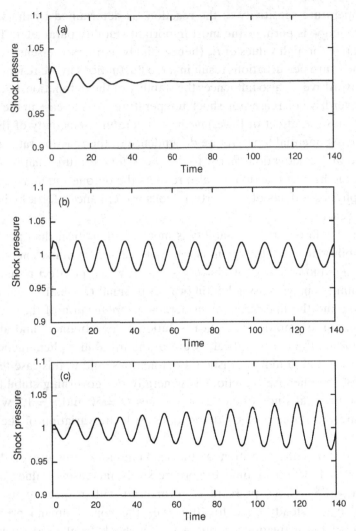

Figure 5.1. Shock pressure as a function of time for increasing values of activation energy, (a) $E_a = 24.00$, (b) $E_a = 24.24$, and (c) $E_a = 25.28$.

Figure 5.2 shows the nonlinear oscillation for increasing values of E_a past the stability limit.

As E_a increases, different modes of oscillation are excited. The oscillation modes are defined by the peak amplitude of the shock oscillation. Thus, a single mode-one oscillation (Fig. 5.2a) bifurcates to a mode-two oscillation (Fig. 5.2b), which can be distinguished by the appearance of two different values of the peak amplitude of the shock oscillation. Successive bifurcations to period four and period eight occur for further increases in the activation energy E_a (Fig. 5.2c and 5.2d). Further bifurcation to period-sixteen oscillation have been obtained by Ng (2005a) for a value $E_a = 27.845$, but it becomes increasingly difficult to distinguish the differences in the peak amplitudes corresponding to the different modes of oscillation.

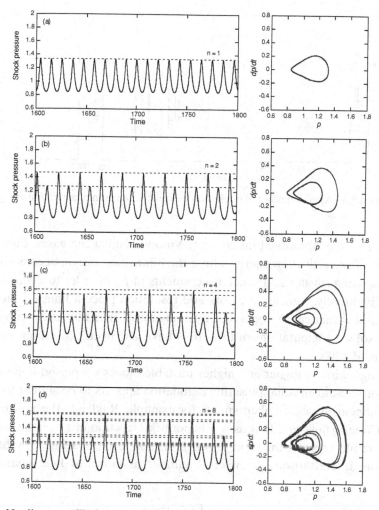

Figure 5.2. Nonlinear oscillations past the stability limit for (a) $E_a = 27.00$, (b) $E_a = 27.40$, (c) $E_a = 27.80$, and (d) $E_a = 27.82$.

The progressive change from single-mode, harmonic oscillation near the stability limit to multimode, nonlinear oscillation via period-doubling bifurcations and eventually to chaos can be conveniently represented in a bifurcation diagram. Figure 5.3 shows a plot of the amplitudes of the oscillation modes as a function of the controlling stability parameter, namely, the activation energy E_a.

For the range of activation energies $25.26 \leq E_a \leq 27.22$, the amplitude of the single-mode oscillation increases steadily with the activation energy. Bifurcation to a two-mode oscillation occurs at around 27.22, as indicated by two distinct peak amplitudes of the two modes (Fig. 5.2b). Bifurcation to a four-mode oscillation occurs at $E_a = 27.71$, with four distinct peak amplitudes for the four oscillation modes (Fig. 5.2c). Bifurcation to higher modes continues as the activation energy further increases. However, it becomes more and more difficult to distinguish the peak

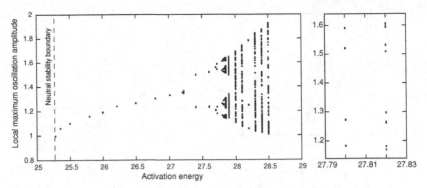

Figure 5.3. Amplitudes of the oscillation modes as a function of activation energy.

amplitudes of the vibrational modes as the mode number increases. Furthermore, the values of the activation energy where the bifurcation occurs become closer and closer, requiring smaller and smaller increments of E_a in order to define the bifurcations. Eventually, truncation errors and resolution place a limit on the number of bifurcations that can be obtained unambiguously in a numerical computation. Highly resolved computations of this effect have also been carried out recently by Henrick *et al.* (2006).

The progression to higher and higher unstable modes via period-doubling bifurcations for one-dimensional pulsating detonation appears to be similar to a host of other nonlinear instability phenomena, for example, Rayleigh–Bernard instability, circular Couette flow, and Belousov-Zhabotinskii reaction in a well-stirred reactor. In all these phenomena, the ratio between the successive spacings of subharmonic bifurcation, δ, tends toward a universal number, as shown by Feigenbaum (1983); namely,

$$\delta = \frac{\mu_i - \mu_{i-1}}{\mu_{i+1} - \mu_i},$$

where μ_i is the location of the bifurcation point. Table 5.1 shows the Feigenbaum number δ computed from the bifurcation limits for one-dimensional pulsating detonations. The results indicate that convergence to the Feigenbaum limit of $\delta = 4.669$ will occur as the number of bifurcations tends to infinity.

Table 5.1. Values of bifurcation limits and Feigenbaum numbers obtained using the one-step Arrhenius kinetic model

i	Activation energy E_a	Oscillation mode	Feigenbaum number
0	25.27	1	–
1	27.22	2	–
2	27.71	4	3.98
3	27.82	8	4.46
4	27.845	16	4.40

The shock pressure fluctuation (i.e., Fig. 5.2) of pulsating detonations is essentially a time series from which the power spectrum density can be computed. This is obtained via a Fourier transform of the autocorrelation sequence of the time series. Figure 5.4 shows the power spectrum density for increasing values of the activation energy. For an activation energy of $E_a = 27.00$, we note that the power spectrum density consists of one large spike corresponding to the fundamental mode, and a small one at exactly twice the frequency, which is the subharmonic of the system. The power spectrum density can be considered also as the distribution of the energy among the various modes of oscillation. As the activation energy increases, the energy is distributed over more modes as the oscillation bifurcates. For a very high activation energy of $E_a = 30$, the power spectral density approaches a noiselike behavior with the energy distributed over a wide range of frequencies.

A more quantitative demonstration of the chaotic nature of the system is to determine its Lyapunov exponent. The Lyapunov exponent measures the exponential behavior between two neighboring trajectories in phase space. A positive Lyapunov exponent indicates a divergence of the phase trajectories, and thus, with time, two neighboring solutions can deviate significantly from each other. In essence, this implies that a chaotic system is extremely sensitive to initial conditions, and a small error in specifying the initial conditions can result in a very large departure of solutions with time. Ng *et al.* (2005b) have demonstrated that for the two activation energies $E_a = 28.17$ and $E_a = 30$, the Lyapunov exponent is positive. Thus, for highly unstable pulsating detonations with high activation energies, it is doubtful that convergence to the same long-time oscillating solution of the detonation is possible, because of the strong dependence on initial conditions and grid-dependent evolution. Sharpe and Falle (2000b) have demonstrated the strong dependence of the long-time behavior of the solution on resolution for very unstable detonations with high activation energy. This raises questions regarding the existence of converged solutions for chaotic systems that are extremely sensitive to initial conditions. It imposes an upper limit on the activation energy (of the order of $E_a = 30$) beyond which numerical simulation of pulsating detonations cannot be considered reliable.

For unstable detonations at high activation energies, the propagation of the detonation resembles a series of failures and reinitiations. Figure 5.5 shows the fluctuation of the leading shock pressure as a function of time for the case of $E_a = 30$. Starting a cycle at the initiation phase, we note that the shock front surges to a peak value in excess of twice the steady CJ shock pressure. The shock then rapidly decays to a value of the order of the steady CJ value and remains at that value for a certain period. It then fails abruptly and drops to a value of about half the steady CJ velocity. At the end of this low-velocity phase, the detonation reaccelerates rapidly to an overdriven state for the next cycle. The duration of the low-velocity phase is found to increase with increasing activation energy. Thus a cycle of pulsation for a very unstable detonation consists of a strong initiation and a failure phase rather than a

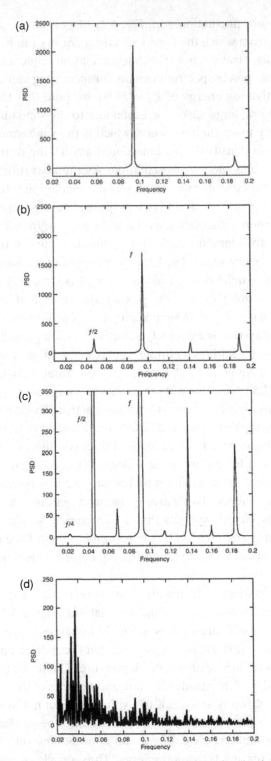

Figure 5.4. Amplitudes of the oscillation modes as a function of activation energy for (a) $E_a = 27.00$, (b) $E_a = 27.40$, (c) $E_a = 27.80$, and (d) $E_a = 30.00$.

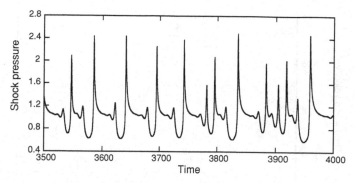

Figure 5.5. Fluctuation of the leading shock pressure as a function of time for $E_a = 30$.

periodic fluctuation about a mean propagation velocity as in the case of stable (low activation energy) detonations.

In the study by He and Lee (1995), it was found that the detonation dies at high activation energies when the low-velocity phase fails to accelerate to an overdriven state to start the next cycle. However, in the later studies of Sharpe and Falle (1999, 2000a, b), they showed that with higher grid resolution, reinitiation was observed, whereas with the coarser grid, reinitiation failed to occur. These authors claimed that with a coarser grid, the series of explosions from the pockets of unburned gases left behind in the wake of the shock front was suppressed. Explosions from these pockets are responsible for the enhancement of the leading shock front, resulting in the reinitiation of the detonation. Similar observations were also observed by Mazaheri (1997) in his numerical study of the direct blast initiation problem, where the series of explosions from unburned pockets in the wake of the shock front leads to the reacceleration of the shock front to an overdriven detonation.

It should also be noted that in a single Arrhenius reaction rate model, the chemical reaction can never be quenched, irrespective of the temperature, because there is no diffusive loss mechanism in the Euler equations. Thus, if one waits long enough, the low-velocity phase will accelerate from the impulses received from the exploding pockets of unburned gases in the wake of the shock front. More realistic models that provide a lower cutoff for the chemical reactions can result in failure. Thus, in one dimension, a highly unstable detonation may fail to propagate due to self-quenching of the reactions during the low-velocity phase of the fluctuating cycle. In two- or three-dimensional detonations, quenching becomes more difficult, as it requires the oscillations in all dimensions to be synchronized so that the low-velocity phase occurs simultaneously. This is highly unlikely, and thus the quenching phenomenon for highly unstable detonations may be characteristic of one space dimension only.

In spite of the large velocity fluctuations of highly unstable pulsating detonations, the averaged velocity (over many cycles) is still found to be remarkably close to the theoretical CJ value. Figure 5.6 shows the averaged detonation velocity as a function of the activation energy. Near the stability limit where the harmonic fluctuations

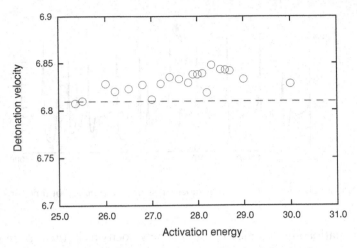

Figure 5.6. Average detonation velocity as a function of activation energy.

are small, the averaged velocity is almost identical to the steady CJ value. However, at very large activation energies where the peak can exceed twice the steady CJ velocity and the low-velocity phase is only half the CJ value, the departure of the averaged velocity from the equilibrium CJ value is increased, but still very small, of the order of 1% (Fig. 5.6). This is due to the fact that without losses, energy is conserved and the averaged detonation velocity depends only on the energetics of the explosive mixture.

In steady CJ theory, the criterion for choosing the tangency solution for the detonation wave is based on the existence of the sonic surface behind the detonation, which essentially prevents the expansion waves in the rear from influencing the detonation front. The existence of a sonic surface is also necessary to match the non-steady solution for the expanding products to the Rankine–Hugoniot solution across the steady detonation front (G.I. Taylor, 1950). Thus it is of interest to see if a sonic plane also exists for the averaged flow of an unstable pulsating detonation.

In a steady ZND detonation, the reaction zone is subsonic relative to the shock front, and thus the $u + c$ characteristics bend toward the shock. At the equilibrium plane at the end of the reaction zone, the flow is sonic relative to the shock, and the $u + c$ characteristic is parallel to the shock trajectory. Downstream of the sonic plane, the characteristics bend away from the shock as the flow becomes supersonic. Figure 5.7 shows the distribution of the flow Mach number, the reaction variable, and the $u + c$ characteristics for such a steady ZND wave. The existence of the sonic plane where the $u + c$ characteristic is parallel to the shock front is evident.

For a nonsteady pulsating detonation, it is of interest to see if a sonic plane corresponding to the chemical equilibrium surface also exists. For nonsteady flow, it is no longer meaningful to speak of sonic conditions; however, we can still determine a limiting $u + c$ characteristic that separates the flow domains (viz., the flow domain where disturbances can catch up with the shock and the domain where they cannot influence the front; Kasimov & Stewart, 2004).

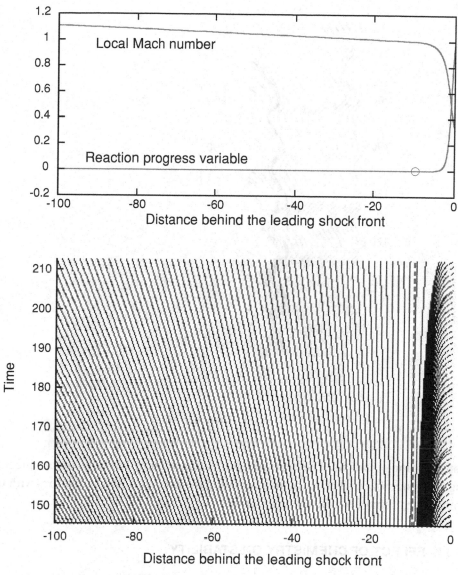

Figure 5.7. Relatively stable detonation with no pulsations.

For an unstable pulsating detonation ($E_a = 27$), the oscillating shock front pressure, the $u + c$ characteristics as well as the boundary of the chemical equilibrium surface ($\lambda \approx 0$) are shown in Fig. 5.8. In a nonsteady detonation, the chemical equilibrium plane no longer coincides with the limiting characteristic of a pulsating detonation.

Although the averaged velocity of a pulsating detonation may be close to the steady CJ value, we cannot determine this averaged velocity by applying steady CJ theory to the averaged flow quantities of a nonsteady detonation wave. In averaging a nonsteady flow, the fluctuations will appear as source terms in the steady

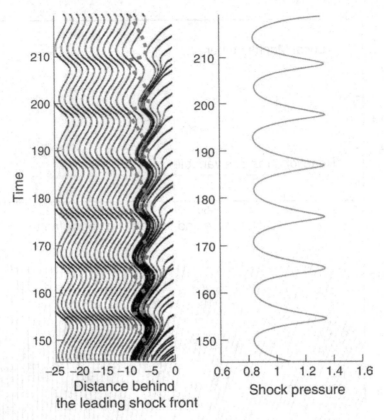

Figure 5.8. Unstable detonation with $E_a = 27$.

conservation equations. The averaged detonation velocity has to be determined using the generalized CJ criterion in integrating the averaged ZND equations with the source terms that describe the fluctuations.

5.11. EFFECT OF CHEMISTRY ON STABILITY

The detailed chemistry of the reactions plays an important role in detonation instability. The results described in the previous section are based on the simple one-step Arrhenius rate model. In real explosive mixtures, the reaction profile generally shows a distinct (approximately thermally neutral) induction zone followed by a rapid exothermic reaction zone where the chemical energy release takes place. The one-step model gives a rather continuous reaction profile without a distinguishable induction zone, especially at low activation energy. Furthermore, a single-step model does not give rise to a quenching of the reaction, irrespective of how low is the shock temperature. In the absence of a loss mechanism, the critical initiation energy, detonation limits, and the like cannot be obtained with a single-step Arrhenius rate model.

The study of unstable one-dimensional pulsating detonation with more realistic chemistry was first attempted by Fickett *et al.* (1972). They introduced a simple one-step representation of a chain-branching kinetic model, which could be tuned to represent two possible reaction-zone structures: one where the exothermic reaction zone is much longer than the induction-zone length. and the other when the reaction-zone length is of the same order as the induction-zone length. They carried out numerical simulations of pulsating detonations for both cases, but it is difficult to identify the key parameter that controls the instability with their chemical model.

Short and Quirk (1997) employed a three-step reaction model and carried out a controlled study of the behavior of pulsating detonations driven by a chain-branching reaction. The model contains a chain-initiation step in which an energetically demanding Arrhenius reaction produces only a small concentration of the radicals from the reactant. This is followed by a lower-activation-energy chain-branching reaction step, also of the Arrhenius form, which then accelerates the growth of the free radicals. The final step is an exothermic chain termination reaction in which the chain carriers are converted to products. It was assumed that the last chain-termination step is independent of temperature. The important contribution of this model is that it gives distinct induction and recombination zones that reproduce the characteristics of chain-branching reactions in real mixtures.

The important parameter in the Short–Quirk model is the chain-branching crossover temperature T_b, in which the chain-branching and chain-termination rates are equal. The crossover temperature also determines the length of the induction zone relative to the recombination zone. Short and Quirk found that the crossover temperature can also serve as a bifurcation parameter, and at low values of T_b, the detonation is stable, whereas for high values of T_b it is unstable. Thus, T_b can be used as the stability parameter that defines the stability boundary.

Short and Quirk carried out both a linear stability analysis and a numerical simulation of the large-amplitude nonlinear oscillations. Their results are qualitatively similar to the single-step model described earlier except for the existence of detonation limits for high values of T_b. For very unstable detonations, the shock temperature can drop below the chain-branching crossover temperature during the low-velocity phase of the pulsation cycle, the chain-branching reaction becomes effectively switched off, and the detonation is then quenched. In contrast, the one-step Arrhenius model has no clear definition for the detonation limit, because there is no sharp cutoff.

The dynamic self-quenching behavior of highly unstable one-dimensional pulsating detonation led Lee and Stewart (1990) to suggest that higher dimensions are required for the existence of highly unstable detonations. Local quenching may not lead to global failure of the detonation wave, because it is highly improbable for all the local fluctuations to be synchronized and in phase. Thus, failure in one local region of the detonation front may be reinitiated by its neighbors, for there is considerable turbulent mixing in a multidimensional unstable reaction zone.

The necessity of transverse waves for the self-sustained propagation of a cellular detonation had been demonstrated by Dupré *et al.* (1988), Teodorczyk and Lee (1995), and Radulescu and Lee (2002). They showed that if the transverse waves are damped out by acoustic absorbing walls to render the detonation one-dimensional, it subsequently failed. Transverse waves play an important role in compressible turbulence, as vorticity is produced through shock–shock, shock–vortex, and shock–density interface interactions.

In a later study, Short and Sharpe (2003) simplified the three-step chain-branching model to a two-step model. The model consists of a thermally neutral chain-branching induction zone governed by an Arrhenius rate law. At the end of the induction zone, a discontinuous transition is assumed, where the reactant is instantaneously converted to chain carriers. An exothermic reaction zone then follows, and it is assumed that this recombination zone is temperature-independent. The extent of the recombination zone is governed by a rate constant k, which in effect determines the ratio of the reaction-zone length to the induction-zone length. Thus, k plays a similar role to that of the crossover temperature in the three-step chain-branching model. In the study by Short and Sharpe (2003), they investigated numerically the nonlinear dynamics of one-dimensional pulsating detonations, and they also compared the numerical results with those for a nonlinear evolution equation derived by Short (2001). The results of the two-step model are qualitatively the same as those of the three-step model. In a recent study by Ng et al. (2005b), that two-step model is modified by using an Arrhenius rate law for the second exothermic reaction rather than assuming it to be temperature-independent as in the study of Short and Sharpe (2003). The result is qualitatively similar.

In all the studies of one-dimensional detonation stability using the chain-branching models, stability is found to depend on the parameters that effectively control the ratio of the induction-zone length to the exothermic recombination-zone length (e.g., the crossover temperature, and the rate constant k of the recombination reaction). In the earlier study of Howe et al. (1976), the length of the induction zone is controlled by delaying the energy release. The energy release rate is assumed to be dependent on the square of the reaction progress variable multiplied by the time rate of change of the reaction variable itself (assumed to be of the Arrhenius form). Therefore, initially when the reaction variable is small, the chemical energy release rate is negligible and only becomes significant when the reaction variable has increased to finite values. Howe et al. also showed that the existence of an induction zone results in instability.

Following the stability studies of previous investigators, Ng et al. (2005b) formally adopted a stability parameter based on the ratio of the induction- to the reaction-zone length. However, they also incorporated the temperature sensitivity of the induction reaction into the definition of the stability parameter χ:

$$\chi = \varepsilon_{\mathrm{I}} \frac{\Delta_{\mathrm{I}}}{\Delta_{\mathrm{R}}},$$

where ε_{I} is essentially the normalized activation energy of the induction reaction with respect to the shock temperature, and Δ_{I} and Δ_{R} denote the induction and reaction lengths, respectively. Because the reaction zone is in general a gradual, slowly

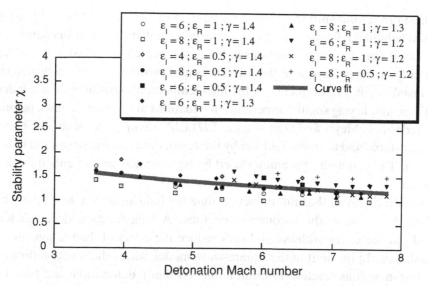

Figure 5.9. Stability parameter as a function of detonation Mach number.

varying function of temperature, it is difficult to define unambiguously a characteristic reaction-zone length. Ng *et al.* proposed that Δ_R be defined as

$$\Delta_R = \frac{u_{CJ}}{\dot{\sigma}_{max}},$$

where u_{CJ} is the particle velocity of a steady CJ detonation in the mixture, and $\dot{\sigma}_{max}$ is the thermicity, defined by Fickett and Davis (1979) as

$$\dot{\sigma}_{max} = (\gamma - 1)\frac{Q}{c_0^2}\frac{d\beta}{dt},$$

where β is the reaction progress variable for the recombination reaction.

With this modified two-step chain-branching reaction model, Ng (2005) investigated numerically the stability of one-dimensional detonations by varying the values of the parameters (γ, Q or M_{CJ}, ε_I, ε_R) to cover a range of the stability parameter χ on both sides of the stability boundary. Figure 5.9 shows the value of χ that defines the stability boundary for a range of the detonation Mach number (or equivalently the heat release Q). The results indicate that the stability boundary corresponds to a single value of χ for the reaction model used. The slight variation of χ may be a consequence of not allowing for the temperature sensitivity of the recombination reaction (i.e., ε_R), which in general is slowly varying. Thus, ignoring ε_R in the definition of the stability parameter has little effect, as indicated by the results shown in Fig. 5.9.

It is evident that stability of the detonation is a consequence of the temperature sensitivity of the chemical reactions. Small fluctuations in the shock temperature result in large fluctuations in the induction delay time as well as (to a lesser extent) the energy release rate of the recombination reactions. Radulescu (2003) and Ng (2005a) attempted to link the stability of detonations to the stability of shock-induced autoignition. As first noted by Voyevodsky and Soloukhin (1965), ignition behind a reflected shock can be unstable in that the onset of chemical reactions can occur at discrete locations in the shock-heated gases instead of occurring uniformly at the plane where the mixture is first processed by the reflected shock. This instability is a consequence of the high temperature sensitivity of the induction reactions, amplifying small temperature fluctuations in the heated gases behind the reflected shock. Meyer and Oppenheim (1971a, b) attempted to define the boundary between stable and unstable ignition by the temperature sensitivity of the induction reaction. The sensitivity parameter used by Ng is similar to Oppenheim's stability criterion.

It should be noted that the induction time (or induction-zone length) should be measured relative to the recombination time. A long reaction time will tend to spread out the energy release and thus reduce the effect of fluctuations in the induction time. In the limit of the square-wave model, where the reaction time is zero (i.e., instantaneous reaction after an induction time), detonations are found to be always unstable. Thus a long reaction time has a stabilizing effect, and this is taken into consideration explicitly in the stability parameter of Ng.

The observation of discrete explosion centers in reflected shock-induced ignition has its equivalent in highly unstable pulsating detonations. He and Lee (1995) and Sharpe and Falle (1999) have also observed discrete unburned pockets of mixture embedded in the combustion products behind the shock front. The subsequent explosion of these pockets of unburned mixtures send shock pulses that catch up with the leading front and thus enhance it. It is also clear that a certain synchronization (or coherence) of these pulses with the leading shock is necessary for the self-sustained propagation of the detonation wave.

Meyer and Oppenheim (1971a, b) have advanced a coherence theory for stable ignition, requiring that the power pulses from neighboring explosion centers be coherent in space and time. Ng pointed out that incoherence in the power pulses from large values of the stability parameter χ leads to gasdynamic instabilities in the reaction zone. It can be concluded that chemistry plays a crucial role in detonation instability.

Instead of using simplified reaction models, numerical simulation of one-dimensional unstable detonations can readily incorporate full chemistry involving several elementary reactions (Yungster & Radhakrishnan, 2005). Such numerical studies are computationally demanding in resolution and time. However, advances in chemical kinetic modeling have led to the development of a number of *reduced mechanisms* where the essential features of the detailed complex reactions can be reproduced accurately. For example, Varatharajan and Williams (2001) have developed a seven-step reduced mechanism for acetylene–oxygen–diluent reactions. A comparison of the reaction-zone profiles for steady ZND detonation between this seven-step reduced mechanism and full detailed chemistry (25 steps) is shown in Fig. 5.10. As indicated, the reduced seven-step mechanism reproduces the temperature and the species distribution of the full detailed 25-step mechanism very well.

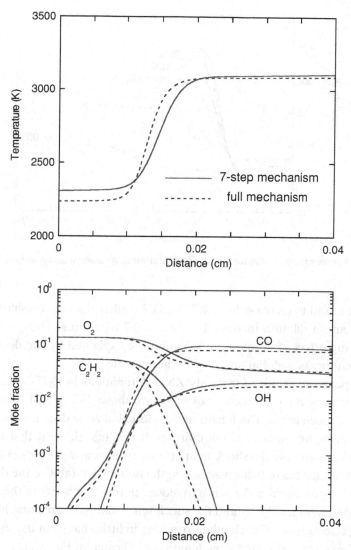

Figure 5.10. Comparison of the ZND structure using a seven-step reduced mechanism and full detailed chemistry for $C_2H_2 + 2.SO_2 + 81\%$ Ar at $\rho_0 = 41.7 kpa$.

   Using the reduced seven-step mechanism, Radulescu *et al.* (2002) carried out a detailed study of pulsating detonations in acetylene–oxygen–argon mixtures. Experimentally, it has been found that high argon dilution tends to stabilize the detonation wave. Attempts to characterize the effect of argon dilution on the detonation stability have not been successful. Since the activation energy (for a one-step Arrhenius rate model) controls the stability, a global activation can be computed for the case of acetylene–oxygen–argon mixtures by fitting a one-step reaction rate law to the reaction-time profiles using the seven-step mechanism (Ng, 2005). However, it was found that the change in the global activation energy with argon dilution is minimal. For stoichiometric mixtures (i.e., $C_2H_2 + 2.5 O_2 + x$ Ar), the global activation

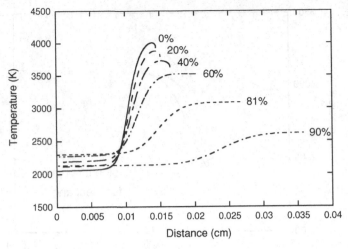

Figure 5.11. Temperature profile for steady ZND detonations using a seven-step reduced mechanism.

energy was found to increase from 4.77 to 5.07 (rather than decreasing) as the percentage of argon dilution increased from $x = 0.7$ to $x = 0.9$. Thus, it is clear that the stabilizing effect of argon dilution cannot be explained by the decrease in the activation energy of a global one-step Arrhenius rate law.

The temperature profiles for steady ZND detonations in $C_2H_2 + 2.5\ O_2 + x$ Ar mixtures for various concentrations of argon are shown in Fig. 5.11 using the seven-step reduced mechanism. The length of the induction zone does not change noticeably for varying amounts of argon dilution. It was first thought that the effect of argon dilution is to raise the shock temperature by increasing the specific heat ratio. This would reduce the induction-zone length and tend to stabilize the detonation.

However, argon dilution also tends to reduce the exothermicity of the mixture and hence the detonation velocity and the shock temperature. It appears that these two competing effects balance each other, resulting in little change in the shock temperature and hence the induction zone length, as indicated in Fig. 5.11. An alternative explanation for the stabilizing effect of argon dilution was proposed by Thibault (1987). He computed the equilibrium Hugoniot curve for different degrees of overdrive and noted the decrease in exothermicity with increasing degrees of overdrive. For H_2–O_2 mixtures with 70% argon dilution, he noted that the equilibrium Hugoniot curve crosses the shock Hugoniot curve at $M_s = 1.41 M_{CJ}$, and the overall reaction becomes endothermic. Thus, in the overdriven phase of the pulsating cycle when $M_s > M_{CJ}$, the exothermicity is reduced and less energy goes to maintain the oscillation. This dampens the amplitude of the oscillation and renders the detonation more stable. However, it is clear from Fig. 5.11 that the temperature as well as the length of the recombination zone increases significantly with increasing argon dilution. With argon dilution, the exothermicity of the mixture is reduced, resulting in a lower temperature rise in the reaction zone. The reaction rates of the exothermic reactions are also reduced, and this increases the reaction time. The examination

Table 5.2. Different thermodynamic states of the quiescent and driver gas used in the computations for varying amounts of argon dilution x in $C_2H_2 + 2.5O_2 + x$ Ar

Argon dilution	P_0 (kPa)	T_0 (K)	P_{vn} (kPa)	T_{vn} (K)	l_{driver} (mm)
90%	100	298	2866	2146	2.2
85%	60	298	1919	2252	1.5
81%	41.7	298	1451	2308	1.5
70%	16.3	298	623	2259	2.0

of the ZND profiles in Fig. 5.11 leads to the conclusion that the stabilizing effect of argon dilution is a result of the increase in the reaction-zone length and a decrease in exothermicity.

Numerical simulations of one-dimensional pulsating detonations in $C_2H_2 + 2.5O_2 + x$ Ar were carried out by Radulescu et al. (2002). For different concentrations of argon dilution, the initial pressure was chosen so that the mixtures had the same sensitivity (i.e., cell size or initiation energy). The properties of the mixture are summarized in Table 5.2. Note that as the percentage of argon dilution is increased, the initial pressure is also increased accordingly to retain the same detonation sensitivity.

Numerical results of the pulsating shock-front pressure for these mixtures are shown in Fig. 5.12. In these simulations, the detonation is initiated by a slug of gas of length l_{driver} brought uniformly to the von Neumann pressure and temperature of the corresponding steady ZND detonation of the mixture. In other words, the initiation process is essentially a shock tube problem. The length of the driver gas slug is also shown in the table. In Fig. 5.12, we note that after an initial transient, the low-amplitude high-frequency mode decays and the low-frequency high-amplitude mode tends to a nonlinear limit cycle oscillation. As the percentage of argon dilution decreases, the amplitude of the oscillation increases (Fig. 5.12b), and a further decrease in argon dilution leads to a multimode irregular oscillation (Fig. 5.12c). At 70% argon dilution (Fig. 5.12d), the pulsating detonation fails after a few cycles of oscillation when the low-velocity phase of the large-amplitude oscillations drops below some critical value where the shock temperature is too low and the resulting induction time is too long for the reaction zone to be able to couple to the shock front. This was found to occur at a shock temperature of about 1300 K when the chain-branching and chain-termination rates balance. This critical temperature is referred to as the crossover temperature, and the present result is similar to the results obtained by Short and Quirk (1997) using a three-step chain-branching model where the detonation limit was observed when the crossover temperature is reached at the low-velocity phase of the pulsation.

Alternatively, Ng et al. (2005b) characterized the stabilizing effect of argon dilution on acetylene–oxygen mixtures by using the stability parameter χ that was defined previously. To compute χ for the various argon-diluted mixtures shown in

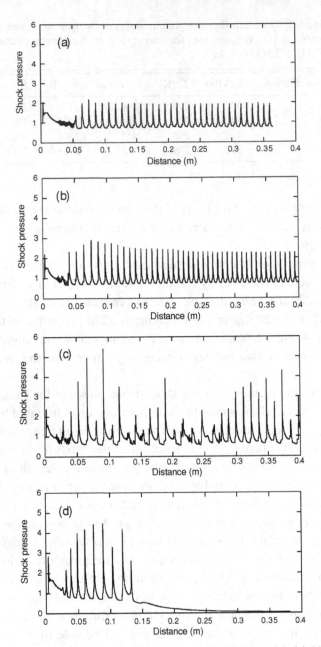

Figure 5.12. Pulsating detonation front pressures for $C_2H_2 + 2.5\ O_2$ with (a) 90%, (b) 85%, (c) 81%, and (d) 70% argon dilution.

Figure 5.12, the steady ZND detonation profile was first determined using the detailed or the reduced kinetic mechanism of the reactions. Figure 5.13 shows the temperature and the thermicity profiles. The induction- and reaction-zone lengths, Δ_I and Δ_R, as well as the peak thermicity rate $\dot{\sigma}_{max}$, can be readily obtained from the ZND profile. To determine the activation energy of the induction reaction, a constant volume explosion was first computed using either the detailed or the reduced

Table 5.3. Values of detonation parameters for varying amounts of argon dilution in C_2H_2–O_2 mixtures in the stability analysis

Argon dilution	P_0 (kPa)	E_a/RT_s	Δ_I(cm)	Δ_R (cm)	χ
90%	100	5.07	1.92×10^{-2}	5.30×10^{-2}	1.83
85%	60	4.86	1.51×10^{-2}	3.41×10^{-2}	2.15
81%	41.7	4.77	1.52×10^{-2}	3.03×10^{-2}	2.39
70%	16	4.77	2.25×10^{-2}	3.32×10^{-2}	3.24

reaction model. The reaction profile was then fitted with a global Arrhenius rate law. Table 5.3 gives the relevant stability parameters for the mixtures investigated in Fig. 5.12; note that the induction-zone length Δ_I, does not change noticeably for varying amounts of argon dilution.

Note that although the changes in the global activation energy are relatively small, the stability parameter χ varies by a significant amount. Previously, the stability boundary has been determined for the two-step chain-branching model (Fig. 5.9). It appears that a value of $\chi \approx 1.5$ can define the stability boundary for a wide range of parameters. It is also of interest to compare the results for the acetylene–oxygen–argon mixture with the two-step chain-branching model. Figure 5.14 shows the four mixtures investigated on a stability map similar to that shown in Fig. 5.9.

In accord with the numerical simulations shown in Fig. 5.14, all four mixtures are unstable and lie above the neutral stability boundary. Also, with decreasing argon dilution, the value of χ departs further from the neutral stability. Thus, it appears that the stability parameter χ used by Ng et al. (2005b) is sufficiently general and can provide a quantitative description of the stability of detonations, even when complex detailed chemistry is used.

It should be noted that the present discussions on one-dimensional instability also apply to two- or three-dimensional cellular instability of detonations. However, the

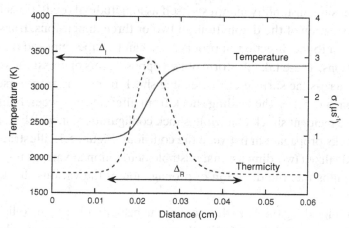

Figure 5.13. Temperature (solid line) and thermicity (dotted line) profiles of the ZND structure for a $C_2H_2 + 2.SO_2 + 78\%$ Ar.

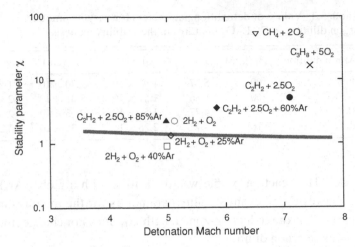

Figure 5.14. Comparison of four argon-diluted acetylene–oxygen mixtures on a stability map.

failure of one-dimensional detonations that occurs when they become too unstable (the amplitude of the oscillations becomes too large during the low-velocity phase, so that the shock strength drops below some critical value that results in the decoupling of the reaction zone from the shock front) is perhaps specific to one-dimensional detonations. If instability is extended to another spatial dimension, the oscillations in the other dimensions may not be synchronized in phase to make all the low-velocity phases occur at the same time and lead to global failure. This is in accord with experimental observations that indicate that when one dimension is suppressed (through damping of the transverse waves by acoustic absorbing walls), the detonation fails (Dupré *et al.*, 1988; Teodorczyk & Lee, 1995).

5.12. TWO-DIMENSIONAL CELLULAR INSTABILITY

In one dimension, instability manifests itself as longitudinal oscillations in the direction of propagation of the detonation. In two or three dimensions, transverse oscillations normal to the direction of propagation can be superimposed on the longitudinal pulsations. These take the form of two opposing sets of pressure (shock) waves that sweep across the surface of the leading shock front. The transverse shocks reflect from each other as the leading shock front alternates between being the Mach stem and the incident shock of a triple-shock configuration, in which transverse (reflected) shocks propagate in between the collisions. Figure 5.15 illustrates the structure of an idealized two-dimensional unstable detonation at various instants in time. The arrows illustrate the direction of propagation of the various shocks in the unstable front.

Consider a local region A on the front at the instant of head-on collision between two transverse waves. Subsequent to the collision, the transverse shocks reflect and propagate away from each other. The leading front is now the Mach stem, with the

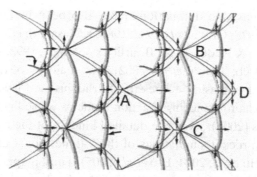

Figure 5.15. Sketch of an idealized two-dimensional unstable detonation at different instants in time.

adjacent portions of the leading front serving as the incident shocks of the triple-shock configuration. At B and C the transverse shocks collide with the neighboring transverse shocks, and the reflected transverse waves now propagate toward each other to collide head on again at D. The leading front of the Mach stem from D to BC now becomes the incident shock to the neighboring Mach stems as the leading front propagates from BC to D. Hence, from A to D the leading front alternates between being a Mach stem and being the incident shock of the three-shock Mach configuration. Upon collision of the transverse waves, the Mach stem is highly over-driven and typically has a velocity of about $1.5V_{CJ}$ (for unstable detonations). It decays subsequently, and before the next collision at D, the leading front can decay to about $0.5V_{CJ}$ before jumping abruptly back to $1.5V_{CJ}$ after the collision of the transverse shocks to start the next cycle. Thus, locally the leading shock pulsates longitudinally between $0.5V_{CJ} < V < 1.5V_{CJ}$ in one cycle of local pulsation. The trajectories of the triple points of the shock interaction are also sketched out in Fig. 5.15, and they represent two sets of diagonally intersecting parallel lines. Associated with a three-shock intersection is also a shear discontinuity, across which pressure is continuous but particle velocity and temperature (hence density) take on a discontinuous jump. The reaction zones behind the shocks are also sketched out in Fig. 5.15.

It should be noted that the structure of the transverse waves can be more complex and can take on a double Mach reflection configuration for very unstable mixtures. Thus, Fig. 5.15 represents only a qualitative picture to illustrate the two-dimensional unstable structure of the detonation. When viewed head on, the leading front takes on a cellular structure with the cell boundaries defined by the intersection of the transverse shock with the leading front. Only one mode of transverse vibration is illustrated in Fig. 5.15. In general, a number of modes can be simultaneously excited, and as in one-dimensional pulsating detonations, there can be other sets of parallel diagonal lines of different spacings superimposed on the fundamental mode to represent the trajectories of the triple points of the higher-order transverse modes.

As in the case of the one-dimensional pulsating longitudinal instability first investigated numerically by Fickett and Wood (1966), two-dimensional as well as three-dimensional unstable detonations have also been investigated numerically (Taki & Fujiwara, 1978; Oran *et al.*, 1982, 1988; Kailasanath *et al.*, 1985; Lefebvre & Oran, 1995; Fujiwara & Reddy, 1989; Bourlioux & Majda, 1992; Williams *et al.*, 1997; Sharpe & Falle, 2000b; Tsuboi *et al.*, 2002). The majority of the two-dimensional studies were carried out using a one-step Arrhenius rate law. However, a more complex four-step chain-branching reaction model was used in the recent study by Liang and Bauwens (2005). Even the detailed kinetics of the hydrogen–oxygen reaction were used in recent simulations of three-dimensional unstable detonations (Oran *et al.*, 1998; Hu *et al.*, 2004; Eto *et al.*, 2005). Thus, it appears that current numerical simulations are no longer restricted to the simple single-step Arrhenius rate reaction when significant computing power is available. However, most numerical simulations are still based on the inviscid reactive Euler equations rather than the Navier–Stokes equations.

In most investigations, the numerical integration of the two-dimensional conservation equations is carried out for a certain computational domain fixed with respect to the propagating detonation wave in the x-direction. Adjustment of the inflow and outflow conditions is effected to maintain the detonation within the computational domain during the initial transient developing stage. To start the numerical integration, different initial data are placed within the computational domain. For example, the steady ZND detonation profile is generally used (e.g., Sharpe & Falle, 2000b; Liang & Bauwens, 2005). In the study of Gamezo *et al.* (1999a, b), a planar shock with a square profile at the CJ detonation pressure was placed near the left-hand boundary of the computational domain as a starting condition. In general, the initial one-dimensional detonation is first allowed to run to steady state before a finite transverse perturbation in the y-direction in the form of a harmonic variation in either the density or the transverse particle velocity is introduced to trigger the transverse instability. However, in the study of Gamezo *et al.* (1999a, b) it was found that numerical noise was sufficient to cause spontaneous onset of the transverse instability. Small initial perturbations grow rapidly via the mechanism of thermal instability (positive feedback between heat release rate and shock intensity; Kuznetsov & Kopotev, 1986) to eventually form a steady cellular detonation. Although the amplitude of the numerical noise is found to influence the transient development of the cellular detonation (i.e., the time to reach steady state), the final steady cellular structure is found to be independent of the characteristics of the numerical noise. As in one-dimensional instability, the finer details of the detonation structure depend on the numerical resolution. Thus, one has to seek a compromise between computational time and the desired resolution of the fine structural details. A demonstration of the dependence of the detonation structure on numerical resolution was given by Sharpe (2001).

For the boundary condition in the transverse y-direction, rigid walls with free-slip conditions compatible with the inviscid Euler equations are used. However, in some numerical studies (e.g., Bourlioux & Madja, 1992) periodic boundary conditions (i.e., the state along one transverse boundary is the same as that at the opposite boundary) are also imposed. The use of periodic boundary conditions can only give an integral number of cells across the width of the channel, because the transverse waves can only reflect from other transverse waves. Thus, the so-called single-headed spin or the zig-zag detonation with only one transverse wave cannot be obtained using periodic boundary conditions. However, when the transverse dimension is very large compared to the cell size, the effect of the transverse boundary condition becomes unimportant. To describe cellular detonations in a channel whose width is of the order of the cell size or less, the real reflecting solid boundary condition must be used, because the structure is strongly influenced by reflection from the solid walls.

The majority of numerical studies of cellular detonations present qualitative pictures of the detonation structure and reproduce the transverse wave trajectories imprinted on smoked foils from experiments. However, in numerical simulation the complete transient flow field of physical and chemical state variables is also computed. Thus, there is a wealth of quantitative information from numerical simulations that can be analyzed (as compared to real experiments, where relatively few parameters can be measured with acceptable accuracies). Perhaps the best attempt to analyze the numerical data obtained was made by Gamezo et al. (1999a, b). We shall present the work of Gamezo to illustrate typical results obtained from numerical simulations of cellular detonations.

A single-step Arrhenius rate law was used by Gamezo, and the activation energy was varied to change the temperature sensitivity of the chemical reactions and to control the stability of the detonation. The preexponential factor, which is a scale factor in the model, was chosen for each case so that the computational domain always contained several equilibrium detonation cells. The other parameters of the system were kept constant (i.e., $\gamma = 1.333$, $Q = 4.867$ MJ/kg, $\rho_0 = 0.493$ kg/m^3, $M = 0.012$ kg/mol), and their values are representative of stoichiometric hydrogen–oxygen mixtures at $p_0 = 1$ bar and $T_0 = 293$ K. Corresponding CJ and ZND parameters are $D_{CJ} = 2845$ m/s, $p_{CJ} = 17.5$ bar, $T_{CJ} = 3007$ K, $p_{ZND} = 34.0$ bar, and $T_{ZND} = T^* = 1709$ K. Changing the activation energy changes the temperature sensitivity of the reaction, but the equilibrium detonation parameters are not affected.

A sketch of the computational domain is shown in Fig. 5.16. With an initial shock of strength $p = p_{ZND} \approx 2p_{CJ}$ placed near the left boundary of the computational domain, the basic conservation equations are then integrated numerically. With the prescribed initial conditions, the detonation is initially overdriven as it propagates from left to right in the computation domain and decays asymptotically with time toward the steady CJ state. The inflow and outflow are adjusted so that the detonation is eventually stabilized near the right boundary, and the computation domain thus moves with the steady detonation wave. The grey scale contour of the maximum pressure computed for each point in space gives a pattern that closely resembles experimental smoked foil records of triple point trajectories due to shock intersections at the leading shock front of a cellular detonation. The evolution of the detonation

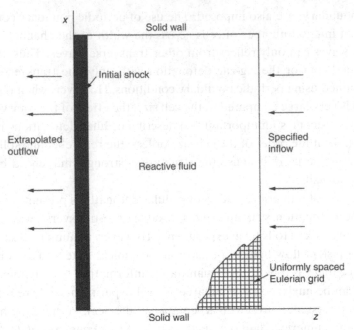

Figure 5.16. Sketch of computational domain.

toward steady state for $E/RT^* = 2.1, 4.9,$ and 7.4 are illustrated in Figs. 5.17 to 5.19, respectively.

The common feature in Figs. 5.17 to 5.19 is the appearance of transverse waves as the detonation decays from its initial overdriven state. The initial transverse waves are weak and propagate along the leading shock front at the acoustic velocity. Thus,

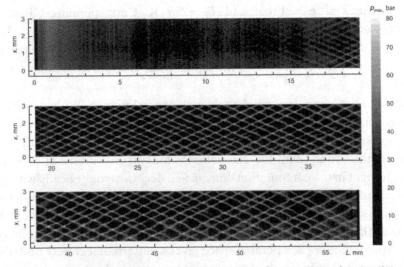

Figure 5.17. Time-integrated maximum-pressure contours of a two-dimensional cellular detonation from numerical simulation ($E_a/RT^* = 2.1$).

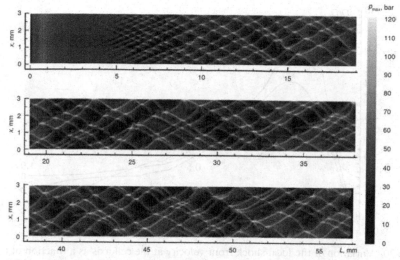

Figure 5.18. Time-integrated maximum-pressure contours of a two-dimensional cellular detonation ($E_a/RT^* = 4.9$).

the initial cell patterns of the overdriven detonations are fairly regular for all three values of the activation energy. As the detonation decays, the transverse waves coalesce, reducing the total number of transverse waves across the width of the channel, which eventually reaches its equilibrium value as the detonation approaches its CJ state. The initial transient phase of the growth of instability is sensitive to the numerical resolution, and thus the results are only correct qualitatively. It should also be noted that, for the channel width chosen, there are always two or three detonation cells present, and the cellular structure is not significantly influenced by the channel width.

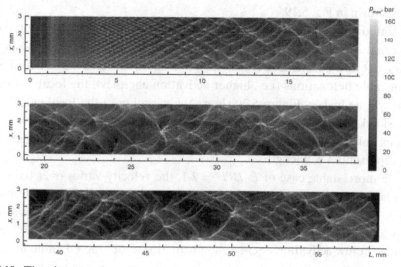

Figure 5.19. Time-integrated maximum-pressure contours of a two-dimensional cellular detonation ($E_a/RT^* = 7.4$).

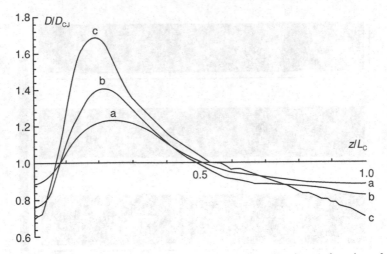

Figure 5.20. Variation of the local shock front velocity at the cell axis as a function of the scaled distance z/L_c, where L_c is the length of the detonation cell (curve a, $E_a/RT^* = 2.1$; curve b, $E_a/RT^* = 4.9$; curve c, $E_a/RT^* = 7.4$).

For low activation energy when the detonation is relatively stable (as in Fig. 5.17), the transverse waves are weak and propagate at more or less the acoustic velocity as the steady CJ state is reached. The cellular pattern is also regular, with only one dominant transverse mode excited. At higher activation energies where the detonation is more unstable (Figs. 5.18 and 5.19), the cellular pattern becomes more irregular, with transverse waves disappearing as they coalesce and spontaneously appearing due to the rapid growth of small perturbations from instabilities. However, the average number of transverse waves across the width of the channel (i.e., the average cell size) remains the same. In fact, it is difficult to point out a dominant and representative cell size from the highly irregular pattern of transverse wave trajectories shown in Fig. 5.19.

As pointed out previously, the local velocity of the leading shock front fluctuates as it alternates between being a Mach stem (upon collision of two opposing transverse waves) and an incident shock (prior to the transverse wave collision). For more unstable detonations (i.e., higher activation energies), the local velocity fluctuations will be higher. Figure 5.20 shows the variation of the local velocity of the leading shock front along a cell length, i.e., one cycle of transverse wave collision. The shock velocity is normalized with respect to the CJ detonation velocity, D/D_{CJ}, and L_c represents a cell length.

For the more stable case of $E_a/RT^* = 2.1$, the velocity varies over $0.6 \leq D/D_{CJ} \leq 1.2$, whereas for the more unstable case of $E_a/RT^* = 7.4$, the fluctuation is $0.7 \leq D/D_{CJ} \leq 1.7$. However, the average velocity of propagation of the detonation wave remains approximately equal to the CJ value, as shown in Fig. 5.21. For a stable detonation at low activation energy, the fluctuation about the CJ value is negligibly small.

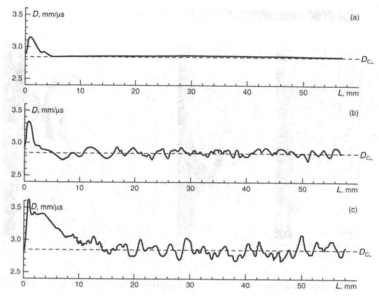

Figure 5.21. Comparison of leading shock velocity fluctuations for stable and unstable detonations: (a) $E_a/RT^* = 2.1$, (b) $E_a/RT^* = 4.9$, (c) $E_a/RT^* = 7.4$.

An interesting numerical experiment is to observe the response of a cellular detonation to perturbations. Because the system is unstable, it should respond to external noise and amplify any perturbation, which can influence the natural unstable cellular structure of the detonation. Gamezo *et al.* introduced a random numerical noise into the heat release term such that the energy release by chemical reactions varied from zero to twice the steady value, that is, from 0 to $2Q$. The cellular pattern for the case of $E_a/RT^* = 4.9$ is shown in Fig. 5.22. In contrast with Fig. 5.18, where no noise is imposed, the calculations indicate that the cellular pattern remains unchanged. Similar results are obtained when the fluctuation parameter is introduced into the reaction rate law to give a random fluctuation in the reaction rate. Therefore, it may be concluded that an unstable cellular detonation structure is extremely robust and is not influenced by external noise. However, comparison between Figs. 5.18 and 5.22 indicates that the initial transient phase of the development of the cellular structure can be influenced by external noise. With external

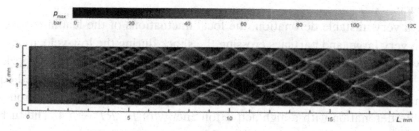

Figure 5.22. Cellular pattern for $E_a/RT^* = 4.9$.

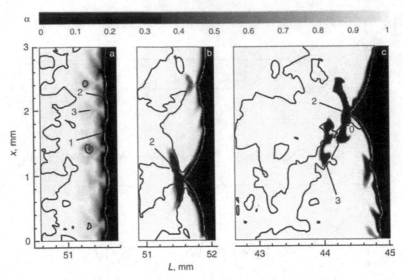

Figure 5.23. Temperature contours for an unstable detonation with $E_\mathrm{a}/RT^* = 7.4$.

perturbations, the growth of the instability is more rapid, and the formation of the cellular pattern is found to occur earlier. It should be noted that for a detonation near the stability limit where perturbation growth and decay rates are small, the imposition of external noise can have a significant influence on the cellular structure, because the disturbances are retained and are not suppressed by the intrinsic instability of the detonation. This is in accord with experiments, where it was found that detonations that are relatively stable are more easily influenced by boundary conditions (Strehlow, 1969; Moen *et al.*, 1986; Dupré *et al.*, 1988).

The insensitivity of a very unstable cellular detonation to external perturbations could be due to the fact that for unstable detonations, transverse waves are continuously being generated and eliminated, hence giving the irregular cell pattern. Thus, the additional random perturbations superimposed on the already unstable structure would not contribute significantly to the ongoing process of growth and decay of transverse waves. Also, in unstable detonations where the local fluctuations are already very large (Fig. 5.20), the fluctuation of the imposed noise is minor compared to the amplitude of the intrinsic instability. For stable detonations, where the intrinsic fluctuations are small, the external noise has a correspondingly stronger influence.

For a very unstable detonation, the local fluctuations of the leading shock front can be very large, and particles crossing the shock during the low-velocity phase of the cycle can have very long induction times. This results in unburned pockets of gases being formed behind the cellular detonation front, as first observed computationally by Oran *et al.* (1982). Figure 5.23 shows the temperature contour of an unstable detonation with a high activation energy of $E_\mathrm{a}/RT^* = 7.4$. The turbulent nature of the reaction zone as indicated by the vortex-like structure can be seen. Since viscosity is absent in the Euler equations, the vorticity is generated by the triple-shock Mach intersections; Kelvin–Helmholtz instability of the shear layers; Taylor, Richtnyer, and Meshkov instabilities of density interfaces; and the baroclinic vorticity generation mechanism due to pressure and density gradients in the reaction zone. It can also be observed that unburned pockets of mixtures are embedded in the products behind the leading front. The size and the depth of penetration of these unreacted pockets increase with increasing activation energy. Thus, for very unstable detonations, combustion of the mixture may not be completed directly behind the leading shock and transverse waves. For more stable detonations, the unburned pockets may eventually react after a long induction time by self-ignition from adiabatic heating due to shock compression as particles cross the leading shock and transverse waves. For higher activation energies, the combustion of the pockets can result from turbulent diffusional transport across the interface of these pockets. Since there is no molecular or turbulent transport term in the Euler equations, the boundaries of these unreacted pockets propagate via numerical diffusion. It should also be noted that in the highly fluctuating turbulent field, the pockets can disintegrate into smaller ones with larger surface areas and hence disappear more rapidly.

The observation of unburned pockets behind highly unstable detonations has an important implication: shock compression alone (by the leading shock and transverse waves) cannot be responsible for the ignition of the mixture. Turbulent transport also plays a role in the combustion process in the detonation wave. Thus, the distinction between the combustion mechanism of detonation (i.e., shock ignition) and deflagration (i.e., heat and mass transport) can no longer be sharply defined.

The existence of unburned pockets that get swept downstream results in an increase in the effective length of the reaction zone of the detonation wave. The delayed energy release of the pockets cannot contribute to the propagation of the detonation if they are behind the sonic plane. In any case, the energy release from these discrete pockets generates pressure waves that result in larger fluctuations in the cellular detonation velocity. Since the entire field data for all the variables are available in a numerical simulation, it is possible to compute the time- and space-averaged profiles of the various parameters (temperature, pressure, reaction progress variables, and reaction rate) for a cellular detonation. Figures 5.24 to 5.26 compare the time- and space-averaged profiles of a cellular detonation (solid line) with the corresponding profiles for a steady ZND detonation (dashed line) for different activation energies ($E_a/RT^* = 2.1$, 4.9, and 7.4, respectively). As can be observed, the differences between the averaged profiles of cellular detonations and the corresponding steady ZND detonations increase with activation energy. The averaged profiles have lower peak values but are more extended, giving a larger effective reaction zone length.

Although three-dimensional unstable cellular detonations have been simulated numerically (Williams et al., 1997; Tsuboi et al., 2000, 2002), the quantitative prediction of the detonation cell size (or the spectrum of cell sizes) is still not possible. This is mainly due to the very high resolution that is required to ensure that

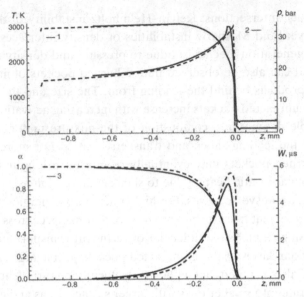

Figure 5.24. Comparison of time- and space-averaged profiles between a cellular and a ZND detonation for $E_a/RT^* = 2.1$.

unstable cellular structures can be described. In an unstable detonation, the strength of the leading shock front can peak at over 1.5 times the steady ZND value. The scale of the reaction length is significantly reduced during these peak excursions, thus requiring extremely fine resolution. Although the effect of numerical noise on the steady cellular structure may be minimal for unstable detonations (as shown in

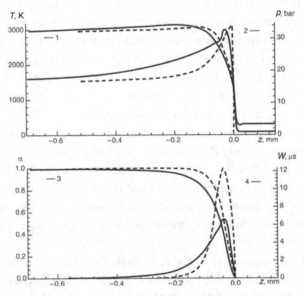

Figure 5.25. Comparison of time- and space-averaged profiles between a cellular and a ZND detonation for $E_a/RT^* = 4.9$.

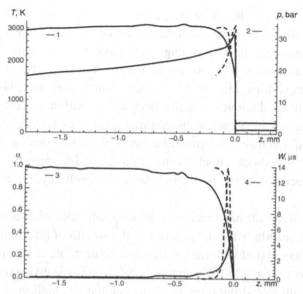

Figure 5.26. Comparison of time- and space-averaged profiles between a cellular and a ZND detonation for $E_a/RT^* = 7.4$.

the study by Gamezo et al. described earlier), it will play a more influential role in more stable detonations. Furthermore, the transient development of the detonation is also highly dependent on the nature of the perturbations initially imposed. Hence, such phenomena cannot be described quantitatively in numerical simulations. As well, when the channel width is small, the perturbations generated by the boundaries will be more significant and can persist for very long times.

It should also be noted that almost all numerical simulations to date are based on the inviscid Euler equations. It has already been discussed that viscosity and transport effects can play important roles in the small-scale phenomena within the structure. Direct numerical simulations using the Navier–Stokes equations have been reported by Oran et al. (1988). Although they claimed the results are similar to those with the inviscid Euler equations, the value of the Reynolds number that was used is far too low to represent real detonations. Thus, on a qualitative basis, numerical simulation serves an extremely useful purpose and can contribute significantly to the understanding of the complex structure of cellular detonations. However, quantitative predictions of the cell size and other dynamic parameters are still not possible.

5.13. CLOSING REMARKS

Numerical simulations of unstable detonations have the advantage that the full nonlinearity of the unstable phenomenon can be described, in contrast to linear stability theory, where only the initial small-amplitude perturbations can be considered. Furthermore, the algebraic complexity of linear stability analysis and the general need to employ a numerical solution of the perturbation equation for the unstable spectrum also fail to make the physical mechanisms of the unstable processes transparent so as to promote the understanding of the physics of the unstable mechanisms.

Numerical simulations thus far are restricted mostly to the integration of the inviscid Euler equations. However, it is clear that turbulence plays an important role in the eventual dissipation of the fluctuations within the cellular detonation structure. There are also numerous complex processes within the reaction zone of the cellular detonation that current numerical simulations cannot resolve adequately (e.g., shock–shock, shock–vortex, and shock–interface turbulence). To describe these processes would be a formidable task well beyond current numerical capability.

A particular area that has received relatively little attention is the development of models and algorithms for the analysis of the wealth of information on the transient flow field associated with the propagation of unstable detonations. The recent work of Radulescu *et al.* (2007), which attempts to statistically analyze the flow field in order to obtain global parameters of the cellular detonation (i.e., the hydrodynamic thickness), is an example of the effort required to fully exploit the results of numerical simulations.

Bibliography

Abouseif, G.E., and T.Y. Toong. 1982. Theory of unstable detonations. *Combust. Flame* 45:67–94.

Abouseif, G.E., and T.Y. Toong. 1986. Theory of unstable two-dimensional detonations: Genesis of the transverse waves. *Combust. Flame* 63:191–207.

Barthel, H.O. 1974. Predicted spacings in hydrogen–oxygen–argon detonations. *Phys. Fluids* 17:1547–1553.

Bdzil, J.B., and D.S. Stewart. 1989. Modeling two-dimensional detonations with detonation shock dynamics. *Phys. Fluids A* 1(7):1261–1267.

Blythe, P.A., and D.G. Crighton. 1989. Shock-generated ignition: The induction zone. *Proc. R. Soc. Lond. A* 426:189–209.

Bourlioux, A. 1991. Numerical study of unstable detonation. Ph.D. thesis, Princeton University, Princeton, NJ.

Bourlioux, A., and A.J. Majda. 1992. Theoretical and numerical structure for unstable two-dimensional detonations. *Combust. Flame* 90:211–229.

Bourlioux, A., and A.J. Majda. 1995. Theoretical and numerical structure of unstable detonations. *Phil. Trans. R. Soc. Lond. A* 350:29–68.

Bourlioux A., A.J. Majda, and V. Roytburd. 1991. Theoretical and numerical structure for unstable one-dimensional detonations. *SIAM J. Appl. Math.* 51:303–343.

Buckmaster, J.D. 1989. A theory for triple point spacing in overdriven detonation waves. *Combust. Flame* 77:219–228.

Buckmaster, J.D., and G.S.S. Ludford. 1987. The effect of structure on the stability of detonations. I – Role of the induction zone. *Proc. Combust. Inst.* 21:1669–1676.

Buckmaster, J.D., and J. Neves. 1988. One-dimensional detonation stability: The spectrum for infinite activation energy. *Phys. Fluids* 31(12):3571–3576.

Clavin, P., and B. Denet. 2002. Diamond patterns in the cellular front of an overdriven detonation. *Phys. Rev. Lett.* 88:044502.

Clavin, P., and L. He. 1996a. Stability and non-linear dynamics of one-dimensional overdriven detonations in gases. *J. Fluid Mech.* 306:353–378.

Clavin, P., and L. He. 1996b. Acoustic effects in the non-linear oscillations of planar detonations. *Phys. Rev. E* 53:4778–4784.

Clavin, P., L. He, and F.A. Williams. 1997. Multidimensional stability analysis of over-driven gaseous detonations. *Phys. Fluids* 9(12):3764–3785.

Clavin, P., and F.A. Williams. 2002. Dynamics of planar gaseous detonations near Chapman–Jouguet conditions for small heat release. *Combust. Theory. Model.* 6:127–139.

Daou, R., and P. Clavin. 2003. Instability threshold of gaseous detonations. *J. Fluid Mech.* 482:181–206.

Dupré, G., O. Peraldi, J.H. Lee, and R. Knystautas. 1988. Propagation of detonation waves in an acoustic absorbing walled tube. *Prog. Astronaut. Aeronaut.* 114:248–263.

Erpenbeck, J.J. 1962. Stability of steady-state equilibrium detonations. *Phys. Fluids* 5:604–614.

Erpenbeck, J.J. 1964. Stability of idealized one-reaction detonations. *Phys. Fluids* 7:684–696.

Erpenbeck, J.J. 1967. Non-linear theory of unstable one-dimensional detonations. *Phys. Fluids* 10:274–288.

Erpenbeck, J.J. 1970. Non-linear theory of unstable two-dimensional detonation. *Phys. Fluids* 13:2007–2026.

Eto, K., N. Tsuboi, and A.K. Hayashi. 2005. Numerical study on three-dimensional C-J detonation waves: Detailed propagating mechanism and existence of OH radical. *Proc. Combust. Inst.* 30:1907–1913.

Fay, J.A. 1962. Detonation wave stability at low pressures. *Propellant and Fuels* 22:15.

Feigenbaum, M.J. 1983. Universal behavior in non-linear systems. *Physica D* 7:16–39.

Fickett, W., and W.C. Davis. 1979. *Detonation*. Berkeley, CA: University of California Press.

Fickett, W., J.D. Jacobson, and G.L. Schott. 1972. Calculated pulsating one-dimensional detonations with induction-zone kinetics. *AIAA J.* 10:514–516.

Fickett, W., and W.W. Wood. 1966. Flow calculations for pulsating one-dimensional detonations. *Phys. Fluids* 9:903–916.

Fujiwara, T., and K.V. Reddy. 1989. Propagation mechanism of detonation – three-dimensional phenomena. *Mem. Fac. Eng. Nagoya Univ.* 41:1–18.

Gamezo, V.N., D. Desbordes, and E.S. Oran. 1999a. Formation and evolution of two-dimensional cellular detonations. *Combust. Flame* 116:154–165.

Gamezo, V.N., D. Desbordes, and E.S. Oran. 1999b. Two-dimensional reactive flow dynamics in cellular detonation. *Shock Waves* 9:11–17.

Gorchkov, V., C.B., Kiyanda, M. Short, and J.J. Quirk. 2007. A detonation stability formulation for arbitrary equations of state and multi-step reaction mechanisms. *Proc. Combust. Inst.* 31:2397.

He, L. 2000. Theory of weakly unstable multi-dimensional detonation. *Combust. Sci. Technol.* 160:65–101.

He, L., and J.H.S. Lee. 1995. The dynamical limit of one-dimensional detonations. *Phys. Fluids* 7(5):1151–1158.

Henrick, A.K., T.D. Aslam, and J.M. Powers. 2006. Simulations of pulsating one-dimensional detonations with true fifth order accuracy. *J. Comput. Phys.* 213:311–329.

Howe, P., R. Frey, and G. Melani. 1976. Observations concerning transverse waves in solid explosives. *Combust. Sci. Technol.* 14:63–64.

Hu, X.Y., B.C. Khoo, D.L. Zhang, and Z.L. Jiang. 2004. The cellular structure of a two-dimensional $H_2/O_2/Ar$ detonation wave. *Combust. Theory Model.* 8:339–359.

Hwang, P., R.P. Fedkiw, B. Merriman, T.D. Aslam, A.R. Karagozian, and S.J. Osher. 2000. Numerical resolution of pulsating detonation waves. *Combust. Theory Model.* 4:217–240.

Kailasanath, K., E.S. Oran, J.P. Boris, and T.R. Young. 1985. Determination of detonation cell size and the role of transverse waves in two-dimensional detonations. *Combust. Flame* 61:199–209.

Kasimov, A.R., and D.S. Stewart. 2002. Spinning instability of gaseous detonations. *J. Fluid Mech.* 466:179–203.

Kasimov, A.R., and D.S. Stewart. 2004. On the dynamics of self-sustained one-dimensional detonations: A numerical study in the shock-attached frame. *Phys. Fluids* 16(10):3566–3578.

Kuznetsov, M.S., and V.A. Kopotev. 1986. Detonation in relaxing gases and the relaxation instability. *Fizo Goreniya Vzryva* 22:75–86.

Lee, H.I., and D.S. Stewart. 1990. Calculation of linear detonation instability: One-dimensional instability of planar detonations. *J. Fluid Mech.* 216:103–132.

Lefebvre, M.H., and E.S. Oran. 1995. Analysis of the shock structures in a regular detonation. *Shock Waves* 4:277–283.

Liang, Z., and L. Bauwens. 2005. Detonation structure with pressure dependent chain-branching kinetics. *Proc. Combust. Inst.* 30:1879–1887.

Liang, Z., B. Khastoo, and L. Bauwens. 2004. Effect of reaction order on stability of planar detonation. *Int. J. Comput. Fluid Dyn.* 19(2):131–142.

Majda, A. 1987. Criteria for regular spacing of reacting Mach stems. *Proc. Nat. Acad. Sci.* 84(17):6011–6014.

Majda, A., and V. Roytburd. 1992. Low-frequency multidimensional instabilities for reacting shock waves. *Stud. Appl. Math.* 87:135–174.

Mazaheri, B.K. 1997. Mechanism of the onset of detonation in blast initiation. Ph.D. thesis, McGill University, Montreal, Canada.

Meyer, J.W., and A.K. Oppenheim. 1971a. Coherence theory of the strong ignition limit. *Combust. Flame* 17:65–68.

Meyer, J.W., and A.K. Oppenheim. 1971b. On the shock-induced ignition of explosive gases. *Proc. Combust. Inst.* 13:1153–1164.

Moen, I.O., A. Sulmistras, G.O. Thomas, D. Bjerketvedt, and P.A. Thibault. 1986. Influence of cellular regularity on the behavior of gaseous detonations. *Prog. Astronaut. Aeronaut.* 106:220–243.

Namah, G.S., C. Brauner, J. Buckmaster, and C. Schmidt-Laine. 1991. Linear stability of one-dimensional detonation. In *Dynamical issues in combustion theory*, ed. P. Fife, A. Linan, and F. Williams, 229–239. Springer-Verlag, New York.

Ng, H.D. 2005. The effect of chemical reaction kinetics on the structure of gaseous detonations. Ph.D. thesis, McGill University, Montreal Canada.

Ng, H.D., M.I. Radulescu, A.J. Higgins, N. Nikiforakis, and J.H.S. Lee. 2005a. Numerical investigation of the instability for one-dimensional Chapman–Jouguet detonations with chain-branching kinetics. *Combust. Theory. Model.* 9:385–401.

Ng, H.D., A.J. Higgins, C.B. Kiyanda, M.I. Radulescu, J.H.S. Lee, K.R. Bates, and N. Nikiforakis. 2005b. Non-linear dynamics and chaos analysis of one-dimensional pulsating detonations. *Combust. Theory. Model.* 9:159–170.

Oran, E.S., K. Kailasanath, and R.H. Guirguis. 1988. Numerical simulations of the development and structure of detonations. *Prog. Astronaut. Aeronaut.* 114:155–169.

Oran, E.S., J.W. Weber, E.I. Stefaniw, M.H. Lefebvre, and J.D. Anderson. 1998. A numerical study of a two-dimensional H_2-O_2-Ar detonation using a detailed chemical reaction model. *Combust. Flame* 113:147–163.

Oran, E.S., T.R. Young, J.P. Boris, J.M. Picone, and D.H. Edwards. 1982. A study of detonation structure: The formation of unreacted gas pockets. *Proc. Combust. Inst.* 19:573–582.

Quirk, J.J. 1991. An adaptive grid algorithm for computational shock hydrodynamics. Ph.D. thesis, Cranfield Institute of Technology, U.K.

Radulescu, M.I. 2003. The propagation and failure mechanism of gaseous detonations: Experiments in porous-walled tubes. Ph.D. thesis, McGill University, Montreal, Canada.

Radulescu, M.I., and J.H.S. Lee. 2002. The failure mechanism of gaseous detonations – experiments in porous wall tubes. *Combust. Flame* 131:29–46.

Radulescu, M.I., H.D. Ng, J.H.S. Lee, and B. Varatharajan. 2002. The effect of argon dilution on the stability of acetylene–oxygen detonations. *Proc. Combust. Inst.* 29:2825–2831.

Radulescu, M.I., G.J. Sharpe, C.K. Law, and J.H.S. Lee. 2007. The hydrodynamic structure of unstable cellular detonations. *J. Fluid Mech.* 580:31–81.

Sharpe, G.J. 1997. Linear stability of idealized detonations. 1997. *Proc. R. Soc. Lond. A* 453:2603–2625.

Sharpe, G.J. 1999. Linear stability of pathological detonations. *J. Fluid Mech.* 401:311–338.

Sharpe, G.J. 2001. Transverse waves in numerical simulations of cellular detonations. *J. Fluid Mech.* 447:31–51.

Sharpe, G.J., and S.A.E.G. Falle. 1999. One-dimensional numerical simulations of idealized detonations. *Proc. R. Soc. Lond. A* 455:1203–1214.

Sharpe, G.J., and S.A.E.G. Falle. 2000a. Two-dimensional numerical simulations of idealized detonations. *Proc. R. Soc. Lond. A* 456:2081–2100.

Sharpe, G.J., and S.A.E.G. Falle. 2000b. Numerical simulations of pulsating detonations: I. Non-linear stability of steady detonations. *Combust. Theory. Model.* 4:557–574.

Short, M. 1996. An asymptotic derivation of the linear stability of the square-wave detonation using the Newtonian limit. *Proc. R. Soc. Lond. A* 452:2203–2224.

Short, M. 1997a. Multidimensional linear stability of a detonation wave at high activation energy. *SIAM J. Appl. Math.* 57:307–326.

Short, M. 1997b. A parabolic linear evolution equation for cellular detonation instability. *Combust. Theory. Model.* 1:313–346.

Short, M. 2001. A non-linear evolution equation for pulsating Chapman–Jouguet detonations with chain-branching kinetics. *J. Fluid Mech.* 430:381–400.

Short, M. 2005. Theory and modeling of detonation wave stability: A brief look at the past and toward the future. In *Proc. 20th ICDERS*, Montreal, Quebec, Canada.

Short, M., and P.A. Blythe. 2002. Structure and stability of weak-heat-release detonations for finite Mach numbers. *Proc. R. Soc. Lond. A* 458:1795–1807.

Short, M., and J.W. Dold. 1996. Linear stability of a detonation wave with a model three-step chain-branching reaction. *Math. Comput. Model.* 24:115–123.

Short, M., and J.J. Quirk. 1997. On the non-linear stability and detonability limit of a detonation wave for a model three-step chain-branching reaction. *J. Fluid Mech.* 339:89–119.

Short, M., and G.J. Sharpe. 2003. Pulsating instability of detonations with a two-step chain-branching reaction model: Theory and numerics. *Combust. Theory Model.* 7:401–416.

Short, M., and D.S. Stewart. 1997. Low-frequency two-dimensional linear instability of plane detonation. *J. Fluid Mech.* 340:249–295.

Short, M., and D.S. Stewart. 1998. Cellular detonation stability. Part 1. A normal-mode linear analysis. *J. Fluid Mech.* 368:229–262.

Short, M., and D.S. Stewart. 1999. The multi-dimensional stability of weak-heat-release detonations. *J. Fluid Mech.* 382:109–135.

Short, M., and D. Wang. 2001. On the dynamics of pulsating detonations. *Combust. Theory. Model.* 5:343–352.

Stewart, D.S., T.D. Aslam, and J. Yao. 1996. On the evolution of cellular detonation. *Proc. Combust. Inst.* 26:2981–2989.

Stewart, D.S., and J.B. Bdzil. 1988. The shock dynamics of stable multidimensional detonation. *Combust. Flame* 72:311–323.

Strehlow, R.A. 1969. the nature of transverse waves in detonations. *Astro. Acta.* 5:539–548.

Strehlow, R.A., and F.D. Fernandes. 1965. Transverse waves in detonations. *Combust. Flame* 9:109–119.

Taki, S., and T. Fujiwara. 1978. Numerical analysis of two-dimensional non-steady detonations. *AIAA J.* 16:73–77.

Taylor, G.I. 1950. The dynamics of the combustion products behind plane and spherical detonation fronts in explosives. *Proc. R. Soc. Lond. A* 200:235–247.

Teodorczyk, A., and J.H.S. Lee. 1995. Detonation attenuation by foams and wire meshes lining the walls. *Shock Waves* 4:225–236.

Thibault P. private communication. See also: Thibault, P.A., and J. E. Shepherd. 1987. Effect of reversible reactions on one-dimensional detonation stability, paper presented at SIAM *Conf. on Numerical Combustion*, San Francisco, March 9–11.

Tsuboi, N., A. K. Hayashi, and Y. Matsumoto. 2000. *Shock Waves.* 10:274–285.

Tsuboi, N., S. Katoh, and A.K. Hayashi. 2002. Three-dimensional numerical simulation for hydrogen/air detonation: Rectangular and diagonal structures. *Proc. Combust. Inst.* 29:2783–2788.

Varatharajan, B., and F.A. Williams. 2001. Chemical-kinetic descriptions of high-temperature ignition and detonation of acetylene–oxygen–diluent systems. *Combust. Flame* 124:624–645.

Voyevodsky, V.V., and R.I. Soloukhin. 1965. On the mechanism and explosion limits of hydrogen–oxygen chain self-ignition in shock waves. *Proc. Combust. Inst.* 10:279–283.

Watt, S.D., and G.J. Sharpe. 2004. One-dimensional linear stability of curved detonations. *Proc. R. Soc. Lond. A* 460:2551–2568.

Williams, D.N., L. Bauwens, and E.S. Oran. 1997. Detailed structure and propagation of three-dimensional detonations. *Proc. Combust. Inst.* 26:2991–2998.

Yao, J., and D.S. Stewart. 1996. On the dynamics of multi-dimensional detonation. *J. Fluid Mech.* 309:225–275.

Yungster, S., and K. Radhakrishnan. 2005. Structure and stability of one-dimensional detonations in ethylene–air mixtures. *Shock Waves* 14:61–72.

Zaidel, R.M. 1961. The stability of detonation waves in gaseous mixtures. *Dokl. Akad. Nauk SSSR* 136:1142–1145.

6 Unstable Detonations: Experimental Observations

6.1. INTRODUCTION

The heart of the detonation phenomenon is the structure where the detailed ignition and combustion processes take place. The Chapman–Jouguet (CJ) theory does not require any knowledge of the structure, as it is based on steady one-dimensional flow across the transition zone with equilibrium conditions on both sides. The CJ theory also assumes that the flow within the reaction zone itself is one-dimensional and steady. The details of the transition zone are provided by the Zeldovich–von Neumann–Döring (ZND) model for the detonation structure. The ZND model is explicitly based on a steady planar structure and is therefore compatible with the one-dimensional gasdynamic CJ theory. The excellent agreement between experimental values of the detonation velocity and the theoretical predictions from the CJ theory fortuitously confirm the validity of the steady one-dimensional assumption of the CJ theory and also, implicitly, the ZND model for the detonation structure.

Early experimental diagnostics lacked the resolution to disprove the steady one-dimensional ZND structure. The state of detonation theory in the early 1960s was best summarized in a remark by Fay (1962): "...the peculiar disadvantage of detonation research is that it was too successful at too early a date. The quantitative explanation of the velocity of such waves by Chapman and Jouguet has perhaps intimidated further enquiry." However, it was during the late 1950s and early 1960s when overwhelming experimental evidence was produced to show that the detonation structure is neither steady nor one-dimensional. Thus, the ZND model cannot describe the structure of an actual detonation wave, for the real structure is transient and three-dimensional. The detonation front is found to be unstable even for readily detonable mixtures well within the detonability limits. The problem of reconciling the steady one-dimensional CJ theory with the transient three-dimensional structure of the detonation front remains unsolved to date.

Instability of the detonation front was actually observed much earlier than the formulation of the ZND model in the early 1940s. However, the instability was thought to be a near-limit phenomenon only, and detonation waves away from the limits were thought to be steady and one-dimensional. The first report of nonsteady detonation waves was by Campbell and Woodhead (1926). They identified the phenomenon of spinning detonations in small-diameter tubes near the detonation limits. Spinning detonations are relatively easy to observe, because the scale of the instability of the detonation front is of the order of the tube diameter (the pitch of the spin is about three times the tube diameter). Thus, in tubes of the order of centimeters, the periodic fluctuation in the speed of the detonation front can readily be observed even with the limited resolution of early streak cameras. Away from the limits, the spin frequency increases and the amplitude of the fluctuations of the frontal instability decreases, making it much more difficult to resolve. Thus, as Fay pointed out, early researchers were reluctant to abandon the steady one-dimensional CJ theory and to pursue a more thorough probe into the detonation structure itself.

In the late 1950s and early 1960s, a number of novel diagnostic techniques were introduced into detonation research (piezoelectric transducers, the soot-foil method, light scattering off the detonation front, high-speed schlieren and interferograms, fully compensated streak photography, thin-film heat transfer gauges, etc.). These techniques provided convincing evidence to demonstrate the universal unstable structure of self-sustained detonation fronts. The transient three-dimensional structure presents formidable experimental difficulties in making detailed measurements within the transition zone. It is only in recent years that planar laser imaging techniques have permitted an instantaneous observation of the cross-section of the three-dimensional unstable detonation structure. Numerical simulations have also permitted the complex phenomenon to be studied in reduced dimensions, with simplified chemistry and in the absence of transport processes (see Chapter 5). In the present chapter, experimental results obtained by the various diagnostic techniques are presented to illustrate the salient features of the unstable detonation front.

6.2. THE SPINNING DETONATION PHENOMENON

The single-headed spinning detonation in a round tube is truly a unique phenomenon. The three-dimensional structure is stationary with respect to a coordinate system fixed to the rotating structure and propagates with the detonation wave. The results of a thorough investigation of the phenomenon were first reported by Campbell and Woodhead (1926), followed by a succession of papers (Campbell & Woodhead, 1927; Campbell & Finch, 1928). It should be noted that Campbell also pointed out that the streak photographs from the earlier study of Dixon (1903) already showed a similar unstable phenomenon, but it went unnoticed because of the uncertainty in the limited resolution of the streak camera Dixon used at that time. To illustrate the early observation of the spin phenomenon, the classical streak photograph taken by Campbell and Woodhead (1927) of a spinning detonation in a $2CO + O_2$ mixture with 2% H_2 in a 15-mm-diameter tube is shown in Fig. 6.1.

It can be observed that the leading edge of the streak trajectory undergoes a regular undulatory behavior, indicating a periodic variation of the detonation front

time ↑

distance ◄

Figure 6.1. Streak photograph of a spinning detonation (Campbell & Woodhead, 1927).

velocity. However, the mean velocity remains constant and corresponds closely to the CJ value for the mixture. Also evident in Fig. 6.1 are the regular horizontal bands of luminosity that extend from the front well into the product gases.

Using various ingenious arrangements of photographic slits and viewing directions of the streak camera, Campbell and co-workers concluded that there is a localized, intense combustion region at the front near the wall of the tube that rotates circumferentially as the detonation propagates, thus tracing out a helical path. The luminous bands behind the front are due to a transverse pressure wave that also propagates circumferentially with the localized, intense region at the detonation front. (Campbell credited his explanation of the results to a Mr. E.F. Greig.) The intense region at the front is referred to as the *spinning head*, and the associated luminous band that extends backward into the product gases is the *tail*. The tail is interpreted as a transverse compression wave that causes a higher luminosity of the product gases due to the adiabatic compression. The circumferential rotation of the spinning head was further confirmed by the cycloidal path obtained when the detonation is viewed end-on with a streak camera looking along the tube axis through a diametral slit. A helical path was also observed on the lead vapor deposit (left over from previous firings) on the wall of a glass section of the detonation tube adjacent to a lead section. This gives further support for the existence of a rotating spinning head.

Initially, Campbell and co-workers proposed that a bulk rotation of the combustion gases is involved in the rotation of the spinning head. However, this was conclusively refuted by the later studies of Bone *et al.* (1935), who found that a longitudinal rib placed inside the tube to impede the gas motion did not influence the spinning phenomenon. Thus, it was concluded that only a transverse pressure wave rotates circumferentially, and the gas particles themselves merely undergo an oscillation about a mean as the wave propagates by.

Perhaps the first to measure the detonation pressure was Gordon (1949), who developed special, fast-response piezoelectric transducers for this purpose. A typical pressure–time record for a spinning detonation in a 48-mm-diameter tube is shown in Fig. 6.2 (Donato, 1982). We first note a double peak at the front, indicating that a weaker shock first reached the transducer, followed by a very sharp spike. The initial weaker signal is from the incident shock, whereas the large spike is due to the Mach stem portion of the folded leading shock front of the spinning detonation. Note that the periodic pressure signal remains for many cycles with only a slight attenuation after the passage of the detonation front, indicating that the transverse pressure wave extends well into the rear of the detonation front. It should also be noted that the amplitudes of the transverse pressure oscillations are not small relative to the pressure at the front. The rotation of the transverse wave is also confirmed by an observation made by Gordon *et al.* (1959), who found that the pressure oscillations are 180° out of phase when registered by two pressure transducers placed directly opposite to each other on the tube wall. From the measurement of the frequencies of the subsequent pressure pulses, Gordon *et al.* also pointed out that the transverse pressure wave need not extend back from the front as a straight line, but can be twisted. The pressure measurements confirmed the presence of a long transverse pressure wave that extends backward from the front, and it is the compression from this transverse wave that gives rise to the luminous bands observed in the streak photograph.

It is interesting to note that as one approaches the detonation limits, the onset of spinning detonation occurs. However, the spinning phenomenon persists for a range of conditions after its first appearance until it finally disappears, and the detonation fails when the limit is reached. In initiation experiments using a powerful initiation source, Mooradian and Gordon (1951) observed that during the transient decay of the initially strongly overdriven detonation in mixtures outside the limits, the spin phenomenon occurs when the detonation reaches the CJ velocity of the mixture and disappears with further decay. Thus, spinning detonations are observed for a small range of velocities around the CJ value during the transient decay of the overdriven wave. However, the spin phenomenon abruptly vanishes with further decay past

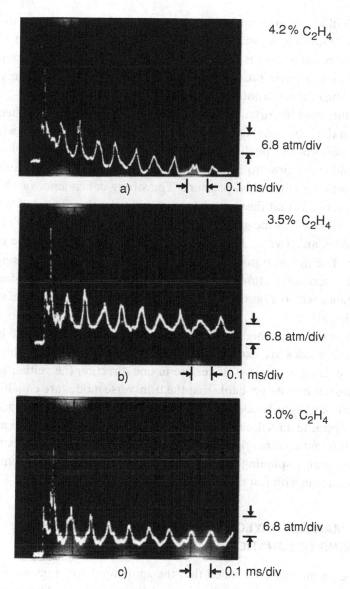

Figure 6.2. Samples of pressure–time histories of a spinning detonation in C_2H_4–air mixtures in a 48-mm-diameter tube (Donato, 1982).

some critical value of the detonation velocity. This indicates that overdriven detonations are stable (or the spin frequency is too high to be resolved) and that spin occurs near conditions of self-sustained propagation. Thus, the spin phenomenon appears to be nature's last resort for maintaining the detonation mode of combustion for most mixtures.

The energy required to sustain the transverse vibration of spinning detonation must be derived from the chemical energy release at the front. In the experiment by Bone *et al.* (1935), the energy released at the front is momentarily interrupted by putting a small gap of inert nitrogen between two columns of the explosive mixture in the tube. When the spinning detonation transits across the inert nitrogen gap where no chemical energy is released, the spin is abruptly suppressed. Thus, the rotation of the transverse pressure wave is driven directly by the energy released at the spinning head at the front.

Perhaps the most important property of spinning detonations is their strong dependence on the tube diameter. In the range of near-limit conditions when spinning occurs, the ratio of the pitch to the diameter of the helical path of the spinning head is found to be close to 3. This ratio is found to be rather insensitive to mixture composition and to initial pressure. The strong dependence on the dimension of the tube suggests that the spin phenomenon is closely associated with the natural acoustic vibration of the gas column behind the detonation where the transverse acoustic modes are governed by the characteristic dimensions of the cross-section of the tube. The lowest transverse acoustic mode corresponds to a single pressure wave rotating around the tube perimeter with a characteristic dimension of πd. This is in accordance with the observations of the pitch-to-diameter ratio of spinning detonation being about 3.

Away from the limits, higher spin frequencies are observed as the higher transverse acoustic modes are excited. At higher spin frequencies, it is highly unlikely that all the excited transverse modes are in one direction (i.e., either left- or right-handed spins). It is more probable that the transverse modes are equally partitioned between left- and right-handed spins. Hence, we have two sets of transverse waves rotating in opposite directions. Because the amplitudes of these waves are not small, the waves interact nonlinearly upon collision with one another. Thus, the structure of higher-frequency spinning detonations can be quite complex as compared to the single-headed spin with just one transverse wave.

6.3. THE MANSON–TAYLOR–FAY–CHU ACOUSTIC THEORY OF SPINNING DETONATION

The early experiments established that the spinning detonation is a manifestation of the vibrations in the gas column behind the detonation. Manson (1945, 1947) was perhaps the first to recognize the direct correspondence of spinning detonations with the transverse acoustic vibration of the gas column behind the detonation. He considered the vibration in a uniform gas column of the product gases and hence ignored the nonsteady expansion in the Taylor wave behind the detonation front. He also assumed that the amplitude of the oscillations is small, and thus the vibration can be described by linear acoustic theory. Manson considered only transverse vibrations and thus ignored the longitudinal oscillations along the axis of the tube. In essence, he applied the solution for the velocity potential, $\phi(r, \theta, t)$, of the two-dimensional linear acoustic equation to describe the transverse oscillation in the cylindrical gas column. The solution for $\phi(r, \theta, t)$ can readily be obtained by the method of separation of variables, and $\phi(r, \theta, t)$ is given as a product of a harmonic

Table 6.1. Values of $k_{n1}R$ for
the circumferential mode

n	$k_{n1}R$
1	1.841
2	3.054
3	4.201
4	5.35
5	6.35

function and a Bessel function of the first kind for cylindrical geometry. The Bessel function of the second kind is not compatible with the boundary condition at the axis $r = 0$, for it is infinite there. Applying the boundary condition at the wall where the particle velocity normal to the wall vanishes gives

$$u = \left(\frac{\partial \phi}{\partial r}\right)_{r=R} = 0,$$

resulting in the first derivative of the Bessel function vanishing at the wall. Thus,

$$J_n'(k_{nm}R) = 0,$$

where $k_{nm}R$ is the zeroth root of the first derivative of the Bessel function. The integers n and m denote the number of circumferential and radial modes, respectively. If we consider only the circumferential mode, we take $m = 1$ and the numerical values of the first few modes of $k_{n1}R$ are given in Table 6.1.

The angular velocity of rotation of the transverse wave can be obtained from the acoustic solution as

$$\omega_n = \frac{k_{nm}c_1}{n},$$

where c_1 is the speed of sound. The linear velocity of the transverse wave at the wall can be written as

$$v_n = \omega_n R = \frac{k_{nm}c_1 R}{n}.$$

Equating the time for the transverse wave to travel around the circumference to the time for the detonation to propagate a distance equal to the pitch, that is,

$$\frac{\pi d}{v_n} = \frac{p_n}{D},$$

the pitch-to-diameter ratio, p_n/d, can be obtained as

$$\frac{p_n}{d} = \frac{n\pi}{k_{nm}R}\left(\frac{D}{c_1}\right),$$

where D is the axial velocity of propagation of the detonation front. For a single transverse mode where $n = 1$, we get

$$\frac{p_1}{d} = \frac{\pi}{1.841}\left(\frac{D}{c_1}\right).$$

From CJ detonation theory, the ratio of the sound speed to the detonation velocity can be written as

$$\frac{c_1}{D} \approx \frac{\rho_0}{\rho_1} \approx \frac{\gamma}{\gamma + 1},$$

where γ is the specific heat ratio of the product gases. Thus, for a typical value of $\gamma \approx 1.2$, we get

$$\frac{p_1}{d} \approx 3.128,$$

which is in good agreement with experimental observations. We also note that the ratio c_1/D is not very sensitive to the mixture composition or to initial pressure. Thus, the pitch-to-diameter ratio for single-headed spinning detonation is generally around 3 for most mixtures.

Extension of the acoustic theory to rectangular tubes can be done readily by writing the acoustic equations in Cartesian coordinates. Accordingly, the solution is given as a combination of harmonic functions and boundary conditions of zero particle velocity normal to the wall, and gives eigenfrequencies of oscillations in the transverse x and y directions (z being the tube axis). Transverse frequencies for other tube geometries can be obtained in a similar fashion.

Away from the limits, Manson found that the acoustic theory can also predict the spin frequencies quite satisfactorily. It can be concluded that the spinning detonations' frequency can adequately be described by considering transverse acoustic vibrations in the gas column behind the detonation. By considering only transverse vibrations, boundary conditions at the detonation front need not be considered, and the only link to the CJ theory is the ratio of the sound speed to the detonation velocity.

It is interesting to note that G.I. Taylor independently formulated a similar acoustic theory for spinning detonations in 1948.* However, when he submitted his results for publication, he was informed by Sir Alfred Eggerton of Manson's earlier publication. He withdrew his paper and instead wrote a letter to Manson, informing him of his work and describing the additional results that he had obtained for the frequencies of vibrations in triangular cross-section tubes. Taylor's work was eventually published in 1958 (Taylor & Tankin, 1958). It is not known if Taylor had carried out a more general three-dimensional theory and included longitudinal oscillations in his study of spinning detonations.

Another independent investigation of spinning detonations was carried out by Fay in his Ph.D. thesis at Cornell University in 1951. Fay (1952) initially attempted

---

* Private communication with N. Manson.

a three-dimensional theory and considered longitudinal as well as transverse vibrations in the gas column behind the detonation. The solution required another boundary condition to be satisfied at the detonation front. This takes the form of a dimensionless quantity, z, called the specific acoustic impedance and is essentially a relationship between the vibrational pressure and the particle velocity. For conditions of practical significance where transverse frequencies are much higher than longitudinal frequencies, the value of z is found to be large. An infinite value of z means no axial velocity, and thus the vibration is only transverse. Hence, Fay showed that the vibration is predominantly transverse, in accord with experiments. For $z \to \infty$, Fay's results are identical to those of Manson.

In Fay's investigation, he also considered the mechanisms by which the vibrations can be sustained. He concluded that there must exist a nonuniform energy release at the detonation front, which couples to the transverse wave motion. This is in accord with the observation of Campbell and his co-workers (Campbell & Woodhead; 1926, 1927; Campbell & Finch, 1928) of an intense reaction region near the wall that rotates circumferentially as the detonation propagates. However, within the framework of the linear acoustic theory, the exact nature of the nonuniform energy release at the front cannot be described.

Perhaps the most complete theoretical analysis of the acoustic vibration in the product gas column behind a detonation wave is that of Boa Teh Chu (1956). He also assumed a uniform gas column, but had to consider a slightly overdriven detonation so that the flow is subsonic behind the detonation, thus permitting the longitudinal waves to communicate between the back boundary and the detonation front. Instead of using the linear acoustic equation in the velocity potential as in Manson's and Fay's analyses, Chu started with the Euler equations. Assuming small perturbations in the flow variables, Chu linearized the Euler equations and obtained an analogous linear wave equation with the perturbation pressure as the dependent variable. For the boundary condition at the detonation front, he perturbed the Rankine–Hugoniot equations and obtained the relationships between the perturbed flow variables at the front and the perturbation of the frontal surface of the detonation wave itself. In contrast to Fay's earlier analysis, Chu obtained an accurate description of the boundary condition at the detonation front.

In both Manson's and Fay's earlier investigations, only the solution of the linear wave equation was subjected to the appropriate boundary conditions to obtain the spin frequency. The mechanism of how the transverse waves propagate was not considered. Chu's analysis differs in that he studied explicitly the generation of pressure waves at the detonation front. In accordance with experimental observations of spinning detonation where there is a rotating, intense combustion region at the detonation front, Chu considered the generation of pressure waves by a rotating heat source. Variation in the rate of heat release in a compressible medium results in the generation of pressure waves. Specifically, Chu chose the variation of heat release of the planar rotating heater to correspond to the form of the solution of the wave equation. In cylindrical coordinates, the solution of the wave equation would be given by a combination of harmonic and Bessel functions. Chu also pointed out that the pressure waves generated by the rotating heater are equivalent to those from a rotating piston with a surface profile corresponding to the solution of the wave equation. In essence, Chu analyzed the generation of pressure waves in the domain $z < 0$ by a planar heater at $z = 0$ rotating at angular velocity ω_0, with a surface distribution of heat release rate corresponding to the solution of the linear wave equation. He obtained the results described in the following paragraphs.

When the driving frequency is greater than the eigenfrequency of the gas column, an undamped helical wave train propagating back from the heater into the region $z < 0$ is generated by the rotating heater. The helical waves rotate about the tube axis at the driving angular velocity ω_0 of the rotating heater. When there are n modes, there will be n helical waves equally spaced around the circumference with a pitch of $2\pi/k$, where k is the wavenumber. As Chu pointed out, the wave train rotates together as if the waves were threads on a helical screw.

When the rotation frequency of the heater approaches the natural transverse eigenfrequency, the pressure waves depend less and less on z and become more and more transverse. When the driving frequency corresponds to the eigenfrequency, the oscillations become purely transverse. The heater is now driving the column of gas at the resonant frequency, the amplitude increases without bound, and the linearized acoustic theory can no longer describe the vibrations.

When the driving frequency is less than the eigenfrequencies, the pressure waves decay rapidly away from the heater. There is also a cutoff frequency below which a particular driving mode fails to propagate out as a wave train. Thus, Chu concluded that the condition for pressure waves generated by a rotating heat source to propagate out as a nondecaying wave train is that the frequency of the pressure fluctuation (at any fixed point in the tube) should not be less than the natural eigenfrequency of the transverse vibration. For tubes of other geometrical cross-sections, the same conclusion applies, except that the surface variation of the heat release rate of the heater must now be driven by some other means.

Chu also analyzed the reflection of longitudinal waves off the detonation front and concluded that the overdriven detonation is stable to all modes of excitation. This result is in accordance with the later result from stability theory that indicates overdriven detonations are, in general, stable. The stability of the detonation wave to small perturbations also implies that there is no natural means to generate and sustain the transverse vibrations. Thus, Chu concluded that his analysis failed to describe the cause of the spinning detonation. However, more detailed stability analyses carried out subsequent to Chu's work all indicate that the planar ZND structure of a detonation is unstable for activation energies of the reactions corresponding to those of practical explosive mixtures.

A more rigorous theory for spinning detonations would have to consider their structure and the non-equilibrium reaction zone. Such an analysis becomes essentially a stability analysis of the detonation structure, but with all the boundary conditions (at the tube wall, the shock front, and the rear boundary of the products) to be

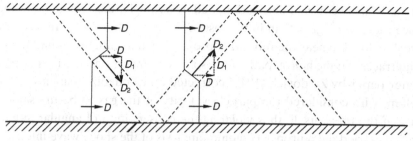

Figure 6.3. Sketch of Shchelkin's model of a spinning detonation front.

satisfied. The nonlinear interaction of the transverse waves at the front should also be properly described. Such an analysis presents formidable difficulties, and a complete numerical simulation that includes the unstable front as well as the vibrations in the gas column in the wake of the detonation has yet to be carried out.

Although the acoustic theory of spinning detonation due to Manson, Taylor, Fay, and Chu failed to explain the origin of spin instability, it does yield a very important result: the strong coupling between the natural transverse eigenmodes and the detonation structure. The transverse vibrations must be sustained by the nonuniform energy release rate at the front. Thus, the frontal structure must adjust itself according to the transverse eigenmode to achieve resonant coupling. Although the energy release at the front drives the transverse vibrations in the product gases in the rear, it is the natural frequency of the transverse vibration that determines the nature of the nonuniform energy release at the front. This is particularly true for low-mode vibrations in near-limit detonations. When the spin frequency is very high, the natural eigenmodes play a lesser role and the transverse vibrations now have to establish resonant coupling with the characteristic chemical reaction rates rather than the characteristic eigenmodes of the tube.

6.4. STRUCTURE OF THE SPINNING DETONATION FRONT

The linear acoustic theory of Manson, Taylor, Fay, and Chu cannot be expected to provide a detailed description of the frontal structure of a spinning detonation. The analysis of Chu only managed to show how the detonation front should deform in accordance with the pressure distribution as obtained from the solution of the wave equation. At the front, the transverse waves are actually shock fronts. Thus, a description of the frontal structure must involve the nonlinear interaction of shock waves. Perhaps the first to suggest that the leading shock front of a spinning detonation consists of a *break* (i.e., an abrupt change in the slope of the shock surface) or a *crease* was Shchelkin (1945). A sketch of Shchelkin's model of the spinning detonation front is shown in Fig. 6.3. The break, traveling with the front, will have the same axial velocity, equal to the detonation velocity D. The tangential velocity of the break is D_1. Thus, the velocity of the break propagating into the unburned mixture is $D_2 = \sqrt{D_1^2 + D^2}$ and is greater than the detonation velocity D. In other words, the break is an overdriven detonation, and accordingly, the pressure and the temperature there are higher than in the rest of the detonation front. In a round tube, the break is located near the tube wall, propagates tangentially around the periphery, and traces out the helical path of the spin head that is observed experimentally.

A later paper by Zeldovich (1946) corrected an erroneous assumption made by Shchelkin, who considered the particle velocity of the gas to be the same as the velocity of the break itself, thus violating the conservation of angular momentum. Zeldovich carried out a proper Hugoniot analysis of the shock wave intersection at the break and managed to compute the pitch angle of the helix in agreement with experimental observation.

The idea of a break or a crease on the leading shock front surface was postulated by Shchelkin. The crease defines the intersection of the shock surfaces on either side of it. To satisfy the boundary condition behind the two intersecting shock waves, it is found that a third shock (a reflected shock) – as well as a slip line, across which the pressure is continuous – is necessary. Thus, the break is essentially a triple-shock Mach intersection, and in a spinning detonation front there is a Mach intersection near the wall formed by the crease in the shock front, but the crease flattens itself out toward the center to give a continuous smooth shock surface. On one side of the crease, the shock is stronger and is called the Mach stem. The weaker shock on the opposite side of the crease is called the *incident* wave. The transverse, or reflected, wave propagates into the shock-compressed but as yet unreacted mixture. Combustion is intense behind the strong Mach stem and the transverse shock. This localized luminous region constitutes the spinning head. However, it was not until the late 1950s and early 1960s that the detailed spinning detonation structure was finally discovered, confirming the concept of the break in the leading shock surface proposed by Shchelkin in 1945.

Although the cycloidal trajectory registered on the end-on streak photograph by Campbell and Woodhead, looking down the axis of the tube, confirmed the circumferential motion of the spinning head, a direct open-shutter photograph of the luminous helical path can also be obtained for certain explosive mixtures. This was first noted by Gordon *et al.* (1959), who photographed the helical luminous path of a spinning detonation in an explosive mixture of ozone, oxygen, and nitrogen pentoxide. A similar open-shutter photograph in $C_2H_2-O_2-Ar$ mixtures in a 25-mm-diameter tube, taken by Schott (1965), is shown in Fig. 6.4. In acetylene–oxygen detonations, the luminosity is highly localized in the nonequilibrium reaction zone. The visible light emission from the product gases is minimal as compared to the reaction zone. The spiral path of the spinning head is evident in Fig. 6.4. The spiral path also shows a lot of fine detailed structure, indicating that the reactions in the spinning head are highly nonuniform.

Perhaps the most useful technique for investigating the structure of unstable detonation fronts is the soot-foil technique. This technique was discovered by Antolik (1875), who first observed that the path of a triple-shock Mach intersection can be recorded as a well-defined thin line on a soot-coated surface. The soot-foil

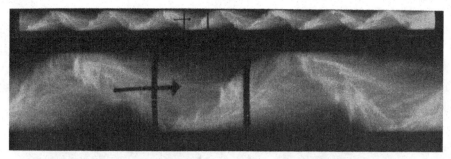

Figure 6.4. Open-shutter photograph of a spinning detonation (Schott, 1965).

technique was subsequently used by Mach and Sommer (1877) in their study of spark discharges and interacting shock waves. It was Denisov and Troshin (1959) who first applied this technique to the study of unstable detonation structure. However, it should be pointed out that Denisov and Troshin were probably inspired by the earlier work on spinning detonation. Campbell and Finch (1928) already noted the spiral path left on lead-vapor-coated tube walls by a spinning detonation. Bone *et al.* (1935) also reported that spiral tracks were etched on the silver-coated inner wall of the detonation tube. In fact, Campbell and Finch tried to coat the inner wall of the tube with various incombustible powders to try to register the helical path of the spinning head. They eventually discovered that a light deposit of French chalk adhered well to the glass surface to give a good recording of the spiral trajectory made by a spinning detonation. The coating of a glass, metal, or Mylar foil with the soot from a wooden match or a richly burning hydrocarbon flame has been established as the standard technique used to record the trajectories of triple-shock Mach intersections of unstable detonations for the past 50 years. Duff (1961) gave a detailed description of the procedure used to obtain uniform soot coatings.

Although the soot-foil technique has been widely used for the past five decades, the precise mechanism by which the soot is removed by the passage of the triple point over the soot-coated surface is still not understood. It is clear that the high temperature and pressure associated with the Mach stem and reflected shock in the vicinity of the triple point cannot inscribe such a sharp, thin line. A suggested mechanism is that the soot is removed by the extremely sharp velocity gradient of the slip line at the triple point. The sharp differential velocity change across the shear discontinuity is thought to provide the scouring action. However, a boundary layer is always present at the wall surface. The no-slip boundary condition at the surface will tend to dissipate the differential velocity gradient across the shear line associated with the triple point. Thus, it is doubtful that the scouring action of shear flows can remove the soot deposit.

On the other hand, if there is a pressure gradient normal to the wall, soot particles can be lifted off the surface effectively. A planar shear layer can be thought of as the superposition of a series of line vortices. Thus, at the triple point, we should have a very concentrated intense vortex tube, and soot particles can be lifted off normal

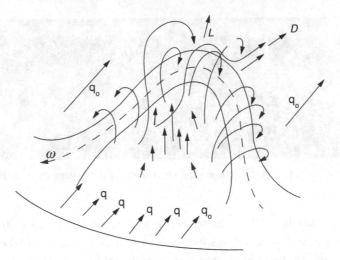

Figure 6.5. Sketch of a hairpin vortex.

to the surface as the triple point (line) sweeps by. In turbulent boundary layers, *hairpin* vortices have been postulated by Theodorsen (1952) and observed experimentally by Head and Bandyopadhyay (1981), among others. Figure 6.5 is a sketch of a hairpin vortex in a turbulent boundary layer as postulated by Theodorsen. The low-pressure core of the vortex element can aspirate the soot particles from the wall. Thus, it is conceivable that a similar mechanism might be present as the shear layer of the triple-shock Mach configuration sweeps across the soot foil.

So far, none of the proposed mechanisms of soot removal has been investigated thoroughly enough to permit any definitive conclusions. However, there is ample experimental proof that the inscription on the soot foil is due to the passage of the triple point of a Mach configuration propagating over the surface.

A typical soot record of the path left behind by a spinning detonation is shown in Fig. 6.6. The Mylar foil covered the entire inner wall of the round tube; thus the vertical dimension of the foil is the circumference, πd, of the tube. The helical path is a wavy band, indicating that the transverse wave itself is not stable. The finite width of the band is due to the transverse shock propagating into the compressed yet unburned mixture in the induction zone behind the incident shock of the Mach intersection. The angle of the helix made with the tube axis is around 45° and is in accordance with the acoustic theory, where

$$\tan \alpha = \frac{\pi d}{p},$$

and the pitch-to-diameter ratio is given by

$$\frac{p}{d} = \frac{\pi}{1.841}\left(\frac{D}{c_1}\right) = \frac{\pi}{1.841}\left(\frac{\gamma+1}{\gamma}\right) = 3.128$$

for $\gamma = 1.2$. Thus, $\alpha = \tan^{-1}\frac{\pi}{3.128} = 45.12°$.

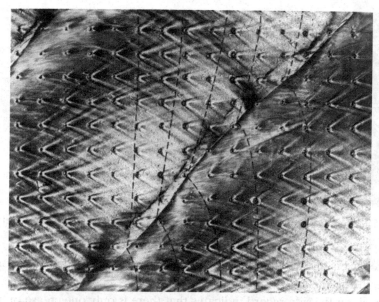

Figure 6.6. Soot-foil record of the path behind a spinning detonation.

The V-tracks shown on the soot record of Fig. 6.6 are due to a network of protrusions made by piercing a Mylar foil with a sharp pin against a soft surface prior to depositing soot on the foil. The passage of the flow over the protrusions creates the V-shaped marker. The perpendicular drawn to the line bisecting the V-marker gives the tangent to the shock surface at the wall. Thus, from the V-markers, the shape of the shock front at the wall of the tube can be constructed. The dashed lines in Fig. 6.6 represent the shock shape at different instants of time. The existence of a crease at the shock front near the wall, as suggested by Shchelkin, is clearly illustrated. Note also that the curvature of the stronger Mach stem is much larger than the incident shock on the opposite side of the crease of the Mach configuration.

A soot foil wrapped around the inner wall of the tube registers only the passage of triple-shock intersections at the outer circumference of the detonation front. The shape of the leading shock front away from the wall toward the axis of the tube can be determined from a sooted surface placed normal to the tube axis so that the detonation front reflects off it. The complex process of how the shock intersections can be recorded on the end-on soot foil from the reflection of the spinning detonation wave is not clear. However, the end-on soot foil does register the shock intersections on the detonation front surface. Figure 6.7 shows simultaneous side-wall and end-on soot records of a spinning detonation. The fine structure due to the instability of the transverse detonation propagating into the shock-compressed reactants behind the incident shock of the triple-shock Mach intersection is clearly evident in the side-wall soot record. The end-on soot record illustrates how the crease extends from the wall into the interior toward the tube axis. The fine structure shown in the

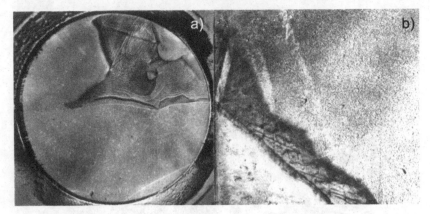

Figure 6.7. Simultaneous (a) end-on and (b) side-wall soot records of a spinning detonation.

end-on record is due to the reflected shock sweeping across the shock-compressed reactants in the induction zone behind the incident shock of the crease. A single helical track on the side record indicates that there is only one crease on the shock surface and that this crease does not extend to the opposite wall. If the crease extends to the opposite wall, then two parallel spiral tracks will be recorded on the side-wall soot foil. This is rarely observed. However, two spiral tracks corresponding to two creases propagating in different directions (i.e., left- and right-handed spins) are commonly observed. The frontal structure for such a double-headed spinning detonation is much more complicated.

Although the side-wall soot record is just a single helical track for single-headed spinning detonations, the frontal structure on the leading shock surface can be quite different. Figure 6.8 illustrates a series of end-on soot records of single-headed

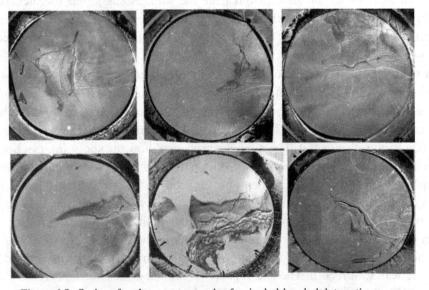

Figure 6.8. Series of end-on soot records of a singled-headed detonation wave.

detonation waves. The side-wall records are all similar in that only one helical path is registered. However, the variations in the interior structure away from the wall indicate that the spinning structure is not stationary in general. Interaction of the radial modes with the circumferential mode can lead to different internal configurations of the structure of the single-headed spinning detonation.

A spinning detonation where a single Mach configuration at the wall rotates circumferentially is, in general, particular to a round tube. In a rectangular tube, the lowest transverse mode will consist of two transverse waves: one propagating in each of the two directions of the tube's rectangular cross-section. However, if one of the dimensions is made sufficiently small, then the transverse mode in that direction can be suppressed, and only one transverse wave propagates in the remaining larger dimension. Such a single-mode detonation is appropriately called a *zigzag* detonation, for the transverse wave sweeps across the leading shock front and reflects off the opposite wall of the rectangular channel. Figure 6.9 shows a simultaneous side-wall and end-on record of a zigzag detonation. The crease extends across the entire height of the channel. Again, the fine structure in the end-on soot record illustrates the instability of the reflected shock as it traverses the shock-compressed mixture of the induction zone behind the incident wave of the Mach configuration of the crease.

The two-dimensional structure of the zigzag detonation was studied by Dove and Wagner (1960) and is simpler than the three-dimensional structure of spinning detonation in a round tube. However, the two-dimensional structure of a zigzag detonation is not stationary, and the shock strengths of the Mach intersection vary cyclically with time upon each reflection of the transverse wave off the tube wall. In contrast, the three-dimensional spinning detonation structure is stationary with respect to a coordinate axis fixed to the rotating detonation front.

The extremely stable and reproducible structure of a spinning detonation in a particular mixture of carbon monoxide and oxygen with a small addition of hydrogen permitted Voitsekhovskii *et al.* (1966) to carry out a thorough study of the spinning detonation structure in the late 1950s. They invented a novel streak photography technique that permitted them to get a detailed description of the spinning detonation structure at the wall of the tube. To understand their *fully compensated* method of streak photography, we first describe the usual streak mode where the film (time axis) moves normal to the direction of propagation. The resulting photograph is a bright line at an angle to the horizontal z-axis of propagation, and the slope is the reciprocal of the wave speed.

In 1949, Shchelkin and Troshin (1965) introduced a variation of the usual streak mode. They used a vertical slit normal to the horizontal z-axis of propagation and arranged to have the plane of the film moving in the same direction of propagation as the detonation wave. When the film speed is adjusted so that the image of the detonation on the film moves at the same speed as the film itself, then the image of the detonation becomes stationary relative to the film. The vertical slit then acts like a focal plane shutter of an ordinary camera, taking a still picture of the detonation.

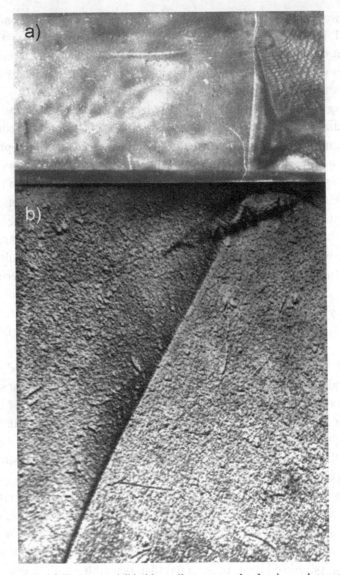

Figure 6.9. (a) End-on and (b) side-wall soot records of a zigzag detonation.

This is called the *compensated* method of streak photography, and Shchelkin and Troshin compensated for the longitudinal velocity of propagation of the detonation wave. In a spinning detonation, the detonation structure rotates as it propagates. Thus, there is an axial as well as an azimuthal velocity component.

Voitsekhovskii *et al.* oriented their streak camera so that the plane of the film moved in the direction tangent to the spiral trajectory of the spinning detonation front. In this manner, both the axial and the circumferential components of the velocity are fully compensated when the film speed matches the velocity of the detonation image tangent to the spiral trajectory. Again, the spinning detonation and

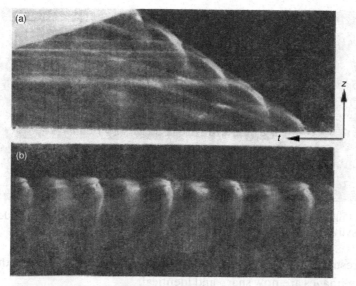

Figure 6.10. (a) Normal and (b) fully compensated streak photographs of a spinning detonation (Voitsekhovskii *et al.*, 1966).

the film can be thought of as stationary relative to each other, and the slit now acts as a focal plane shutter rotating, with the film, around the spinning detonation and photographing the luminosity pattern around the wall. An alternative explanation of this fully compensated streak technique was given by Voitsekhovskii as follows: consider the film and the detonation to be stationary relative to each other. Assume that the wall of the detonation tube is coated with a paint that varies in accordance to the distribution of luminosity of the spinning detonation front at the wall. Then, if the detonation tube is rolled on the film surface like a printer drum, the imprint on the film will correspond to the fully compensated streak photograph of Voitsekhovskii.

Figure 6.10 shows both a normal and a fully compensated streak photograph of a spinning detonation. In the fully compensated photograph shown in Fig. 6.10b, there appear two alternating images of the spinning head: one clear and one blurred. This is due to the fact that the detonation head appears twice at the slit in one revolution – when it is near the slit (closer to the camera), and again when it is at 180° on the far side, opposite the slit. When it is on the side closer to the camera, the motion of the film is in the same direction as the spinning head and full compensation is achieved. When it is on the opposite side to the slit, the film and detonation move in opposite directions and there is no compensation. The image on the opposite side of the slit also appears diffused because it is slightly out of focus. Light emission within the structure is also registered.

To eliminate the image on the far side of the slit, Voitsekhovskii *et al.* placed a small rod on the tube axis and thus blocked the image from the far side. Figure 6.11

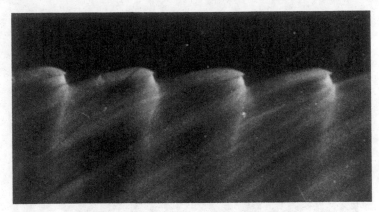

Figure 6.11. Fully compensated photograph of the spinning head on the near side of the tube (Voitsekhovskii *et al.*, 1966).

shows the resulting sequence of images of the spinning head, only from the side near the slit. The images are now sharp and identical.

To obtain the image of the leading shock front as well, Voitsekhovskii *et al.* placed the detonation tube in the path of the beam of a Toepler schlieren system.

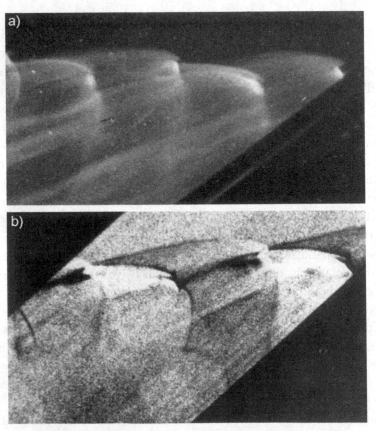

Figure 6.12. (a) Self-luminous and (b) schlieren photographs of a spinning head (Voitsekhovskii *et al.*, 1966).

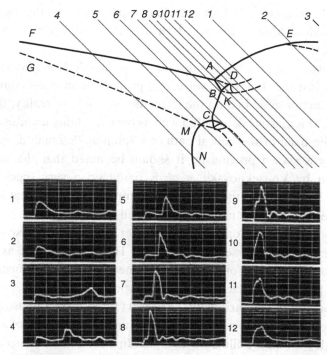

Figure 6.13. Detailed structure of a spinning detonation front and pressure histories across various sections of the spinning front (Voitsekhovskii *et al.*, 1966).

They also oriented the tube axis in such a manner that there was no superposition of the contours of the leading shock front and of the phenomenon occurring at the rear wall of the tube. They managed to obtain a full picture of the shocks and reaction zone of the spinning head. Figure 6.12 shows fully compensated streak photographs of the spinning head of both self-luminosity (Fig. 6.12a) and schlieren (Fig. 6.12b).

By placing miniature piezoelectric transducers on the wall of the detonation tube, Voitsekhovskii *et al.* managed to obtain the pressure–time histories along different cross-sections of the spinning head. The locations of the pressure transducers are identified from a simultaneous fully compensated streak photograph. Figure 6.13 gives the detailed structure of the spinning detonation front and the pressure–time histories along different cross-sections. The numbered parallel lines indicate the duration of propagation and the corresponding pressure–time histories along the various lines.

The crease of the leading shock front is located at A, and the Mach intersection consists of a stronger Mach stem AE on the right side of the crease, and a weaker incident shock AF on the left side. The strength of the curved Mach stem AE decreases away from the triple point A as indicated by pressure traces 10, 11, 12, 1, 2, and 3. The strength of the weaker incident shock AF is relatively constant, as indicated by the first peak in pressure traces 4 to 8. The transverse or reflected shock of the Mach configuration is *AB*. However, in the spinning head, the transverse shock consists of a double Mach stem in order to match the flow conditions behind the various waves.

For simplicity, we shall first consider the entire double Mach interaction system *ABDKCM* as just the transverse wave that propagates into the compressed mixture in the induction zone behind the incident shock *AF*. In reality, the transverse wave *ABDKCM* is a strong detonation with its own instability manifested as the fine structure in the band of the helical path of a spinning detonation, as indicated in the soot records shown previously. It should be noted that the detailed structure described by Voitsekhovskii *et al.* is based on a very special mixture of $2\,CO + O_2 + 3\%\,H_2$ where the spinning structure is extremely stable, and reproducible. In general, the spinning structure itself is unstable, and its velocity also fluctuates as it propagates along the tube. Higher harmonics are also superimposed on the fundamental transverse spinning mode. Thus, the structure as illustrated in Fig. 6.13 is not typical, and for this reason we need not be too concerned with the finer details of the double Mach interactions of the transverse wave.

In an independent study carried out by Schott (1965), fast-response platinum heat transfer gauges were used to map out the shock and reaction front of a spinning detonation in C_2H_2–O_2 highly diluted with argon. In this mixture, the spinning detonation is also relatively stable. By placing four heat transfer gauges around the tube periphery and using a smoke foil to identify the location of the gauge relative to the spinning head, Schott was able to map out the structure of the spinning detonation. The heat transfer gauges can respond to the shock front because the resistance of the film increases from the heating by the shock. When the reaction front arrives at the gauge, the gauge is shunted by the ionized gases. Thus, it is possible to determine relative positions of the shock and reaction front at the various gauge positions, so that, with the help of the soot record of the spin trajectory, the detailed structure of the spinning front at the wall can be determined.

Figure 6.14 shows the instantaneous positions of the shock and reaction fronts around the circumference of the tube. The results are constructed from five experiments, which is only possible when the spin structure is highly reproducible from shot to shot. In the stronger Mach stem and the transverse wave near the triple point, the separation between the shock and reaction front cannot be resolved. As can be observed, the spinning structure, as determined by Schott using a different technique, is almost identical to that obtained by Voitsekhovskii *et al.* It should be emphasized that the detailed structure obtained by Voitsekhovskii *et al.* and by Schott are based on mixtures that give unusually stable spinning detonations. In general, the fundamental spin mode itself is unstable and higher harmonics are excited in the vicinity of the spin head, because both the Mach stem and the transverse shock are themselves unstable detonations with their own unstable structure. The higher harmonics perturb the dominant spin mode and cause it to fluctuate as it propagates.

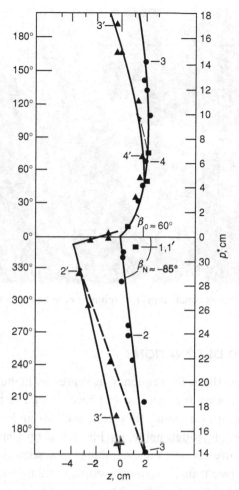

Figure 6.14. Instantaneous positions of the shock and the reaction front around the circumference of the tube (Schott, 1965).

The detailed structure obtained by Voitsekhovskii *et al.* and by Schott describes the phenomenon near the wall. Away from the wall, the three-dimensional structure changes. As indicated from the end-on soot record, the crease disappears as the Mach intersection alternates toward the tube axis and the leading shock surface becomes continuous. To achieve a better understanding of the entire three-dimensional structure of spinning detonation, a Plexiglas model was constructed by Schott. Figure 6.15 shows the three-dimensional model of a spinning detonation structure. The twisted band rotates the model as it slides along the axis. As pointed out previously, the single-headed spinning structure is a stationary structure with respect to a coordinate system fixed to it. Higher spin modes, and the single transverse zigzag detonation in a thin rectangular channel, do not have a stationary structure. Even for a coordinate system fixed to the transverse wave in a zigzag detonation, the strengths of the incident shock and Mach stem vary with time.

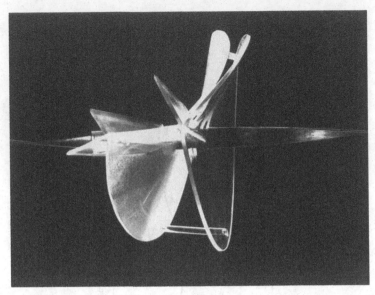

Figure 6.15. Three-dimensional model of a spinning detonation (courtesy of G. Schott).

6.5. MULTIHEADED DETONATIONS

Away from the limits, the spin frequency increases as higher acoustic modes are excited. Figure 6.16 shows the smoked-foil records of the progressive evolution of a single-headed spin in a round tube to a double- and a four-headed spin as the mixture becomes more detonable. In Fig. 6.16a of a single-headed spinning detonation, we can already see the beginning of a second transverse mode. In Fig. 6.16b, there are two transverse modes corresponding to a right-handed and a left-handed spin. In Fig. 6.16c, there are two transverse waves in each of the spin directions.

Because the writing on the soot foil is due to the passage of a triple-shock Mach intersection, in a high-frequency spinning detonation there will be a number of creases, or Mach configurations, at the wall, distributed around the periphery of the detonation front, usually with equal numbers of left- and right-handed spins. Since each Mach configuration corresponds to a spin head, high-frequency spinning detonations are generally referred to as *multiheaded* detonations. Figure 6.17 shows the soot record of a multiheaded detonation with V-markers to determine the shape of the front.

The multiple creases of the shock front at the wall are clearly illustrated. The Mach stems propagating circumferentially around the wall collide and reflect off one another. Unlike linear acoustic waves, where we can superimpose two sets of noninteracting transverse waves, the interaction of the colliding Mach stems is nonlinear. The Mach stems extending radially toward the tube axis also interact with one another. Thus, the structure of a multiheaded spinning detonation will involve a continuous nonlinear interaction of transverse waves. The pattern of shock

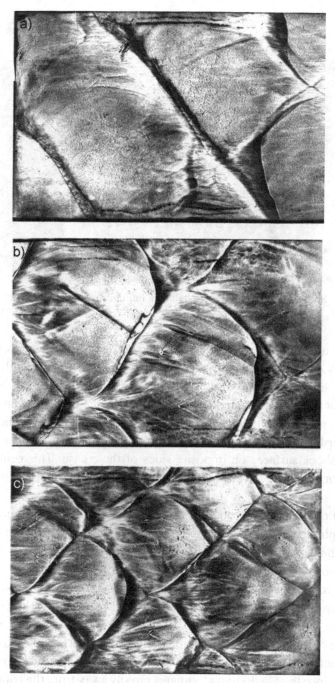

Figure 6.16. (a) Single-headed, (b) double-headed, and (c) four-headed spin.

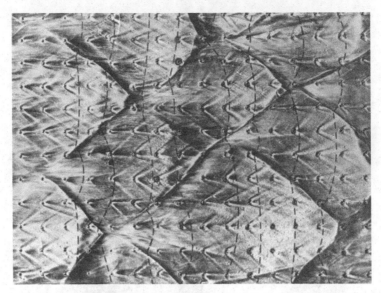

Figure 6.17. Soot-foil record of the path behind a multiheaded detonation (Lee *et al.*, 1969).

interactions across the surface of a multiheaded detonation front is shown in the simultaneous side-wall and end-on soot records of Fig. 6.18. Although the side-wall soot record shows a relatively regular pattern of two sets of transverse waves giving the characteristic diamond, or fish-scale, pattern, the corresponding end-on soot record indicates a rather complex pattern of shock intersections. The lines on the end-on record are the creases in the leading shock front formed by the intersection of the shock surfaces on opposite sides of the crease. This results in a third transverse, or reflected, shock that extends back from the crease into the product gases.

Thus, in a multiheaded detonation front, we may think of a system of transverse waves sweeping across the leading front. The boundaries of the intersection of the transverse shock with the leading front form the pattern of lines shown in the end-on soot record. As in the case of the single-headed spin structure, one side of the shock intersection boundary is the stronger Mach stem, where intense chemical reactions occur almost instantaneously behind the Mach stem with little delay. The opposite side of the shock intersection is the weaker incident wave, where the induction time before the onset of reaction is much longer. Thus, the transverse shock propagates into this induction zone of the incident shock and is a strong detonation wave itself. Since the Mach stem attenuates rapidly away from the triple point (line) of shock intersection, the intense chemical reaction is localized at the boundaries of the triple-shock-wave intersections where the Mach stem and strong detonation meet.

Figure 6.19 shows the compensated streak photographs of a multiheaded detonation in $2\,C_2H_2 + 5O_2$ mixtures at various initial pressures (Voitsekhovskii *et al.*,

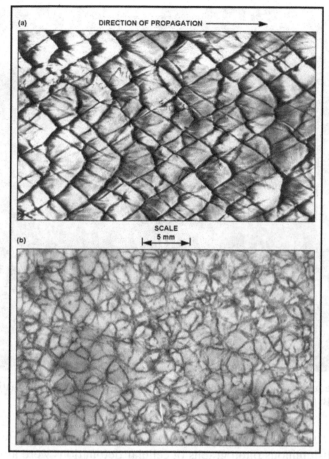

Figure 6.18. (a) Side-wall and (b) end-on soot records of a multiheaded detonation (courtesy of S.B. Murray).

1966). The streak photographs in the top row of the figure were taken with the axis of the camera normal to the direction of propagation, whereas the photographs in the bottom row were taken with the axis of the camera at 45° to the tube axis, thus presenting an oblique view of the circular cross-section of the detonation front.

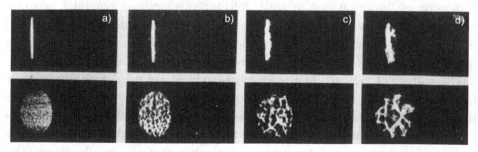

Figure 6.19. Compensated streak photographs of a multiheaded detonation (Voitsekhovskii *et al.*, 1966).

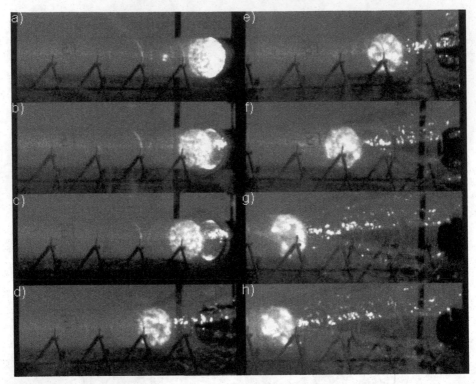

Figure 6.20. Sequence of frames from a high-speed movie of a cellular detonation in C_2H_4–air (courtesy of S.B. Murray).

Similar self-luminous photographs of cellular detonations were also obtained by Murray (1984) using a high-speed movie camera. Figure 6.20 shows a sequence of frames of a high-speed movie of a cellular detonation in C_2H_2–air mixtures (3.85% C_2H_2). The detonation was first initiated in a rigid steel tube 0.89 m in diameter and then exited into a 0.25-mm-thick plastic tube containing the same mixture. The frames are 0.2 ms apart, and the sequence goes from left to right, top to bottom. The cellular pattern in the detonation front is evidenced by the nonuniform luminosity pattern. As the detonation emerges from the rigid tube into the plastic tube, the expansion of the detonation products into the surroundings generates expansion waves that converge toward the tube axis. This attenuates the detonation, and in the case shown in Fig. 6.20, the detonation eventually fails. The progressive increase in the cell size as the detonation attenuates can also be noted in the figure.

The pattern of localized luminosity distributed across the detonation front surface that is shown in Fig. 6.19 corresponds to the boundaries of the shock intersections, identical to those shown on the end-on soot records of Fig. 6.18. At higher frequencies (at higher initial pressures), the nonuniform luminosity pattern at the front gives it a grainy, cellular structure not unlike that of a cellular flame. Thus, multiheaded spinning detonations are commonly referred to as *cellular* detonations,

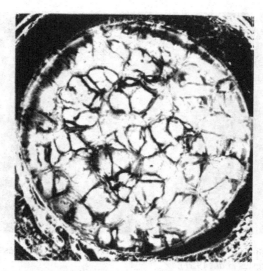

Figure 6.21. End-on soot-foil record of a higher-frequency spinning detonation (Lee, 1984).

which is a more appropriate term. For high-frequency spinning detonations, there are no identifiable, individual spin heads that rotate around the tube periphery as in a single-headed spin. Rather, the intense combustion regions correspond to the boundaries of the shock intersections and are distributed over the surface of the front to give the appearance of a cellular pattern.

Figure 6.21 shows an end-on soot record of the cellular pattern of a higher-frequency spinning detonation. The pattern itself looks chaotic and changes with time as the transverse waves sweep back and forth across the detonation front. However, the side-wall soot record will still show a relatively orderly fish-scale pattern of two sets of opposing transverse waves.

Since the cell boundaries correspond to Mach intersections and the Mach stems are much stronger than the incident shock on the opposite side of the boundary, the pressure distribution across the front surface should also be nonuniform. Applying the impedance mirror technique used in the study of condensed-phase detonation, Presles *et al.* (1987) obtained a cellular pattern of the pressure distribution across the detonation surface. In place of a soot foil, Presles mounted a thin Mylar diaphragm at the end of the detonation tube. The outside surface of the diaphragm was aluminized and reflective. Upon reflection of the detonation from the thin diaphragm, its surface deformed according to the nonuniform pressure distribution on the detonation front. A short-duration flash was timed to illuminate the reflective aluminized surface, and an instantaneous photograph of the deformed diaphragm was obtained. Figure 6.22 shows the nonuniform pressure distribution on the detonation front as recorded by the deformation of the diaphragm. The cellular pattern is identical to that obtained from the luminosity distribution (Fig. 6.19) as well as the end-on soot foil (Fig. 6.21). This high-frequency spinning detonation front has a

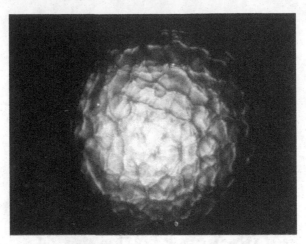

Figure 6.22. Mylar diaphragm deformed as a result of the nonuniform pressure distribution across a detonation front (Presles *et al.*, 1987).

characteristic cellular pattern defined by the boundaries of the shock intersections. Intense combustion and high pressures are also localized at the shock intersection boundaries that define the cells.

For higher-frequency spinning detonations, the motion of the transverse waves that give rise to the time-dependent cellular structure of the detonation front is not evident. For the lower-mode spinning detonations, it is interesting to construct possible patterns of transverse wave motion. Figure 6.23 shows the transverse wave motion for a single-, a double-, and a four-headed spinning detonation as postulated by Voitsekhovskii *et al.* In the double-headed spin pattern shown in Fig. 6.23b, the side-wall soot pattern consists of a left-handed and a right-handed transverse wave trajectory, giving a single diamond across the width of the tube circumference. In the four-headed spin, the side-wall soot pattern would indicate two left-handed and two right-handed transverse waves, giving two diamonds across the width of the tube circumference. A more illustrative three-dimensional sketch of the front surface of a four-headed spinning detonation, as given by Denisov and Troshin (1960), is shown in Fig. 6.23d. It is clear that for higher-frequency spinning detonations, it would be very difficult to describe the transverse wave motion across the detonation front.

The collisions of finite-amplitude transverse shocks lead to the periodic strengthening and decay of the detonation front, giving it a pulsating motion. Denisov and Troshin called these cellular detonations "pulsating detonations" in view of the periodic fluctuations of the local velocity of the detonation front. However, Duff (1961) adopted a different interpretation and argued that the linear acoustic theory can be extended to describe the higher-frequency cellular detonations as well. Duff focused on the interpretation of the side-wall soot pattern of the transverse wave trajectories and considered that the fish-scale pattern is the result of the superposition of two sets of opposite-spinning transverse waves. He then measured the angles that

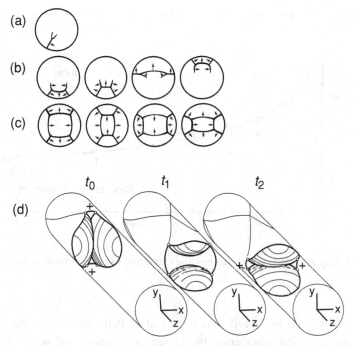

Figure 6.23. Sketch of (a) single-, (b) double-, and (c) four-headed spin (Voitsekhovskii *et al.*, 1966); (d) three-dimensional sketch of a four-headed spinning detonation (Denisov & Troshin, 1960).

the transverse waves made with the tube generatrix and compared them with the prediction from the acoustic theory. The pitch-to-diameter ratio from acoustic theory is given by

$$\frac{p}{d} = \frac{n\pi}{k_{nm}R}\left(\frac{D}{c_1}\right),$$

where n is the circumferential and m is the radial mode (i.e., $n = m = 1$ for single-headed spin). The ratio $D/c_1 \approx (\gamma + 1)/\gamma$ and the angle α that the helical trajectory makes with the tube axis are given by

$$\tan\alpha = \frac{\pi d}{p_n} = \left(\frac{k_{nm}R}{n}\right)\left(\frac{c_1}{D}\right).$$

Assuming $m = 1$, the various values of $k_{nm}R$ can be obtained from the properties of the Bessel function (see Table 6.1 in Section 6.3), and $\tan\alpha$ can then be computed (assuming a typical value of $\gamma = 1.2$).

Figure 6.24 shows a comparison between values of α measured and calculated from acoustic theory for various values of n. The agreement is fairly good especially for the higher frequencies, where the transverse waves are weaker and closer to acoustic waves. Thus, Duff concluded that the postulation of a new kind of wave by Denisov and Troshin is not necessary and that the acoustic theory is adequate to describe high-frequency spinning detonations.

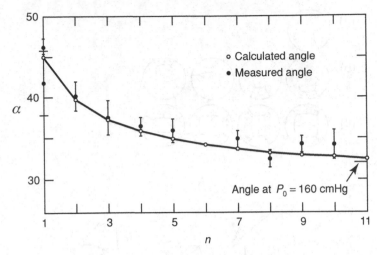

Figure 6.24. Comparison between measured and computed values of α from acoustic theory (Duff, 1961).

However, it should be noted that the local velocity of the detonation front indeed pulsates, strengthening upon the collision of transverse waves and decaying continuously afterward until the next collision. Thus, Denisov and Troshin's definition of a pulsating detonation refers to the local fluctuation of the detonation front. Duff, however, was concerned with the transverse wave trajectories and the transverse vibration of the product gases. Thus, the agreement of the trajectory angle α with acoustic theory does not negate the description of the pulsating motion of the front. The acoustic theory of Manson does not describe the detonation front. For high-frequency spinning detonations, the cell size or the transverse wave spacing will be small compared to the tube diameter. In that case, the transverse wave pattern becomes independent of tube geometry; thus the acoustic theory that links the transverse vibration to the eigenmodes of the tube becomes invalid. The transverse vibration must now be coupled to another characteristic length of the detonation, which should come from the chemical length scales of the reactions.

In comparing the acoustic theory with the wave interactions at the front, Duff pointed out an important paradox. The agreement between acoustic theory and the wave interactions at the front indicates a close coupling between the perturbation in the product gases and the detonation front. Yet, the CJ theory states that the detonation zone is independent of the dynamics of the products. Thus, there arises a contradiction between the acoustic theory for the vibrations of the product gases behind the detonation front and the CJ theory.

6.6. CELLULAR STRUCTURE IN OTHER GEOMETRIES

Thus far, we have mostly been concerned with single and multiheaded spinning detonations in round tubes. In tubes with other geometrical cross-sections, the

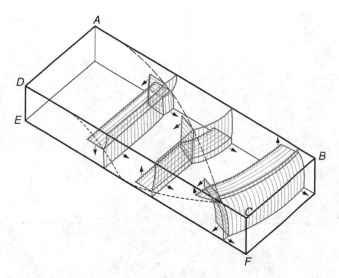

Figure 6.25. Sketch of a detonation in a rectangular channel (courtesy of A.K. Oppenheim).

transverse modes are much more complicated. Consider a rectangular channel where there are now two characteristic dimensions, x and y (z is along the tube axis of propagation). The lowest transverse mode would correspond to one transverse wave sweeping across the front in each of the x and y directions as illustrated in the sketch shown in Fig. 6.25. One of the transverse modes can be suppressed if the channel dimension is made sufficiently small, say in the vertical (y) direction. The result will be a single transverse wave in the longer x-direction only, analogous to a single-headed spin in a round tube, as shown previously in Fig. 6.9. In both square and triangular tubes, Bone *et al.* (1935) reported single-headed spinning detonations at the limits. This is perhaps due to the fact that the various characteristic dimensions are not too different and permit the cross-sectional area to be approximated by a circle. Thus, at the very limit where detonation propagation must be coupled to the lowest transverse vibrational mode in order to sustain itself, the largest characteristic dimension of the cross-section (i.e., the perimeter) provides the lowest eigenfrequency.

Figure 6.26 shows the soot record for a single-headed spinning detonation in a square channel. The soot from the four side walls of the 19-by-19-mm square channel are unfolded to give the periphery of the cross-section. The single helical trajectory of the spin head, similar to that in a round tube, is evident. However, we can also see that other transverse harmonics are excited and superimposed on the spinning head, perturbing the circumferential rotary motion. Since a spinning detonation is the "last resort" for nature to sustain the propagation of a combustion wave in the detonation mode, it is understandable that the lowest transverse mode, corresponding to the largest characteristic length scale of the tube geometry (i.e., the perimeter), would be excited near the limit. Thus, a resonant coupling between the transverse vibration and the reaction at the front permits the self-sustained propagation of the detonation under limiting conditions.

Figure 6.26. Soot-foil record for a single-headed detonation in a square channel (Lee *et al.*, 1969).

Away from the limits, higher transverse modes in both the x and y directions are excited. Figure 6.27 shows the end-on reflection soot records for multimode detonations in a rectangular channel. The mixture is $C_2H_4 + 3\,O_2 + 10\,Ar$ at various initial pressures. In Fig. 6.27a, the initial pressure is $p_0 = 150$ Torr. The cell size (transverse wave spacing) is small relative to the tube dimensions. We can see that the two transverse vibrational modes in the x and y directions are fairly stable and the transverse waves intersect to give a fairly regular pattern of rectangular cells. In Fig. 6.27b and 6.27c, the initial pressures are 120 and 100 Torr, respectively. The transverse waves are now stronger, and their nonlinear interactions result in a pattern of more irregular cells. At higher frequencies, the transverse waves are weaker and tend to be closer to weak acoustic waves, and thus the transverse modes can be superimposed on one another as governed by linear acoustic theory.

Simultaneous side-wall soot records and laser schlieren photographs of a detonation wave in $4\,H_2 + 3\,O_2$ at 87.3 Torr in a rectangular channel 25.4 by 38.1 mm are shown in Fig. 6.28. The side-wall record shows a relatively regular pattern of interacting transverse waves in one direction. However, the transverse waves in the other dimension are highly irregular, as indicated by the vertical lines made by the transverse waves (propagating normal to the plane of the soot foil and reflecting off it). The schlieren photographs indicate a highly irregular turbulent structure because they integrate the density gradient in the direction of the schlieren beam.

If the height of the channel, h, is made sufficiently small compared to the cell size λ, then the transverse waves in the y-direction can be suppressed, and we obtain a two-dimensional detonation with transverse waves only in the x-direction. Voitsekhovskii *et al.* (1966) reported a value of $\lambda/h \approx 6$ to 10 for suppressing the mode

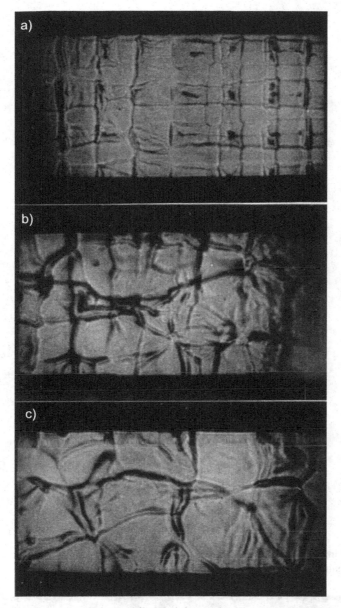

Figure 6.27. End-on soot records for multiheaded detonations in a rectangular channel (courtesy of P. Van Tiggelen).

in the y-direction. Figure 6.29 shows instantaneous schlieren photographs of two-dimensional detonations in a thin channel where transverse waves (normal to the plane of the photograph) have been suppressed. The mixture is H_2–O_2–40% Ar, where it is known that the transverse wave pattern is relatively regular. One can observe a series of transverse waves intersecting with the leading front, giving a regular cell spacing at the front. The transverse waves extend far back into the product gases, attenuating to weak acoustic waves. At a lower initial pressure (Fig. 6.29b),

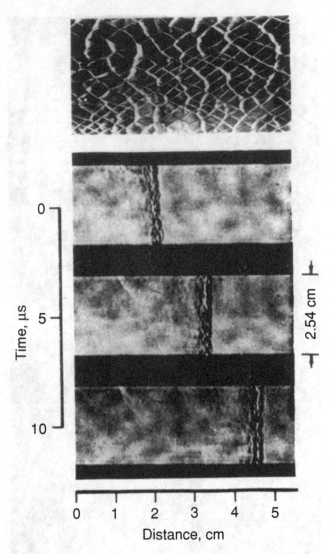

Figure 6.28. Simultaneous soot record and laser schlieren photographs of a detonation wave in a rectangular channel (Oppenheim, 1985).

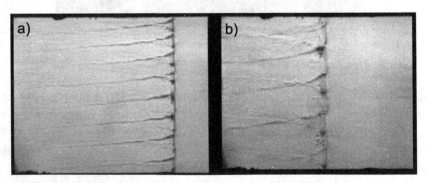

Figure 6.29. Schlieren photographs of multiheaded detonations in a thin channel at (a) $P_0 = 13$ kPa and (b) $P_0 = 8$ kPa (courtesy of M. Radulescu).

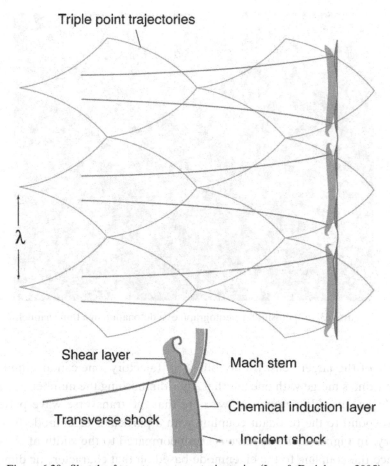

Figure 6.30. Sketch of transverse wave trajectories (Lee & Radulescu, 2005).

the front structure is better resolved, showing the reaction front and shear layers corresponding to the Mach intersections. A schematic diagram illustrating the trajectories of the transverse waves as registered on a soot film on the side wall, together with an enlarged sketch showing the detailed triple-shock Mach interactions, is given in Fig. 6.30.

If the detonation is unstable, the soot-foil pattern of transverse waves becomes irregular. Figure 6.31 shows an open-shutter photograph of the transverse wave trajectories for $C_2H_2 + 2.5\,O_2$ in a thin channel. It is clear that the pattern is irregular, with the continuous growth and decay of transverse waves. The weak waves within a cell are called the *substructure*, and their scale corresponds to the higher harmonics. The top left-hand corner of Fig. 6.31 shows a fairly regular pattern of interacting transverse waves. However, in the lower half of Fig. 6.31, the transverse wave pattern is much more irregular, as evidenced by the fluctuations in the transverse wave spacings. A large variation in the strength of the transverse waves, as indicated by the luminosity of the transverse wave trajectories, can also be observed. Growth of transverse waves, as indicated by the appearance of fine transverse wave trajectories within a so-called detonation cell, can be observed in

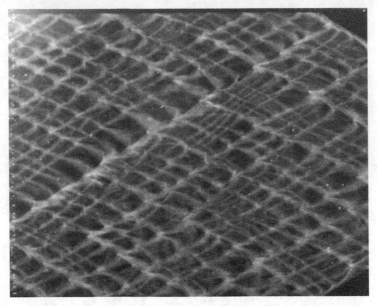

Figure 6.31. Open-shutter photograph of a detonation in a thin channel.

a number of the larger cells. If one follows a trajectory, one can also find that the cells sometimes merge with one another, thus diminishing the number of transverse waves. In an unstable detonation, we note that the transverse wave pattern does not correspond to the resonant coupling with a particular eigenmode from acoustic theory. In Fig. 6.31, the cell size is small compared to the width of the channel, and hence the coupling to the eigenmode based on that characteristic dimension is weak.

A highly resolved schlieren photograph illustrating the detailed structure of the reaction zone just prior to the collision of two transverse waves is shown in Fig. 6.32. The detonation wave is propagating in a thin channel, so that it is two-dimensional, and the mixture is relatively unstable with an irregular transverse wave pattern. A sketch defining the various features of the schlieren photograph is also given in Fig. 6.32. Of particular interest are the turbulent nature of the reaction zone and the presence of unburned gas pockets that are shed off the front and are convected downstream into the wake of the detonation front. These unburned gas pockets, being surrounded by hot product gases, will eventually be consumed. However, the energy released is no longer able to influence the propagation of the detonation front. Thus, in highly unstable mixtures, the turbulent fluctuations at the front, due to strong transverse wave interactions, may result in a reduction of the chemical energy release that goes to support the propagation of the detonation front. The presence of unreacted gas pockets that eventually burn via turbulent diffusion at the interface also implies that the combustion process in a cellular detonation may not be entirely due to autoignition via shock compression, as assumed in the ZND model.

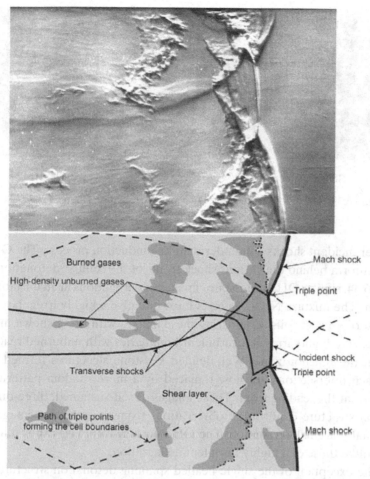

Figure 6.32. Schlieren photograph of a detonation propagating in a thin channel, with an accompanying sketch (Radulescu *et al.*, 2007).

The results from two-dimensional detonations in thin channels may not be indicative of a three-dimensional detonation, for, in suppressing the transverse vibration in one direction, we essentially force the detonation to be outside the limits. Thus, it is important to be able to observe the internal structure of a self-sustained, three-dimensional detonation. Because optical diagnostics integrate the flow field along the line of sight, the three-dimensional structure remained unresolved until the development of planar laser imaging techniques. In general, the fluorescence image of the OH radical is obtained by illuminating a planar cross-section of the three-dimensional structure with a thin laser light sheet. The intensity of the fluorescence image gives the concentration of the particular excited species.

Figure 6.33 shows a sequence of OH fluorescence images in a relatively stable mixture of $2 H_2 + O_2 + 17 Ar$ at $p_0 = 20$ kPa. The typical keystone pattern is due to the triple-shock Mach interactions, where the reaction is most intense at the Mach stem and transverse shock. The indentations are as yet unburned reactants behind

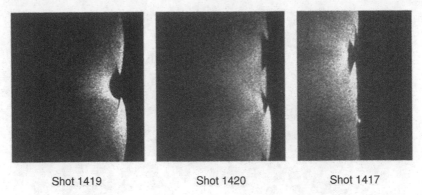

Shot 1419   Shot 1420   Shot 1417

Figure 6.33. OH fluorescence images of a stable detonation (Pintgen *et al.*, 2003a).

the weaker incident shocks still undergoing the induction process. The OH pattern is fairly uniform behind the various shock fronts of the cellular detonation.

In Fig. 6.34, similar OH fluorescence images of detonations in an unstable mixture are shown. The mixture N_2O–H_2–N_2 is known to give a highly irregular transverse wave pattern on soot foils. Comparing these images with those shown in Fig. 6.33, we can observe highly irregular turbulent boundaries with unburned reactants embedded in the reaction zone. Well-defined keystone structures formed by triple-shock Mach intersections are now replaced by a more random pattern of shock interactions at the leading front. These images of self-sustained, three-dimensional detonation structure confirm the observations from the schlieren images of two-dimensional marginal detonation. The OH fluorescence images shown in Fig. 6.34 are not unlike those of a highly turbulent deflagration.

With the exception of the single-headed spinning detonation structure, which is stationary, cellular detonations are unstable in general. Between the collision of two transverse waves, the leading shock front decays. In a thin channel where the detonation can be made two-dimensional, Strehlow and Crooker (1974) measured the

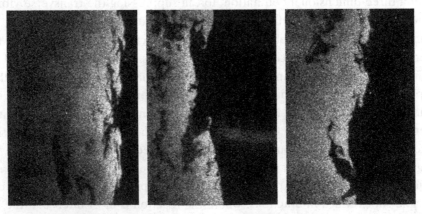

Figure 6.34. OH fluorescence images of an unstable detonation (Pintgen *et al.*, 2003b).

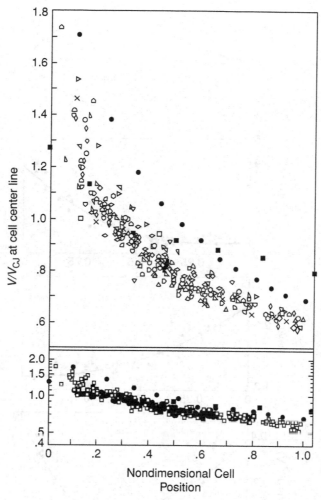

Figure 6.35. Variation of the leading shock velocity in a detonation cell (Strehlow & Crooker, 1974).

variation of the leading shock-front velocity using laser streak schlieren photography. They studied a mixture of $2H_2 + O_2 + 3$ Ar at $p_0 = 58$ Torr. The results from a number of experiments are plotted together in Fig. 6.35. The results indicate that, at the beginning of the cell immediately after the collision of two transverse waves, the leading shock-front velocity can be as high as 1.8 times the average CJ velocity. The front velocity decays rapidly to the CJ value at a distance of about a third of the cell length. Subsequently, it decays to an even lower velocity, and at the end of the cell just prior to the collision of the transverse waves, the shock-front velocity has decayed to approximately 0.6 of the CJ velocity. Note that the magnitude of the fluctuation along a cell length depends on the stability of the mixture as well as how far the condition is from the limits. For relatively stable mixtures well within the limits, the fluctuation is smaller, as the transverse waves themselves are weak, approaching linear acoustic waves.

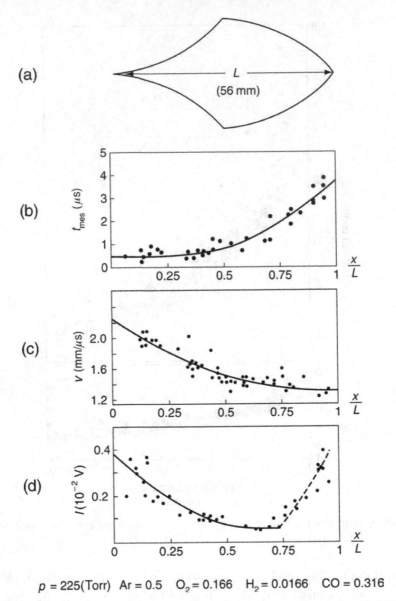

$$p = 225(\text{Torr}) \quad \text{Ar} = 0.5 \quad \text{O}_2 = 0.166 \quad \text{H}_2 = 0.0166 \quad \text{CO} = 0.316$$

Figure 6.36. Variation of measured parameters ((b) induction time, (c) velocity, (d) OH emission) within a detonation cell (Dormal *et al.*, 1979).

The velocity measurements within a cell were repeated by Dormal *et al.* (1979), who also measured the induction time, pressure, and OH emission along the length of the cell for marginal detonations in a thin channel. Their results, shown in Fig. 6.36, confirm those of Strehlow and Crooker. It is interesting to note that in the last quarter of the cell, the OH intensity (as well as the pressure) increases sharply just prior to the collision of the transverse waves to start the next cycle. This has not been investigated thoroughly as yet and is probably due to the complex config-uration of the approaching transverse shocks that cut off a volume of the mixture that becomes compressed by advancing shock waves on all sides, as in an implosion process. With direct measurements of the fluctuation along a cell length, we note that the shock front propagates in a pulsating manner as described by Denisov and Troshin (1960).

In unconfined cylindrical and spherical geometries, there is no characteristic length scale for the transverse vibrations. The circumference of the wave front increases with radius as a diverging detonation expands. Thus, we may not identify the transverse vibration with one of the natural eigenmodes as in the case of detonations in confined tubes. The cellular structure of diverging detonations will then be dependent on the characteristic length scale derived from the chemical rate processes (e.g., the ZND reaction-zone length). Since the chemical length scale is generally small compared to the characteristic dimensions of a tube in the case of confined detonations, we cannot expect to find diverging cellular CJ detonations until the radius is large compared to the cell size. Thus, diverging detonations must be supported, or overdriven, using a powerful ignition source until the radius of the detonation becomes sufficiently large. This is in accordance with experimental observations, which indicate that diverging detonations require a large amount of energy for initiation and that self-sustained propagation is only possible after the detonation is overdriven to a sufficiently large radius.

Figure 6.37 shows an open-shutter photograph of a diverging cylindrical detonation initiated by a powerful exploding ignition source. The trajectories of the transverse waves now show two intersecting sets of logarithmic spirals. Furthermore, if

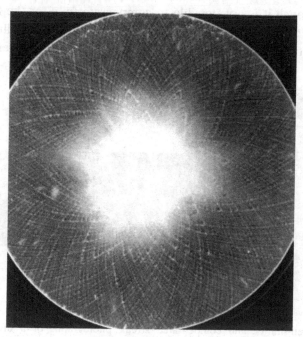

Figure 6.37. Open-shutter photograph of a diverging cylindrical detonation.

the diverging detonation is to propagate at a constant CJ value, the average dimension of the detonation cell (or transverse wave spacing) must be a constant governed by the mixture composition and its initial conditions. Thus, the average number of transverse waves per unit length along the circumference of the detonation front must be constant. This requires that the growth rate of transverse waves must be able to cope with the rate of increase in the surface area of the detonation front. The growth of new transverse waves must originate from the instability at the detonation front itself. Note that in the Mach stem and transverse shock, one can observe a fine structure corresponding to an overdriven detonation in the vicinity of the triple point. Such fine structures manifest themselves as the fine diamond pattern in the band associated with the spiral path of the single-headed spinning detonation (Fig. 6.7). It is these transverse waves from the fine structure that amplify to form new transverse waves as the diverging front expands and as the cell becomes bigger when the time between collisions of the transverse waves increases. In spherical detonations, the growth rate is more rapid, because the surface area is proportional to the square of the radius. Accordingly, the rate for new transverse waves to appear must also be higher than for cylindrical detonations of the same radius. Thus, the existence of spherical detonations requires more unstable mixtures where the growth of instabilities is more rapid.

When the growth rate of new transverse waves is insufficient to cope with the rate of increase of the surface area, the cells get bigger and the detonation quenches. Figure 6.38 shows simultaneous open-shutter and compensated streak records for a failing diverging detonation. A number of horizontal slits are used (which can be seen in the open-shutter photograph), and the streak photograph shows the reaction-zone structure. Although it is not possible to compensate fully for the film and image speed of the detonation because its velocity varies, a qualitative illustration of the reaction-zone structure of the failing wave can nevertheless be observed.

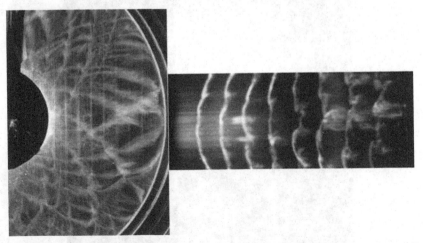

Figure 6.38. Open-shutter and compensated streak photographs of a diverging cylindrical detonation.

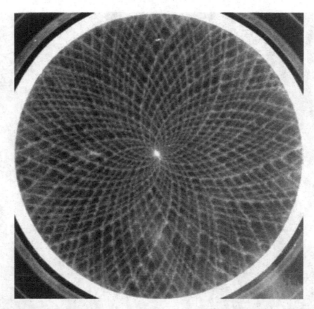

Figure 6.39. Open-shutter photograph of a converging cylindrical detonation.

In contrast to a diverging detonation front, the surface area decreases in a converging detonation. Figure 6.39 shows the open-shutter photograph of a converging cylindrical detonation in a thin channel. Unlike a diverging detonation, which propagates at the constant CJ velocity, an imploding detonation amplifies due to area convergence and becomes progressively more overdriven as it propagates toward the center of symmetry (Knystautas & Lee, 1971). Figure 6.39 again shows two sets of intersecting logarithmic spirals, as in the case of diverging detonations. However, as the detonation converges, the average cell size, or transverse wave spacing, decreases. The converging detonation is progressively being overdriven, and the cell size becomes smaller according to the degree of overdrive. Eventually, as the overdriven detonation approaches a strong shock, the front becomes stable and the transverse waves decay to weak acoustic waves.

Early investigators doubted the existence of steady diverging waves, but they based their consideration on a one-dimensional ZND structure (Jouguet, 1917; Courant & Friedrich, 1950). However, if one considers instabilities as a required condition for the self-sustained propagation of detonation waves, then the existence of diverging detonations depends on a sufficiently fast growth rate of the instabilities to maintain a constant average cell dimension. Figure 6.40 shows an instantaneous (highly magnified) image converter photograph of a spherical detonation in a highly sensitive C_2H_2–O_2 mixture at an initial pressure $p_0 = 300$ Torr. The magnification is 500 times. The cellular structure of the front surface is also quite evident, even though the cell dimension is exceedingly small in the C_2H_2–O_2 mixture at this initial pressure. End-on reflected soot records of spherical detonation obtained by

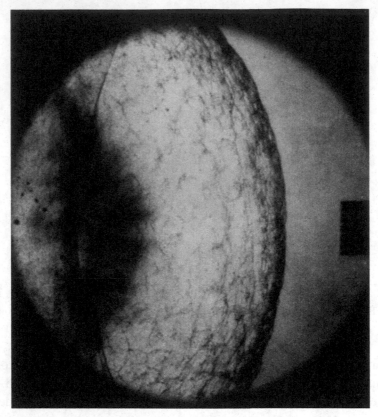

Figure 6.40. Instantaneous image converter photograph of a spherical detonation (courtesy of M. Held).

Shchelkin and Troshin (1965) and Duff and Finger (1965) show identical cellular structures to planar waves for the same mixture.

Figure 6.41 shows a simultaneous spark schlieren photograph and a soot-foil record of a spherical detonation in $H_2 + Cl_2$ at $p_0 = 100$ Torr. The turbulent cellular structure is quite apparent. The soot-foil record registers the end-on reflection of the spherical detonation, and the cellular pattern is similar to the end-on soot record of a planar detonation propagating in a tube (e.g., Fig. 6.21). The reflected spherical wave on the soot plate becomes a Mach stem as the wave expands. Thus, the cell sizes are smaller, as the Mach stem is an overdriven detonation. As the spherical wave expands further, the Mach stem becomes an incident wave, and the cell size returns to the normal cell size of the mixture.

We may conclude that detonations of all geometries are unstable and that the front has the universal cellular structure formed by the interaction of transverse waves with the leading front. Close to the limits, the geometry of the confinement controls the detonation front structure, because the nonuniform heat release at the front must adjust itself so that resonant coupling with the transverse vibrational modes can be achieved. Away from the limits, where the cell size is small compared

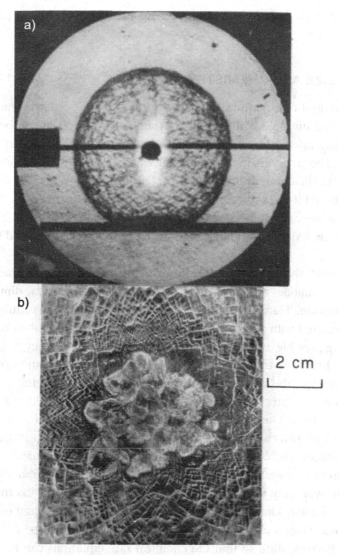

Figure 6.41. Simultaneous (a) spark schlieren and (b) soot-foil records, showing the multiheaded structure of a spherical detonation (Lee *et al.*, 1972).

to the characteristic dimension of the tube geometry, the front structure becomes independent of the tube geometry. Thus, the detonation structure is the same for all geometries. The cell size is now governed by the length scales of the chemical reaction. For diverging detonations, where the surface area of the detonation front increases with radius, new cells must be formed at a rate to match the rate of increase in the surface area so as to maintain a constant average cell size. Without the formation of new cells, the existing cell size would increase. Then the leading shock front would decay as the time between collisions of transverse waves increased and the detonation failed. It appears that the key to self-sustained propagation of detonation is the cellular structure. A steady, one-dimensional ZND model for the structure appears to lack the essential mechanism responsible for the propagation of detonation waves.

6.7. CELL SIZE AND CHEMISTRY

The experimental observations described in this chapter indicate that, in general, detonations are unstable with a cellular frontal structure. The cell boundaries are defined by the intersection of an ensemble of transverse waves with the leading shock front. The transverse waves themselves are sustained by the nonuniform energy release at the detonation front. Thus, the morphology of the frontal structure must be such that it gives the appropriate nonuniform energy release distribution at the front to achieve resonant coupling with transverse vibrations. Hence, contrary to CJ theory, there exists an intimate relationship between the front and the processes behind it.

Near the limit, the transverse vibration corresponds to the lowest acoustic eigenmode. The eigenmodes are, in turn, governed by the characteristic dimension of the tube cross-section. Thus, the characteristic length scale of the cellular detonation front is correlated with the characteristic length scale of the tube cross section at the limits (e.g., see Fig. 6.22). Away from the limits, the frequency of the transverse vibrations is high, and the corresponding cell size in the detonation front becomes small compared to the characteristic dimensions of the tube. Hence, the tube geometry no longer controls the cell size, and the detonation structure is the same for all tube geometries. The cell size now depends on the properties of the explosive mixture. For resonant coupling, the cell size must be a consequence of the nonlinear feedback between gasdynamic and chemical processes. In a chemical reaction, there is a spectrum of time scales corresponding to the rates of the various elementary reactions. However, only the few dominant ones that control the exothermic energy release are relevant. The chemical length scales are also dependent on the thermodynamic state. Thus, a characteristic gasdynamic process must be chosen to define the thermodynamic state so that the chemical rate equations can be integrated to yield the desired chemical length scale.

Experiments have all shown that, even at the limit, the detonation propagates at some mean velocity very close to the theoretical CJ value. Within the structure itself, where the local velocity of the front can vary from $1.8V_{CJ}$ to $0.6V_{CJ}$, the mean velocity over a cell length (i.e., cycle of fluctuation) is still about the theoretical CJ value for the mixture. Thus, it appears that the CJ velocity can be chosen to provide the initial thermodynamic state for the integration of the kinetic rate equation. The resulting ZND reaction-zone length will then serve as an appropriate chemical length scale to characterize the transverse vibrations.

Perhaps the first to correlate the detonation cell size λ with the ZND reaction-zone length l were Shchelkin and Troshin (1965), who assumed a simple linear dependence: $\lambda = Al$, where A is a constant. The ZND reaction-zone length l can be readily determined using CJ theory, first to compute the von Neumann state behind the leading shock front, followed by the integration of the kinetic rate equations for the elementary reactions (along the Rayleigh line). The constant A has to be determined by matching the calculated value to one experimental data point. In general, the smallest cell size corresponding to stoichiometric composition (or equivalence ratio $\phi = 1$) is chosen to evaluate the constant A. Once A is found, the cell size for other compositions can be obtained when the ZND reaction-zone length is computed. The computation of the ZND reaction-zone length and the use of it for the estimation of the detonation cell size, using detailed chemical-kinetic data, was pioneered by Westbrook and Urtiew (1982) and by Shepherd *et al.* (1986). Shepherd *et al.* integrated the basic conservation equation for the ZND reaction zone, following a Rayleigh line. Westbrook and Urtiew, in contrast, computed the reaction time based on a constant volume explosion path, and thus could obtain the reaction-zone length by multiplying the reaction time by the particle velocity behind the shock. In view of the assumption of the simple linear dependence between the ZND reaction-zone length and the cell size, the difference between Shepherd *et al.* 's exact Rayleigh-line integration and Westbrook and Urtiew's constant volume path calculation is not important.

It is perhaps relevant to discuss the method of measurement of the detonation cell size. Shchelkin and Troshin used the reflected end-on soot record (e.g., Fig. 6.21), which gave the cell pattern of the front (at the instant of time of reflection from the soot foil). They drew a line across the tube's cross-section, counted the intersections of the cell boundaries with the line, and divided the length of the line by the number of intersections to obtain the average cell dimension. The standard technique is to use the side-wall soot record, which registers the trajectories of the triple points of Mach intersections at the wall. The side-wall soot record covers a certain distance of travel by the detonation front (i.e., the length of the soot foil, rather than one instant of time as represented by the end-on soot record). It has been found that the two methods give similar results, but the use of the side-wall record of the fish-scale pattern of transverse wave trajectories is more accurate, especially when the pattern is irregular. Observing the detonation over a longer distance of travel greatly facilitates the identification of the dominant cell size.

To obtain a single representative value for the cell size from an irregular soot foil pattern is not an easy task (e.g., Figs. 6.18 and 6.28). There exists a distribution of cell sizes, further complicated by the presence of fine structures of the higher harmonics that are embedded within the cells. The higher harmonics grow to become new cells, and they can disappear when they merge with existing cells upon the collision of the transverse waves. To obtain consistent results, the novice observer must obtain guidelines from more experienced researchers. Longer foils that permit the detonation structure to be observed over a longer period are of value. Use of more than one foil sample, either from the same experiment with foils placed at different locations along the tube or from different experiments, also helps to provide a reliable value for the cell size that characterizes a particular mixture.

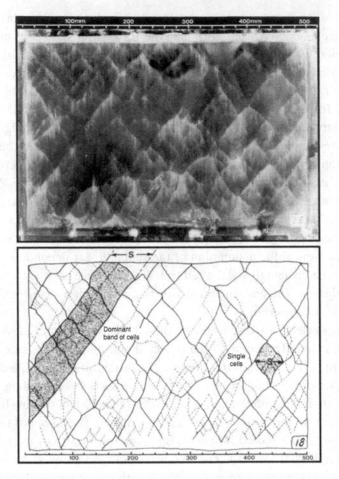

Figure 6.42. A typical soot record of a detonation with an accompanying hand tracing of the transverse wave trajectories (Moen *et al.*, 1982).

As an example in foil interpretation, Fig. 6.42 shows a typical soot record for C_2H_4–air detonation in a large 0.89-m-diameter tube. A hand tracing of the soot foil is also shown in Fig. 6.42 to give a clearer picture of the transverse wave trajectories. As can be observed, there is a large spectrum of cell sizes, as well as substructure, within the cells. However, study of the foil for a certain period of time led to the choice of the shaded row of cells as the dominant band from which a characteristic cell size (or transverse wave spacing) can be obtained.

In an attempt to render soot-foil interpretation more objective, Shepherd *et al.* (1986) and Lee *et al.* (1993) applied a digital image-processing technique to analyze soot records. The first difficulty that had to be overcome was in achieving an even deposit of soot on the foil. A large variation in the soot deposit results in a large fluctuation in the background grey scale and in poor resolution of the scanned image for digitizing the data. In the analysis of Lee *et al.*, the actual soot record is first traced by hand prior to scanning, which is a tedious task. In the work of Shepherd *et al.*, the relatively regular cell pattern of C_2H_2–O_2 with high argon dilution (80%) permitted a fairly well-defined single peak in the cell size spectrum to be obtained. However, with a lower percentage of argon, the distribution is rather flat and no single value for the dominant cell size can be obtained from the spectrum. Similar results were reported by J.J. Lee *et al.* (1995), who obtained a histogram of cell sizes as well as the autocorrelation function of the digitized data of the soot pattern, which provide a measure of the irregularity of the soot pattern. The conclusion was that the cell size from digital processing agreed with that estimated by the human eye for the more regular pattern, but digital image processing techniques could potentially yield much more information about the soot pattern. However, we have yet to learn how to use the additional information, and currently just a single value of the dominant cell size for the mixture suffices. The trained eye remains the standard method used to determine a representative value for the cell size from a soot record of the transverse wave pattern. Cell size data for most common fuels (with air or oxygen and inert diluents) at various initial conditions have been collected and are available at the Caltech Explosion Dynamics Laboratory website of J.E. Shepherd (Kaneshige & Shepherd, 1997).

A comparison between experimental and computed values of the cell size based on the ZND reaction-zone length and the linear dependence $\lambda = Al$ proposed by Shchelkin and Troshin for common fuel–air mixtures (at standard initial conditions $p_0 = 1$ atm, $T_0 = 298$ K) is shown in Fig. 6.43. The constant A is determined by fitting experimental data at stoichiometric composition (i.e., equivalence ratio $\phi = 1$). Note that, for each fuel, a different value of A is obtained by fitting with the cell size value at $\phi - 1$ for the particular fuel. The agreement is quite good on a qualitative basis. However, on a quantitative basis, deviation from the linear fit can be as much as an order of magnitude, especially near the lean limit. The agreement is always adequate near stoichiometric conditions, because A is determined by fitting to experimental data at $\phi = 1$.

Attempts have been made to improve the fit by assuming A to be dependent on the mixture composition. Although a slight improvement was obtained, the factor A is still not universally valid. A more recent attempt was made by Gavrikov *et al.* (2000), who assumed A to be dependent on two stability parameters. The first is the activation energy that measures the temperature sensitivity of the reaction, and the second is based on the relationship between the chemical energy release and the so-called critical internal energy. It is known that exothermicity of the mixture also influences the stability of the detonation. Furthermore, instead of choosing the von Neumann state of a CJ detonation, Gavrikov *et al.* chose a higher shock strength to compute the ZND reaction-zone length, to reflect the fluctuation of a pulsating front during a cell cycle. These modifications appear to give an improved correlation over the linear law of Shchelkin and Troshin, but the functional form of the fitting constant is now much more complex.

Another recent attempt to obtain a better empirical relation between λ and l was made by Ng (2005). He also used a stability parameter involving the product of the

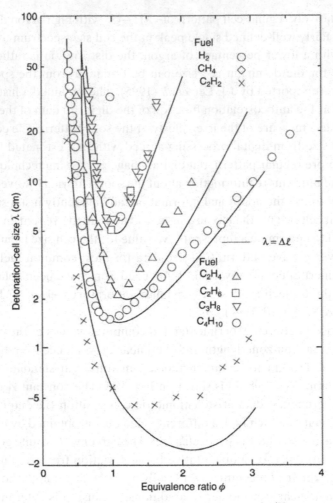

Figure 6.43. Detonation-cell size as a function of equivalence ratio (Knystautas *et al.*, 1985).

activation energy with the ZND induction-to-reaction-zone length ratio. Ng also assumed a polynomial fit in his stability parameter, hence increasing the number of fitting constants to be determined. However, he claimed that once these constants in the polynomial expression are determined, they are valid for different mixtures. Hence, they need to be determined first. He obtained a much better correlation with experimental data than Gavrikov.

Thus far, no theory exists for the prediction of cell size from the fundamental properties of the explosive mixture. Although various empirical relationships can be proposed to improve the correlation of the ZND reaction-zone length (which can be computed) with the measured cell size data, these empirical relationships are essentially an exercise in curve fitting and do not provide any information about the physical relationship of the cell size to the characteristic chemical length scale of a ZND detonation. It should be noted that it is commonly accepted that experimental estimates of the dominant or representative cell size from soot foils can vary by a factor of two.

6.8. CLOSING REMARKS

Experimental investigations over the past 50 years have conclusively demonstrated that self-propagating (not overdriven) detonation waves are unstable in almost all common explosive mixtures. Thus, the classical ZND model for the detonation structure cannot describe real detonations. In other words, the ignition and subsequent reactions are not governed by the adiabatic heating of the leading shock front. In an unstable detonation front, reactants entering into the reaction zone are undoubtedly heated first by various shock waves (incident shock, Mach stem, and transverse shocks), depending on their location relative to the cellular front. However, they are also immediately subjected to adiabatic expansions in the flow field in the wake of these nonsteady shock waves, and the induction reactions may even be quenched in certain cases. In the reaction zone itself, there exist various density interfaces, transverse pressure waves, reaction fronts, shear layers, and vortex structures. Thus, the combustion process of the reactant particles in the reaction zone cannot be ascribed to a single mechanism. Accordingly, the development of a theory for the qualitative description of the structure of an unstable cellular detonation (or to simulate the processes in the reaction zone numerically) presents a formidable task.

How chemical reactions are initiated and the subsequent combustion processes occur in the turbulent reaction zone of a cellular detonation can be perhaps described by the development of better diagnostic and numerical techniques. However, why nature prefers the detonation structure to be so complex instead of the simple steady one-dimensional ZND model is more difficult to answer. Experiments indicate that instability is more pronounced in near-limit conditions when the detonation finds it difficult to maintain self-sustained propagation. Instabilities also set in when the temperature sensitivity of the chemical reaction rate is high. Hence, instability must serve to provide a means to avoid quenching of the reactions from any small temperature fluctuation that the detonation may encounter. Instability may therefore be looked upon as nature's way of bringing in more physical processes to keep the detonation from failing.

One obvious effect of instability is to increase the dimension of the phenomenon. From steady (or stationary) to nonsteady brings in time as an added dimension. Instabilities also increase the spatial dimension from one to two or three dimensions. Increasing the number of dimensions permits a nonuniform local concentration of energy. For example, the Mach stem from shock collisions results in much higher local temperatures. Thus, when the temperature behind a planar shock falls below the auto-ignition limit, ignition can readily be effected if the one-dimensional planar shock is replaced by a three-shock Mach configuration via instability. Additional vorticity-generating mechanisms result in higher dimensions, and thus turbulence is brought into the reaction zone. Turbulence promotes local transport rates, thereby facilitating the chemical reactions in the reaction zone. The detailed description of the various processes within the reaction zone of a cellular detonation illustrates the advantages of an unstable front in maintaining self-sustained propagation. Thus, if a detonation is the mode of combustion that nature attempts to achieve, then instability is the means that permit detonations to be self-sustained.

One may also ask why a self-propagating deflagration wave tends to accelerate and transit to a detonation wave when the initial and boundary conditions permit it to do so. This question is perhaps analogous to asking why laminar flow in a pipe transits to a turbulent flow when the Reynolds number exceeds a certain critical value. Questions of this nature are universal and serve as the fundamental motivation to understand the physical world around us.

Bibliography

Antolik, K. 1875. *Ann. Phys. Lpz.* 230(1):14–37.

Bone, W.A., R.P. Fraser, and W.H. Wheeler. 1935. *Phil. Trans. R. Soc. Lond. A* 235: 29.

Campbell, C., and A.C. Finch. 1928. *J. Chem. Soc.* 131:2094.

Campbell, C., and D.W. Woodhead. 1926. *J. Chem. Soc.* 129:3010.

Campbell, C., and D.W. Woodhead. 1927. *J. Chem. Soc.* 130:1572.

Chu, B.T. 1956. In *Gasdynamics Symp. on Aerothermochemistry*, 95–111. Evanston, IL: Northwest University Press.

Courant, R., and K.O. Friedrich. 1950. *Supersonic flow and shock waves*, 430. Interscience.

Denisov, Yu. N., and Ya. K. Troshin. 1959. *Dokl. Akad. Nauk SSSR* 125:110.

Denisov, Yu. N., and Ya. K. Troshin. 1960. In *8th Int. Symp. on Combustion*, 600.

Dixon, H.B. 1903. *Phil. Trans. A* 200:315.

Donato, M. 1982. The influence of confinement on the propagation of near limit detonation. Ph.D. thesis, McGill University, Montreal.

Dormal, M., J.C. Libouton, and P. Van Tiggelen. 1979. *Acta Astronaut.* 6:875–884.

Dove, J., and H. Wagner. 1960. In *8th Int. Symp. on Combustion*, 589.

Duff, R. 1961. *Phys. Fluids* 4(11):1427.

Duff, R. and M. Finger. 1965. *Phys. Fluids* 8:764.

Fay, J.A. 1952. *J. Chem. Phys.* 10(6).

Fay, J.A. 1962. In *8th Int. Symp. on Combustion*, 30.

Gavrikov, A.I., A.A. Efimenko, and S.B. Dorofeev. 2000. *Combust. Flame* 120:19–33.

Gordon, W. 1949. In *3rd Int. Symp. on Combustion*, 579.

Gordon, W., A.J. Mooradian, and S.A. Harper. 1959. In *7th Int. Symp. on Combust.* 752.

Head, M.R., and P. Bandyopadhyay. 1981. New aspects of turbulent boundary-layer structure. *J Fluid Mech.* 107:297–338.

Jouguet, E. 1917. *Mécanique des explosifs*, 359–366. Ed. Doin.

Kaneshige, M., and J.E. Shepherd. 1997. Detonation database. GALCIT Tech. Rept. FM 97. (Web page, http://www.galcit.caltech.edu/detn_db/html/db.html).

Knystautas, R., C. Guirao, J.H.S. Lee, and A. Sulmistras. 1985. In *Dynamics of shock waves, explosions, and detonations*, ed. J.R. Bowen, N. Manson, A.K. Oppenheim, and R.I. Soloukhin.

Knystautas, R., and J.H.S. Lee. 1971. *Combust. Flame* 16:61–73.

Lee, J.H.S. 1984. Dynamic parameters of detonations. *Ann. Rev. Fluid Mech.* 16:311.

Lee, J.J., D. Frost, J.H.S. Lee, and R. Knystautas. 1993. In *Dynamics of gaseaus Combustion*, ed. A.L. Kuhl, 182–202. AIAA.

Lee, J.J., D. Garinis, D. Frost, J.H.S. Lee, and R. Knystautas. 1995. *Shock Waves.* 5:169–174.

Lee, J.H.S., R. Knystautas, C. Guirao, A. Bekesy, S. Sabbagh. 1972. *Combust. Flame.* 18:321–325.

Lee, J.H.S., and M.I. Radulescu. 2005. *Combust. Explos. Shock Waves* 41:745–765.

Lee, J.H.S., R. Soloukhin, and A.K. Oppenheim. 1969. *Acta Astronaut.* 14:565–584.

Mach, E., and J. Sommer. 1877. *Sitzungsber. Akad. Wien* 75.

Manson, N. 1945. *Ann Mines*, 2éme livre, 203.

Manson, N. 1947. *Propagation des détonations et des deflagrations dans les melanges gazeux*. Paris: L'Office National d'Etudes et des Recherches Aéronautique and L'Institut Franais des Pétroles.

Moen, I.O., S.B. Murray, A. Bjerketvedt, R. Rinnan, R. Knystautas, and J.H.S. Lee. 1982. *Proc. Combust. Inst.* 19:635.

Mooradian, A.J., and W.E. Gordon. 1951. *Chem. Phys.* 19(9):66.

Murray, S.B. 1984. The influence of initial and boundary conditions on gaseous detonation waves. Ph.D. thesis, McGill University, Montreal.

Ng, H.D. 2005. The effect of chemical reaction kinetics on the structure of gaseous detonations. Ph.D. thesis, McGill University, Montreal.

Oppenheim, A.K. 1985. Dynamics features of combustion. *Phil. Trans. R. Soc. Lond. A* 315:471.

Pintgen, F., C.A. Eckett, J.M. Austin, and J.E. Shepherd. 2003a. *Combust. Flame* 133(3):211–220.

Pintgen, F., J.M. Austin, and J.E. Shepherd. 2003b. In G.D. Roy, S.M. Frolov, R.J. Santoro, S.A. Tsyganov (Eds.), *Confined detonations and pulse detonation engines*. (pp. 105–116) Moscow: Torus Press.

Presles, A.N., D. Desbordes, and P. Baner. 1987. *Combust. and Flame* 70:207–213.

Radulescu, M.I., G.J. Sharpe, C.K. Law, and J.H.S. Lee. 2007. *J. Fluid Mech.* 580:31–81.

Schott, G.L. 1965. *Phys. Fluids* 8(1):850.

Shchelkin, K.I. 1945. *C. R. Acad. Sci. U.R.S.S.* 47:482.

Shchelkin, K.I., and Ya. K. Troshin. 1965. *Gasdynamics of combustion*. Baltimore: Mono Book Corp.

Shepherd, J.E., I. Moen, S.B. Murray, and P.A. Thibault. 1986. *Proc. Combust. Inst.* 21:1649–1658.

Strehlow, R., and A. Crooker. 1974. *Acta Astronaut.* 1:303–315.

Taylor, G.I., and R.S. Tankin. 1958. Gasdynamic aspects of detonations. In *High speed aeronautics and jet propulsion*, Vol. 3, ed. H. Emmons. Princeton University Press.

Theodorsen, T. 1952. Mechanism of turbulence. In *Proc. Second Midwestern Conf. on Fluid Mechanics*, Vol. 21, No. 3, 1–18.

Voitsekovskii, B.V., V.V. Mitrofanov, and M. Ye. Topchiyan. 1966. "The structure of a detonation front in gases." English translation, Wright Patterson Air Force Base Report FTD-MT-64-527 (AD-633,821). This monograph conveniently summarizes all the work of the Novosibirsk group of the late 1950s and early 1960s.

Westbrook, C.K., and P.A. Urtiew. 1982. Chemical kinetic prediction of critical parameters in gaseous detonations. *Proc. Combust. Inst.* 19:615–623.

Zeldovich, Y.B. 1946. *C. R. Acad. Sci. U.R.S.S.* 52:147.

7 Influence of Boundary Conditions

7.1. INTRODUCTION

The detonation velocity from the Chapman–Jouguet (CJ) theory is independent of initial and boundary conditions and depends only on the thermodynamic properties of the explosive mixture. Experimentally, however, it is found that initial and boundary conditions can have strong influences on the propagation of the detonation wave. It is the finite thickness of the reaction zone that renders the detonation vulnerable to boundary effects. Thus, the influence of the boundary can only enter into a theory that considers a finite reaction-zone thickness. In the original paper by Zeldovich (1940) on the detonation structure, he already attempted to include the effect of heat and momentum losses on the propagation of the detonation wave. However, in the one-dimensional model of Zeldovich, the two-dimensional effect of losses at the boundary cannot be treated properly. It was Fay (1959) who later modeled boundary layer effects more accurately as a divergence of the flow in the reaction zone behind the shock front, resulting in a curved detonation front. Thus, heat and momentum losses become similar to curvature effects due to the lateral expansion of the detonation products in a two- or three-dimensional detonation. However, for small curvature this can be modeled within the framework of a quasi-one-dimensional theory.

Perhaps a more important effect of the boundary is on the instability of the detonation front. In general, detonations are unstable and have a cellular structure. It is the instability itself that facilitates the propagation of the detonation by providing local high temperatures via shock interactions to effect ignition. Near the limits, when the cell size is of the same order as the dimension of the tube cross-section, a strong coupling between the unstable structure and the boundaries occurs. Thus, boundary conditions play an important role in the propagation of the detonation, particularly near the limits. For example, a spinning detonation will fail if the tube diameter is significantly reduced, because the characteristic length scale of the boundary is now too small to permit coupling with the chemical length scale of the detonation.

The nature of the boundary surface itself is also important. For example, if the boundary attenuates pressure waves that are incident upon it, the transverse waves of the cellular detonation will be dampened out by the absorbing wall, and without cells, the detonation will fail. Very thin, supple walls not only deform readily under the detonation pressure, resulting in the lateral expansion of the products to cause flow divergence and wave curvature, but can also dampen out the transverse waves as they reflect from the yielding boundary.

Thus, in practice, the propagation of a detonation is strongly dependent on boundary conditions, in contrast to CJ theory. In general, the effect of the boundary is to reduce the propagation velocity of the detonation front below that predicted by CJ theory (i.e., to result in a velocity deficit), causing the detonation to attenuate and fail. On the other hand, in very rough tubes, the boundary can also provide a means for shock reflections from the surface protrusions. This generates transverse waves, local high temperatures from shock reflections, and turbulence to promote mixing within the reaction zone that enhances the chemical reaction rates. Although the detonation velocity may be decreased, the higher local temperatures from shock reflections and the turbulence generated by the rough wall can actually render the detonation more robust.

With the boundary playing various positive and negative roles on the mechanism of propagation, it is clear that detonation limits are also strongly dependent on boundary conditions. We shall discuss the effect of boundary conditions on velocity deficits, limits, stability of the structure, and ignition mechanisms in this chapter.

7.2. VELOCITY DEFICIT

Advances in electronic diagnostics permitted precise measurements of the detonation velocity to be made in the early 1950s. Kistiakowsky *et al.* (1952, 1955) obtained quantitative data to show the dependence of the detonation velocity on the tube diameter. Figure 7.1 shows their results for the variation of the detonation velocity with the tube diameter for various concentrations of C_2H_2 in C_2H_2–O_2 mixtures. As expected, the detonation velocity decreases with decreasing tube diameter, as wall effects become more dominant.

It was also found that, in general, the detonation velocity depends almost linearly on the inverse tube diameter. Figure 7.2 shows the results obtained by Brochet (1966) for $C_3H_8 + 5\,O_2 + z\,N_2$ for various concentrations z of the inert diluent nitrogen. Two grades of propane were used, and the differences in the results are due to other hydrocarbon impurities in commercial-grade propane. As illustrated in Fig. 7.2, the dependence of the detonation velocity on the inverse tube diameter is almost linear. This permits an interpolation of the straight line to infinite tube diameter (i.e., $1/d \to 0$) to obtain a value of the detonation velocity, D_∞, which was thought to be independent of the tube diameter. Thus D_∞ can be used for comparison with the theoretical CJ value to assess the validity of the equilibrium assumption

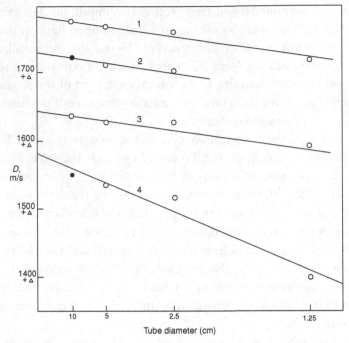

Figure 7.1. Detonation velocity as a function of tube diameter: (1) 7.5%, (2) 53%, (3) 69.6%, and (4) 75% C_2H_2 in C_2H_2–O_2 mixtures (Kistiakowsky *et al.*, 1955).

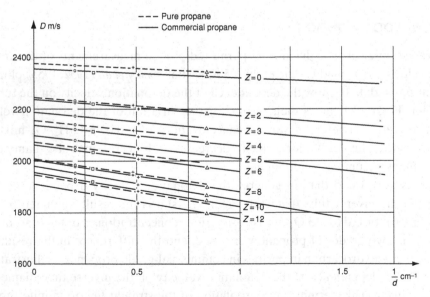

Figure 7.2. Detonation velocity as a function of the reciprocal of tube diameter (Brochet, 1966).

of the CJ theory. However, it was found that D_∞ can sometimes be even greater than, but is usually less than, the theoretical CJ value, indicating that the procedure of extrapolating to infinite tube diameter may not be valid. In general, the behavior of the detonation velocity in large-diameter tubes may not follow the $1/d$ dependence when the boundary layer thickness becomes negligible compared to the tube diameter. The work by Renault (1972) for different surface finishes also shows that the velocities in small tubes were most affected by the surface finish. Hence, the small-tube data controlled the slope, and therefore the intercept as $1/d \to 0$.

From D_∞, we may also define a velocity deficit as

$$\frac{\Delta D}{D_\infty} = \frac{D_\infty - D(d)}{D_\infty}, \tag{7.1}$$

which measures the deviation of the detonation velocity from an ideal value that is independent of the tube diameter. Quite often the theoretical CJ velocity itself is used instead of D_∞ to provide a more standard reference velocity. Like the detonation velocity $D(d)$, the velocity deficit also varies linearly with the inverse tube diameter.

The dependence of the detonation velocity on tube diameter is credited to losses to the tube wall. The effect of heat and momentum losses to the wall was first investigated by Zeldovich (1950). Because viscous drag and heat transfer are proportional to the *wetted area* (i.e., the circumference multiplied by the length of the reaction zone) whereas the total momentum associated with the detonation wave depends on the volume (i.e., the tube area times the reaction-zone thickness), we see that the velocity deficit depends on the ratio of surface area to volume, and thus on $1/d$.

Manson and Guénoche (1956) proposed an alternative mechanism to account for the velocity deficit. They considered that, as a result of heat losses, a layer of mixture adjacent to the wall of the tube becomes *quenched*: its reaction rate decreases significantly. Hence, the total chemical energy that goes to support the detonation will be less by the amount in the quenched layer. Again, this mechanism also produces a dependence on the surface area to volume ratio. Thus, even though the quenching and loss mechanisms differ, the qualitative dependence on tube diameter is the same.

Since heat and momentum losses are confined to the very thin viscous and thermal boundary layers, the core flow structure outside the boundary layer can only be influenced by the pressure waves (generated in the boundary layer) that penetrate into the reaction zone. Zeldovich considered a one-dimensional model for the flow in the reaction zone, and replaced the two-dimensional effect of viscous shear force at the wall with a body drag force in the momentum equation. This creates the problem that work done by this drag body force should also appear in the energy equation. However, in a two-dimensional model no work is done by the viscous drag force, because the particle velocity also vanishes at the wall due to the

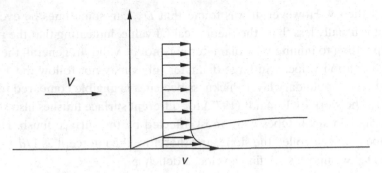

Figure 7.3. Sketch illustrating flow divergence.

no-slip boundary condition. A further study of non-ideal detonations in rough tubes using a similar quasi-one-dimensional model was later reported (Zeldovich *et al.*, 1988).

A more accurate treatment of the effect of a two-dimensional boundary layer on the propagation of the detonation wave was presented later by Fay (1959), who argued that the physical picture of the flow proposed by Zeldovich is incorrect. In Fay's theory, the boundary layer causes the streamlines in the reaction zone to diverge and thereby reduces the propagation velocity. The flow divergence is due to the negative boundary layer displacement thickness of the flow with respect to a reference coordinate system fixed to the propagation shock front, as shown in Fig. 7.3.

In a reference frame fixed to the shock front, both the flow ahead and the wall move at the shock velocity *V*. Due to the higher flow velocity in the boundary layer, streamlines will be deflected toward the boundary layer. Heat loss to the wall also results in a higher density in the boundary layer. Thus, the flow in the core must diverge to accommodate the higher mass flow rate in the boundary layer. Divergence of the streamlines in the reaction zone results in a curved detonation front where the curvature is proportional to the rate of increase of the flow area away from the shock front. The velocity deficit due to the boundary layer thus becomes identical to that of a curved detonation front resulting from the lateral expansion of the products in a weakly confined cylindrical explosive charge.

The theory for the propagation of a curved detonation front in a finite diameter of explosive charge was developed by Jones (1947), Wood and Kirkwood (1954), and Eyring *et al.* (1949). However, if the area divergence is small, the flow in the reaction zone can be approximated as quasi-one-dimensional, and the conservation equation can be written as

$$d(\rho u A) = 0, \tag{7.2}$$

$$\rho u A \, du + A \, dp = 0, \tag{7.3}$$

$$d(h' + u^2/2) = 0, \tag{7.4}$$

where h' includes the chemical energy Q. Integrating the above equations between the shock and the CJ plane gives

$$\rho_1 u_1 = \rho_2 u_2 (1 + \xi), \tag{7.5}$$

$$p_1 + \rho_1 u_1^2 = (p_2 + \rho_2 u_2^2)(1 + \xi) - \int_0^\xi p \, d\xi, \tag{7.6}$$

$$h_1 + Q + \frac{u_1^2}{2} = h_2 + \frac{u_2^2}{2}, \tag{7.7}$$

where

$$h = \frac{\gamma p}{\rho(\gamma - 1)} = \frac{c^2}{\gamma - 1}, \tag{7.8}$$

$$\xi = \frac{A_2 - A_1}{A_1} = \frac{A_2}{A_1} - 1. \tag{7.9}$$

Following Fay, we can write the integral in Eq. 7.6 as

$$\int_0^\xi p \, d\xi = p_2 \epsilon \xi, \tag{7.10}$$

where ϵ has to be determined by integrating the conservation equations across the reaction zone for a given chemical reaction rate law (as discussed previously in Chapter 4). However, the range of values for ϵ can be obtained as follows. At the CJ plane, the pressure is p_2, whereas at the shock front, the pressure equals the von Neumann pressure, which is approximately twice the CJ pressure p_2. Hence, $1 < \epsilon < 2$, and since the pressure drops rapidly away from the shock front with the onset of chemical reactions and asymptotes toward the CJ plane, the value of ϵ should be closer to unity.

From the conservation equations, the pressure, density, temperature, and downstream Mach number M_2 can be solved for in terms of u, γ_1, γ_2, and Q. Noting that at the CJ plane ($M_2 = 1$) an expression for M_1, as obtained by Dabora et al. (1965) and Murray (1984), is given by

$$M_1 = K \sqrt{\frac{1}{1 + \gamma_2^2 \psi}}, \tag{7.11}$$

where

$$K = \sqrt{\left(\frac{Q}{c_1^2}(\gamma_1 - 1) - \frac{\gamma_1 - \gamma_2}{\gamma_1(\gamma_2 - 1)} \right)^2 \frac{\gamma_2^2 - 1}{\gamma_1 - 1}}, \tag{7.12}$$

$$\psi = \left(\frac{1}{1 - v} \right)^2 - 1, \tag{7.13}$$

$$v = \frac{\epsilon \xi}{(\gamma_2 + 1)(1 + \xi)}. \tag{7.14}$$

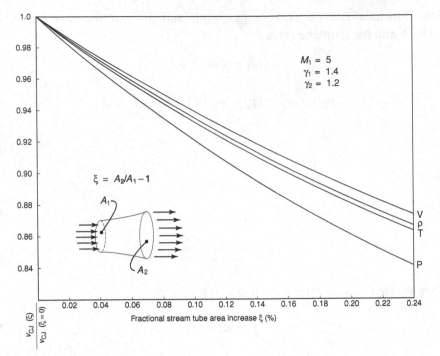

Figure 7.4. Variation of the detonation velocity as a function of area divergence (Murray, 1984).

In Eq. (7.11) we have assumed that since $M_1 \gg 1$ in general, $1/M^2 \ll 1$ and can be neglected. For most practical gaseous explosive mixtures, $M_1 \geq 5$.

An expression for the velocity deficit can now be formed from Eq. 7.11, namely,

$$\frac{\Delta V_1}{V_1} = \frac{\Delta M_1}{M_1} = \frac{M_1(\xi = 0) - M_1(\xi)}{M_1(\xi = 0)}, \tag{7.15}$$

where $\xi = 0$ corresponds to the ideal CJ value where there is no boundary layer. From Eq. (7.11), we obtain

$$\frac{\Delta V_1}{V_1} = \frac{\Delta M_1}{M_1} = 1 - \left\{ \frac{(1-\nu)^2}{(1-\nu)^2 + \gamma_2^2(2\nu - \gamma^2)} \right\}^{1/2}, \tag{7.16}$$

where ν is given by Eq. 7.14. Taking an approximate value of $\epsilon = 1$ and typical values of $\gamma_1 = 1.4$, $\gamma_2 = 1.2$, and $M_1 = 5$, the variation of $M_1(\xi)/M_1(0)$, the density, the temperature, and the pressure ratio as functions of the area divergence ξ is shown in Fig. 7.4.

Because ξ is in general quite small, we may neglect second-order terms in ξ^2, and Eq. 7.5 reduces to

$$\frac{\Delta V_1}{V_1} \approx \frac{\gamma_2^2}{\gamma_2 + 1}\xi. \tag{7.17}$$

The area divergence can be expressed in terms of the boundary layer displacement thickness δ^* as

$$\xi = \frac{A_2}{A_1} - 1 = \frac{\pi(R + \delta^*)^2}{\pi R^2} - 1 \approx \frac{2\delta^*}{R} = \frac{4\delta^*}{d} \tag{7.18}$$

for a round tube of radius R. Thus the velocity deficit may be written as

$$\frac{\Delta V_1}{V_1} \approx \frac{4\gamma_2^2}{\gamma_2 + 1} \frac{\delta^*}{d} \approx 2.6 \frac{\delta^*}{d}. \tag{7.19}$$

For $\gamma_2 = 1.2$, Eq. 7.19 gives the velocity deficit as inversely proportional to the tube diameter, which is in accord with experimental observations. For smooth tubes, the boundary layer displacement thickness has been determined in shock tube experiments by Gooderum (1958) as

$$\delta^* = 0.22x^{0.8} \left(\frac{\mu_e}{\rho_1 u_1}\right)^{0.2}, \tag{7.20}$$

where x is the distance from the shock front, μ_e is the viscosity of the gas in the reaction zone, and ρ_1, u_1 are the density and the velocity in front of the shock (in the shock-fixed coordinate system).

To compute the detonation velocity (or velocity deficit) for a given mixture and a given tube diameter, the reaction-zone thickness has to be known. The reaction-zone thickness is a function of the detonation velocity. The exact solution requires the integration of the conservation equations for the flow in the reaction zone, as well as an iteration for the detonation velocity. However, for small velocity deficits we may take, as a first approximation, the reaction-zone thickness to be that of an ideal ZND detonation with no losses, propagating at the CJ velocity. The theoretical reaction-zone thickness for a ZND laminar detonation with no losses is at least two orders of magnitude less than the experimental value based on schlieren photographs or inferred from measurements of the relaxation time (Kistiakowsky & Kydd, 1956). Since real detonations are unstable, the detonation cell size λ rather than the ZND reaction-zone length, provides a more appropriate length scale to characterize the thickness of a cellular detonation. The cell length $L_c \approx 1.5\lambda$ is more representative of the reaction-zone thickness since it measures the length of a pulsation cycle of a detonation in the direction of propagation. Extensive experimental data for the detonation cell size λ are available for most of the common fuel–air and fuel–oxygen mixtures that permit the velocity deficit in smooth tubes due to boundary layer growth to be determined.

Alternatively, the critical tube diameter d_c has also been chosen as a length scale, rather than the cell size, to characterize a real detonation front (Moen et al., 1985). The critical tube diameter is the minimum diameter through which a planar detonation can emerge into open space and continue to propagate as a spherical detonation. The critical diameter is chosen instead of the cell size because it can be determined less ambiguously. There exists a correlation of $d_c \approx 13\lambda$ that is valid for most explosive mixtures. Thus, $\lambda \approx d_c/13 = 0.077d_c$, and the cell length $L_c \approx 1.5\lambda \approx 0.11d_c$. Using the cell length (based on experimental values of either the cell size or the critical tube diameter) as the reaction-zone thickness for a real detonation, the area divergence ξ can be determined from the displacement thickness δ^*. Thus the detonation velocity (or the velocity deficit) in various-diameter tubes can be obtained.

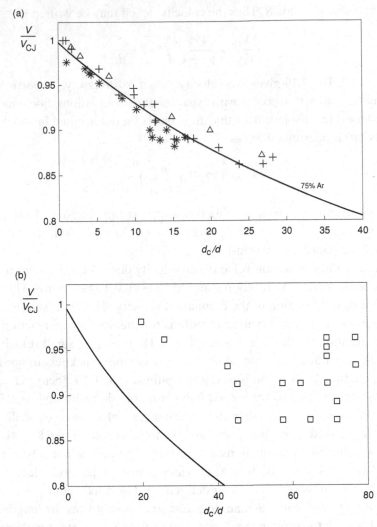

Figure 7.5. Detonation velocity deficit in (a) stable mixtures of $C_2H_2 + 2.5O_2 + x$ Ar ($+$ $x = 70\%$, $*$ $x = 75\%$, and $\triangle$ $x = 80\%$) and (b) an unstable mixture of $\square$ $0.5(C_2H_2+5\ N_2O)+0.5$ Ar; solid line is comparison with Fay's theory (Laberge *et al.*, 1993).

Figure 7.5a shows the results for $C_2H_2 + 2.5\ O_2$ with 70%, 75%, and 80% argon dilution (Laberge *et al.*, 1993). Stoichiometric acetylene–oxygen mixtures with high concentrations of argon dilution have been shown to be so-called stable mixtures where the transverse waves are very weak, the cell pattern is regular, and the propagation mechanism corresponds closely to the classical ZND model. As shown in Fig. 7.5a, the agreement between experiments and Fay's theory is quite good. For these argon-diluted mixtures, the maximum velocity deficit is about 15%. Figure 7.5b shows the results for $C_2H_2 + 2.5\ O_2 + 0.5$ Ar mixtures. This mixture is known to be less stable (i.e., the transverse waves are strong and the cell patterns are more irregular). As can be observed in the figure, the experimental results do not agree with the quasi-one-dimensional theory of Fay. This indicates that unstable

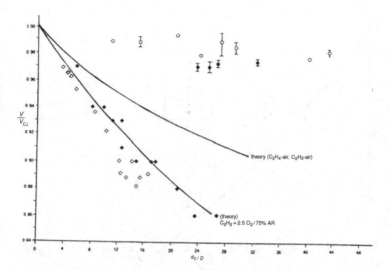

Figure 7.6. Detonation velocity deficit as a function of tube diameter for stoichiometric acetylene–oxygen with ◆ 70% Ar (D = 10.5 mm); ◊ 75% Ar (D = 12.7 mm); ● ethylene–air (D = 12.7 mm), and ○ acetylene–air (D = 12.7 mm) (Moen *et al.*, 1985).

detonations are less influenced by boundary layer effects, because the propagation mechanism is dominated by instability in the detonation structure.

Similar results are obtained for fuel–air mixtures, where again the cell pattern is more irregular and the transverse waves are stronger. Figure 7.6 illustrates the tube-diameter dependence of the velocity deficit for fuel–air mixtures. Whereas $C_2H_2 + 2.5 \, O_2 + 70\%$ Ar mixtures are adequately described by Fay's theory, the velocity deficits in fuel–air mixtures do not agree with the prediction of Fay's theory. Thus, instability mechanisms rather than the shock ignition mechanism of the classical ZND model control the propagation of fuel–air mixtures.

Since Fay's theory essentially accounts for the effect of flow divergence in the re-action zone on the propagation of the detonation wave, it should be applicable to the propagation of a detonation in a tube with weak confinement (i.e., whose walls readily yield under detonation pressures). This is the case for condensed explosives, where the detonation pressure is so high that it is practically impossible to con-fine the detonation. Interesting experiments have been carried out by Sommers and Morrison (1962) and by Dabora *et al.* (1965) where gaseous detonations propagate in a channel with one wall made of a very thin film (250 Å) that yields readily under detonation pressures. Murray and Lee (1986) have also carried out experiments in a square channel where one, two, or three of the tube walls can be replaced by thin inert films of different thicknesses and different materials. In this manner, the de-gree of flow divergence due to the lateral expansion of the detonation gases can be controlled. The flow divergence can be determined by assuming the wall to be a thin piston driven by the expanding detonation products into the inert gas outside. Fig-ure 7.7 shows the velocity deficit as a function of the divergence of the stream tube in the reaction zone (over a distance equal to the cell length of the mixture). Fairly

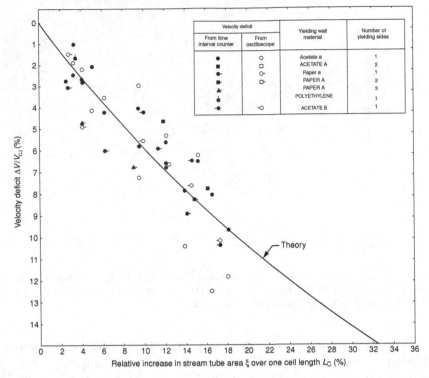

Figure 7.7. Detonation velocity deficit as a function of stream tube divergence (Murray & Lee, 1986).

good agreement with Fay's theory is observed, indicating that the area divergence concept can be applied to a wider range of situations other than just boundary layer effects in rigid tubes.

7.3. DETONATIONS IN ROUGH-WALLED TUBES

In general, in smooth-walled tubes, the velocity deficit seldom exceeds about 15% when the detonation fails. However, if the wall of the detonation tube is made very rough, a steady self-sustained detonation can be observed with a velocity deficit of over 50%. This raises an interesting paradox: on one hand, wall roughness can cause a drastic decrease in the averaged detonation velocity, but on the other hand, wall roughness can also provide a mechanism for the self-sustained propagation of low-velocity detonations. The typical averaged shock strength of a low-velocity detonation is about $M \approx 2.5$. The temperature behind such a shock wave is only of the order of 630 K, which is far too low for autoignition to occur. Thus, low-velocity detonation in tubes with very rough walls requires new ignition and combustion mechanisms that differ from that of the classical one-dimensional ZND theory.

As early as 1923, Laffite (1923) demonstrated that, by depositing a strip of coarse sand on the tube wall, the deflagration-to-detonation transition distance can be

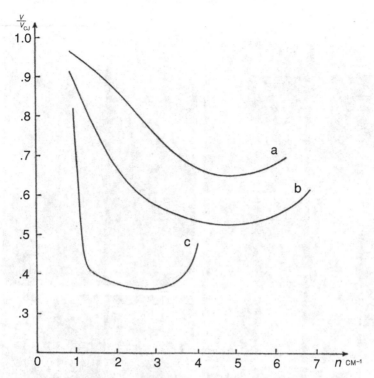

Figure 7.8. Velocity deficit in a rough-walled tube as a function of the number of turns of the Shchelkin spiral (Guénoche, 1949). (a) Tube diameter = 9 mm; spiral diameter = 1 mm; (b) tube diameter = 7 mm; spiral diameter = 1 mm; (c) tube diameter = 7 mm; spiral diameter = 2 mm.

shortened considerably. Chapman and Wheeler (1926) also showed that by putting obstacles in a tube, flame speeds of the order of a few hundred meters per second can be achieved even in relatively insensitive mixtures such as methane and air. Shchelkin (1940) inserted a long length of spiral coiled wire into a tube to increase the wall roughness and observed transition distances two orders of magnitude less than that in a smooth tube. He also observed quasi-steady detonation wave speeds as low as 50% of the CJ value. Guénoche (1949) carried out a more extensive study by measuring the detonation velocity in C_2H_2–O_2 mixtures in different tube diameters and with spirals of different wire diameter and pitch. Figure 7.8 shows the results obtained by Guénoche, where V/V_{CJ} denotes the ratio of the detonation velocity in the rough tube to that in the smooth tube, and n denotes the number of turns of the spiral per centimeter. As can be observed, the velocity decreases with increasing roughness d/D (i.e., ratio of wire diameter to tube diameter) and increasing frequency of the repeated obstacle. Quasi-steady detonation velocities as low as 40% of that in a smooth tube were observed.

Brochet (1966) later carried out similar studies in C_3H_8–O_2–N_2 mixtures and obtained streak schlieren photographs of the detonation wave in a rough-walled tube with a spiral coil. Figure 7.9 shows a series of streak schlieren photographs in $C_3H_8 + 5\,O_2 + z\,N_2$ for various amounts of nitrogen dilution z. In order to understand these results, it is best to compare them with the corresponding streak

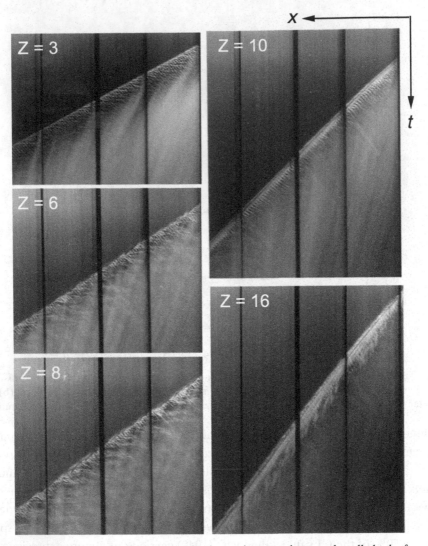

Figure 7.9. Streak schlieren photographs of a detonation wave in a rough-walled tube for varying amounts of nitrogen dilution in stoichiometric propane–oxygen–nitrogen (Brochet, 1966).

schlieren photographs of a detonation in a smooth tube (without the spiral) shown in Fig. 7.10.

In a smooth tube, the detonation for low nitrogen dilution ($z \leq 8$) has the usual high-frequency, multiheaded spin structure, as indicated by the horizontal striae. For high nitrogen dilution ($z \approx 16$) in the smooth tube, the frequency decreases toward the fundamental frequency of a single-headed spin as the limit is approached. In the rough tube with the spiral wire helix (Fig. 7.9), we observe striae at the detonation front that correspond to the frequency of the transverse perturbation generated by the spiral. The frequency of the striae now corresponds to the pitch of the spiral and thus does not change with the nitrogen concentration as in a smooth tube.

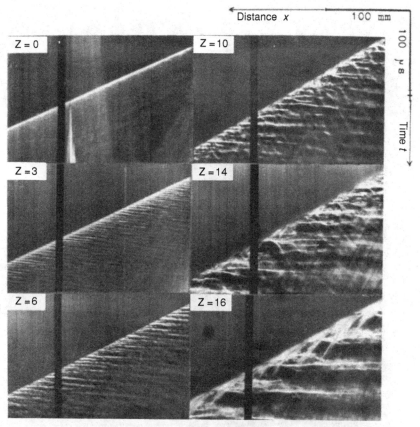

Figure 7.10. Streak schlieren photographs of a detonation wave in a smooth tube (Brochet, 1966).

Large velocity deficits are also observed. For higher nitrogen dilution ($z = 8$) in the rough tube, there is indication that low-frequency striae, corresponding to the fundamental mode of a single-headed spin, begin to appear. The onset of single-headed spin does not appear until $z \geq 14$ in a smooth tube. Thus, the spiral appears to have an adverse effect on the propagation of the detonation and hastens its approach toward the detonation limits and failure.

At a value of $z = 10$ in the rough tube, the shock is observed to separate from the reaction front; however, the propagation velocity remains constant. The reaction front is seen to propagate at the same speed as the shock front, trailing behind it by as much as 6 mm. The striae corresponding to the pitch of the spiral are still clearly evident in the reaction front. In a smooth tube, the reaction front velocity is less than that of the shock front, and as the detonation fails, the separation increases with time as in a normal deflagration wave (Fig. 7.11). For higher nitrogen dilution ($z = 16$) in the rough tube, the two fronts are still observed to propagate together at about the same speed, even though the separation distance increases now to about 10 mm. The velocity is reduced to about 40% to 50% of the corresponding value in the smooth tube.

Figure 7.11. (a) Streak schlieren and (b) self-luminous streak photographs of a detonation wave in a smooth tube in a mixture of $C_3H_8 + 5\,O_2 + 16\,N_2$ (Brochet, 1966).

These results indicate that, for readily detonable mixtures, the wire spiral has an adverse effect on the propagation of the detonation by increasing the velocity deficit and hastening the onset of single-headed spin. However, for limiting mixtures, where the detonation in a smooth tube fails with the decoupling of the shock front from the reaction front, wall roughness now tends to have a positive effect on the propagation by maintaining the coupling between the shock front and the reaction zone, even though the two are separated by a relatively large distance. Thus, in a rough tube, we have observed a new phenomenon of low-velocity detonation

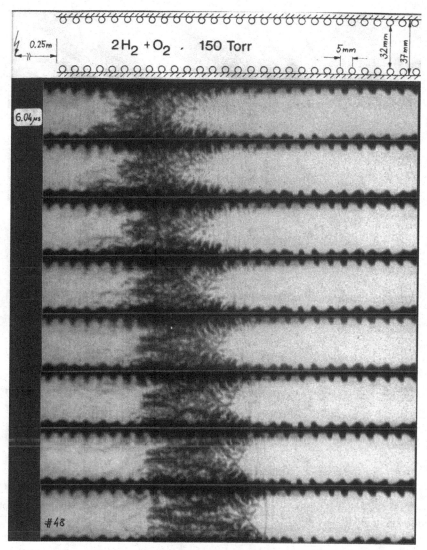

Figure 7.12. Schlieren photographs of a low-velocity detonation (or quasi-detonation) in a rough-walled tube (Teodorczyk *et al.*, 1991).

where the detonation speed can be as low as half the normal CJ value, and yet the shock and reaction front are still coupled and propagate at the same velocity.

For these detonations in rough tubes, where the velocities are substantially below the normal CJ values, two questions need to be answered. The first is the ignition mechanism, and the second is the mechanism that is responsible for the low wave velocity itself. Zeldovich and Kompaneets (1960) proposed that shock reflections from the obstacles are responsible for ignition. Although the incident shock strength is too weak to result in autoignition, Zeldovich proposed that the reflection of the shock from the obstacle gives rise to higher local temperatures to effect ignition. Once ignited, the reaction front then spreads out from the wall toward the tube axis, giving it an inverted conical shape, as shown in Fig. 7.12. However, the wall

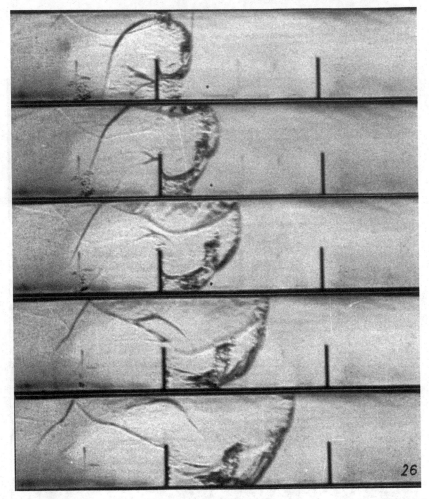

Figure 7.13. Sequence of schlieren photographs of a detonation wave propagating past an obstacle (Teodorczyk *et al.*, 1991).

roughness also generates transverse shock waves that traverse the turbulent reaction zone and enhance the burning rate.

A photographic study of the propagation of detonations in a rough tube with obstacles was carried out by Teodorczyk *et al.* (1988, 1991). Figure 7.13 shows a sequence of high-speed schlieren photographs of a detonation propagating past an obstruction. As the detonation diffracts around the obstacle, the expansion waves cause the detonation to fail, with the reaction front decoupling from the shock. However, as the shock reflects from the tube wall, the initial regular reflection changes to a Mach reflection when the shock angle exceeds the critical value. The stronger Mach stem is capable of autoignition, and the Mach stem becomes an over-driven detonation, which expands subsequently to engulf the entire decaying shock-reaction front.

In Fig. 7.14, the sequence of schlieren photographs illustrates the initiation of the detonation from the reflected shock wave from the obstacle itself. This reflected

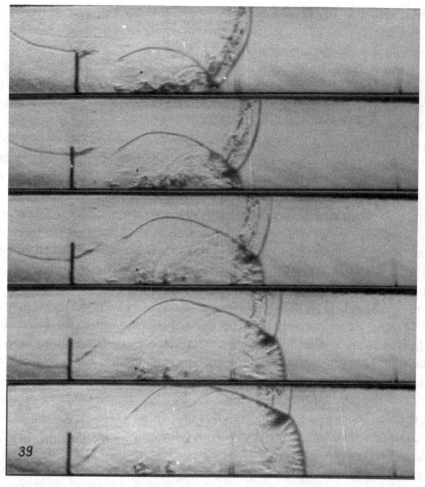

Figure 7.14. Sequence of schlieren photographs illustrating the initiation of a detonation due to shock reflection from an obstacle (Teodorczyk *et al.*, 1991).

shock deflects around the obstacle to merge with the leading shock front and to form a detonation. The detonation then propagates downward to engulf the mixture behind the diffracted shock. Thus, it is possible to cause local reinitiation of the failed detonation front via various shock reflection configurations. In general, the obstacle also scatters the incident shock, generating transverse shocks that interact to produce Mach stems and local initiation of a detonation. This is not unlike the ignition mechanism in a normal cellular detonation. In a smooth tube, transverse waves and their interactions can only come from instability. However, in a rough tube, the obstacles play an active role in the scattering of the incident shock front to create strong transverse shocks. Thus, detonations in rough tubes can be said to be more robust and can maintain steady propagation, even for high velocity deficits where detonations in a smooth tube normally fail.

Regarding the mechanism responsible for the propagation velocity in rough tubes, we cannot consider a steady one-dimensional theory, where the detonation velocity is obtained as $V \approx \sqrt{Q}$ according to CJ theory. If the velocity is half the CJ value, then the effective energy release would be only a quarter of the normal value. Losses and incomplete combustion cannot account for the loss of 75% of the chemical energy of the mixture. It has been proposed that a detonation in the obstacle-filled tube fails and reinitiates periodically (Teodorczyk *et al.*, 1991). Failure is due to diffraction around the obstacles, and reinitiation is due to shock reflections. Thus, the propagation velocity has to be averaged over the duration of failure and reinitiation. The obstacles also serve to scatter the wave in three dimensions, but the propagation speed is taken as that of propagation along the tube axis. A quantitative theory would have to take the multidimensional transient processes of the interaction of the detonation front with the obstacles into consideration.

For the low-velocity waves ($\approx$500 m/s), where even the reflected shock temperature is too low to cause autoignition, turbulent mixing between products and reactants could be the mechanism that permits the reaction front to be coupled to the shock, as shown in Fig. 7.12. It is clear that the spiral obstacle generates a train of strong transverse pressure waves that criss-crosses the reaction zone. The strong density-gradient field in the reaction zone can interact with the pressure-gradient field to generate strong vorticity from the baroclinic mechanism. Thus, vorticity production is not limited to the usual shear velocity-gradient field in the vicinity of the wall. That sufficiently rapid mixing between products and reactants can lead to autoignition and to detonation initiation has been demonstrated by Knystautas *et al.* (1979) in their study of direct initiation by a turbulent jet of hot products.

A coupled shock and reaction front that propagates at the same velocity also requires an explanation, for, according to one-dimensional theory, such steady self-sustained waves can only exist as CJ detonation waves. For steady deflagration waves, the precursor shock progressively separates from the deflagration front, since the shock and deflagration propagate at different velocities. However, if one considers a two-dimensional model of the flow behind the leading precursor shock, then it is possible to obtain a coupled shock–reaction-front complex in which the reaction front propagates via the usual transport mechanism of heat and mass diffusion of a deflagration. In Fay's model of the velocity deficit due to boundary layer effects, we see that the flow diverges behind the shock due to the negative boundary layer displacement thickness (with respect to a shock-fixed coordinate system). The flow is subsonic behind the shock, and area divergence tends to slow the flow as it moves downstream. At some distance downstream from the shock, the flow velocity will decelerate to a sufficiently low velocity so that a flame propagating against the flow can be stationary relative to the shock front. The flame propagates by the usual diffusion transport mechanism and is not autoignited by shock compression. Thus, it is possible to have a *coupled* shock–reaction-front complex propagating at the same velocity if one considers two-dimensional effects. This mechanism for low-velocity detonation was suggested by Manzhalei (1992) in his investigation of detonation propagation in capillary tubes. The various mechanisms of detonation propagation, with velocities that span a wide range of values from the CJ value to as low as 30%

of the CJ value due to the influence of boundary conditions, were also discussed by Mitrofanov (1997). In smooth tubes, large velocity deficits can only be realized in small-diameter tubes where the viscous boundary layer can exert a significant effect on the flow. However, in rough tubes with obstacles such as a wire spiral or regularly spaced orifice plates, the *effective* viscous layer near the wall can be much larger (of the order of the wall roughness or obstacle height). Thus, the low-velocity detonation regime can be realized even in large-diameter tubes.

An extensive study of the propagation of deflagrations and detonations in rough tubes for fuel–air mixtures has been carried out at McGill (Lee, 1986). Since fuel–air mixtures are much less sensitive than fuel–oxygen mixtures, much larger-diameter tubes (up to 30-cm diameter) have to be used. For large-diameter tubes, it is not convenient to use a coiled wire spiral for wall roughness, and regularly spaced orifice plates (generally spaced one tube diameter apart) were used instead. The roughness is characterized by the blockage ratio $BR = 1 - (d/D)^2$, where d and D denote the orifice and tube diameters, respectively. This is similar to the usual way in which roughness is characterized by the ratio d/D.

For a given tube diameter and blockage ratio, it is found that different propagation regimes correspond to different mixture sensitivities. To characterize the mixture sensitivity, the detonation cell size λ can be used. Figure 7.15 shows the detonation velocity as a function of hydrogen concentration for tubes 5, 15, and 30 cm in diameter. For sensitive mixtures, where the cell size is small compared to the orifice diameter, the detonation velocity was found to be close to the CJ value. The wall roughness does not appear to have much influence on the propagation of the detonation. A large velocity deficit can occur, but the propagation mechanism remains that of a normal cellular detonation. Due to the large velocity deficit, this regime is referred to as the *quasi-detonation* regime.

For smaller tube diameters, where the cell size is of the order of the orifice opening (i.e., $\lambda \approx d$), a sudden transition from the quasi-detonation regime to a lower-velocity regime is observed. The lower-velocity regime is referred to as the *choked* or *sonic* regime, because the detonation wave speed is found to correspond closely to the sound speed of the combustion products. In Fig. 7.15, the sound speed of the products corresponding to a constant pressure combustion process of the mixture (hence called the isobaric sound speed) is plotted for comparison.

Figure 7.16 shows a streak photograph of the combustion wave propagation in a methane–air mixture in a tube with periodic orifice plate obstacles. The luminosity pattern reveals two sets of pressure waves, corresponding to the two sets of characteristics of the flow. The velocity of the combustion front can be seen to be the same as the characteristics, propagating in the same direction. Therefore, this regime is called the sonic regime. The coupling between the reaction zone and the leading (scattered) shock front is due to intense turbulence generated by the obstacles. The structure is illustrated in the schlieren photographs in Fig. 7.12.

In Fig. 7.15, we note that a further decrease in the sensitivity of the mixture results in another abrupt downward transition to a *slow deflagration* regime where the

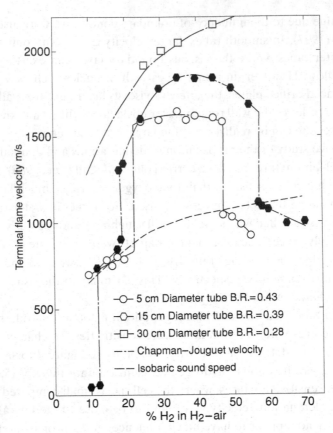

Figure 7.15. Steady flame velocities as a function of H_2 in H_2–air (Peraldi *et al.*, 1986).

propagation velocity is of the order of tens of meters per second. In this regime, normal turbulent diffusion mechanisms are responsible for the propagation of the combustion wave. At such a slow propagation speed, pressure waves play no role in vorticity generation and in the turbulence mechanisms.

The propagation of a detonation in a porous medium is very similar to its propagation in a rough, obstacle-filled tube. Detailed studies of detonation propagation in a porous medium consisting of a densely packed bed of solid particles have been carried out by Lyamin *et al.* (1991) and Makris *et al.* (1993, 1995). Figure 7.17 shows the variation in detonation velocity (V/V_{CJ}) with the equivalence ratio ϕ for a number of fuel–oxygen and fuel–air mixtures. The porous medium is a packed bed of ceramic spheres 12.7 mm in diameter. For a fixed porous bed, the detonation velocity is seen to vary continuously with mixture sensitivity, without the abrupt jump at certain mixture sensitivities as in the case of the orifice plate obstacles (as shown in Fig. 7.15). For very sensitive mixtures, where the detonation cell size is small compared to the characteristic dimension of the void space between the solid spheres, the detonation velocity is found to be close to the theoretical CJ value. As in the obstacle-filled tubes, the detonation velocity decreases with decreasing mixture sensitivity. However, there is no abrupt transition

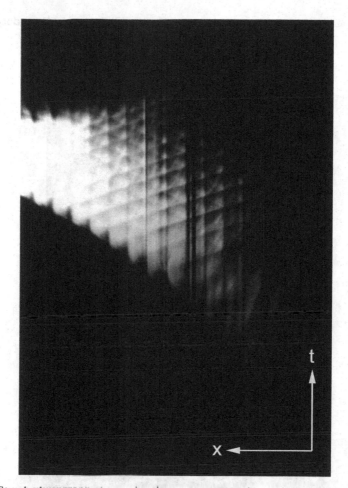

Figure 7.16. Streak photograph of a combustion wave propagating at the sound speed of the combustion products (courtesy of H. Gg. Wagner).

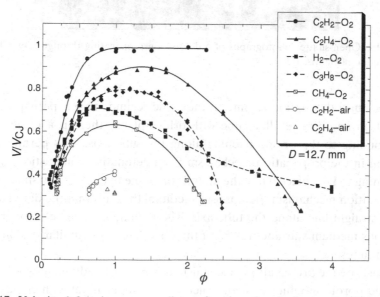

Figure 7.17. Velocity deficits in porous media as a function of equivalence ratio (Makris, 1993).

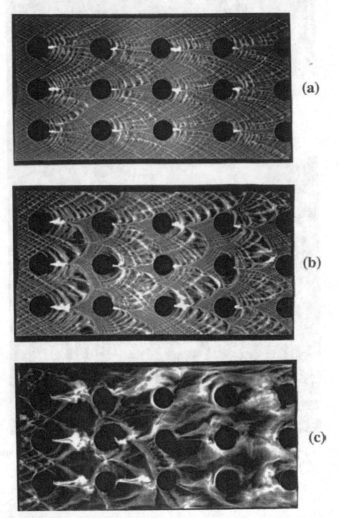

Figure 7.18. Open-shutter photographs of a detonation propagating through a two-dimensional porous medium (Makris, 1993).

from the quasi-detonation regime to the sonic regime. This is perhaps due to the fact that, in an obstacle-filled tube with orifice plates, there is a *see-through* area corresponding to the orifice opening along the tube axis. This permits a distinct transition in the propagation mechanism from detonative combustion to deflagrative burning to take place when the dimension of the opening is too small, inhibiting the detonation mechanism. In a porous medium, the path that the detonation takes is not a straight line along the tube axis. The minimum wave velocity observed in the porous medium was about 30% of the CJ value and is similar to that observed in rough tubes.

To illustrate the propagation mechanism in a porous medium, a two-dimensional simulated porous medium was constructed in a thin channel with circular disks as obstacles. The obstacles are symmetrically arranged to permit clear observation of the phenomenon. In Fig. 7.18a, the detonation propagates at almost the CJ velocity through the obstacles. Diffraction around the obstacles and reinitiation at the Mach stems in the wake of the obstacles can be observed. When the sensitivity is decreased (Fig. 7.18b), diffraction around the obstacles causes local failure, but reinitiation still occurs regularly in the wake of each obstacle. In Fig. 7.18c, reinitiation of the failed detonation is not observed, and combustion proceeds as a turbulent deflagration. Velocity deficits depend on the average duration of the failed detonation and on the frequency of reinitiation due to shock interactions.

We may summarize by saying that, for stable detonations in smooth tubes, boundary layer effects manifest themselves as flow divergence and hence a curved detonation front. Velocity deficits are consequences of wave curvature, and failure occurs at some critical value of this curvature (or equivalently, the velocity deficit). For unstable detonations, where instabilities control the propagation mechanism, velocity deficits are not due to flow divergence. There is little or no correlation of the velocity with tube diameter. Near the limit, the instability mechanism itself becomes strongly dependent on tube diameter and geometry, and failure occurs when the boundary conditions inhibit the development of the unstable detonation structure.

Rough tubes or obstacle-filled tubes have both positive and negative effects on the propagation of detonations. For readily detonable mixtures, wall roughness tends to have an adverse effect by introducing large velocity deficits. However, for less detonable mixtures, wall roughness promotes the continued propagation of supersonic combustion waves. Shock reflections generate hot spots, and obstacles facilitate the same cellular detonation mechanism by providing surfaces for shock reflections and scattering. Where autoignition is no longer possible from shock reflections, the presence of strong transverse shock waves from scattering by the obstacles provides strong vorticity production, leading to high turbulent burning rates within the reaction zone. This permits the reaction front to be coupled to the leading shock even though the shock is too weak to effect autoignition. Thus, boundary effects are seen to play profound roles in the propagation of detonation waves.

7.4. ACOUSTICALLY ABSORBING WALLS

Of particular interest is the effect of an acoustically absorbing wall on the propagation of a detonation wave. Evans *et al.* (1955) were the first to investigate the effectiveness of acoustically absorbing wall material in delaying the onset of detonation. They found that the detonation induction distance (the distance a flame travels before detonation onset) can be increased by more than three times when the wall of the detonation tube is lined with an acoustically attenuating material such as porous sintered bronze. The propagation of self-sustained detonations in porous-walled tubes was later investigated by Dupré *et al.* (1988), Teodorczyk and Lee (1995), and Radulescu and Lee (2002). It was found that upon entering the section of the tube with the acoustically absorbing liner, the detonation attenuates rapidly and propagates in the absorbing wall section with a lower velocity. For decreasing sensitivity of the mixture, the velocity deficit increases until quenching of the detonation occurs at a critical value for a given absorbing wall material. Figure 7.19

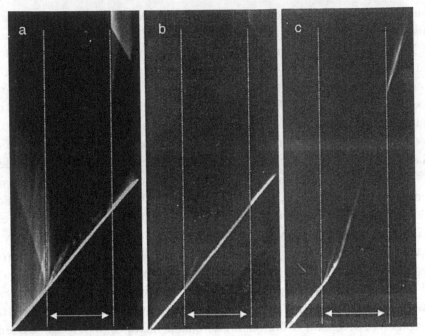

Figure 7.19. Self-luminous streak photographs of a CJ detonation propagating into a porous-walled tube section in stoichiometric C_2H_2–O_2 at (a) $p_0 = 4$ kPa, (b) $p_0 = 2.5$ kPa, and (c) $p_0 = 2.2$ kPa. Horizontal arrows denote length of damping section (Radulescu & Lee, 2002).

shows typical streak photographs of detonations propagating from a rigid tube into a porous-walled section for mixtures of decreasing sensitivity (i.e., decreasing initial pressure). A decrease in the detonation velocity is observed, and the velocity deficit increases with decreasing sensitivity of the mixture. In Fig. 7.19c, the detonation fails after propagating a short distance into the absorbing wall section.

The effect of the acoustic absorbing wall on the attenuation of the detonation is illustrated in the sequence of open-shutter photographs in Fig. 7.20. The two-dimensional channel is divided into three sections (each section is 25.4 mm high and 4 mm thick). The detonation propagates from a solid wall section prior to entering a porous wall section. Figure 7.20a, 7.20b, and 7.20c are in decreasing order of sensitivity, which is governed by the initial pressure of the mixture. From Fig. 7.20b, we can observe the attenuation of the detonation from the enlargement of the cells (or increase in the transverse wave spacing) as the detonation enters the acoustically absorbing wall section. Growth of new cells within an enlarged cell can be observed to maintain the self-sustained propagation of the detonation. In Fig. 7.20c, the detonation fails after propagating a short distance into the acoustic absorbing section, as the rate of cell regeneration cannot cope with the rate of cell destruction. A deflagration with a decoupled leading shock followed by a reaction front results when the detonation fails.

The velocity deficit as a function of the sensitivity of the mixture, as measured by the detonation cell size λ, is shown in Fig. 7.21 for three fuel–oxygen mixtures.

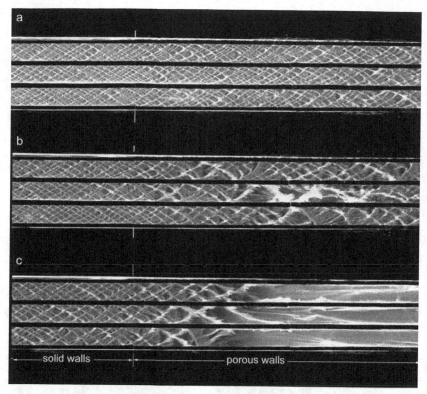

Figure 7.20. Open-shutter photographs of a detonation in stoichiometric C_2H_2–O_2 at (a) $p_0 =$ 3.6 kPa, (b) $p_0 = 3.2$ kPa, and (c) $p_0 = 2.7$ kPa, entering a porous-walled channel section (Radulescu & Lee, 2002).

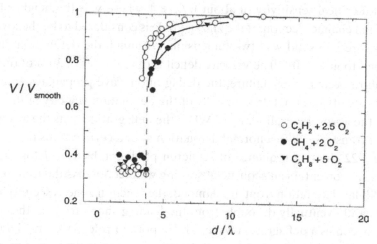

Figure 7.21. Velocity deficit in the porous-walled section for unstable mixtures (Radulescu & Lee, 2002).

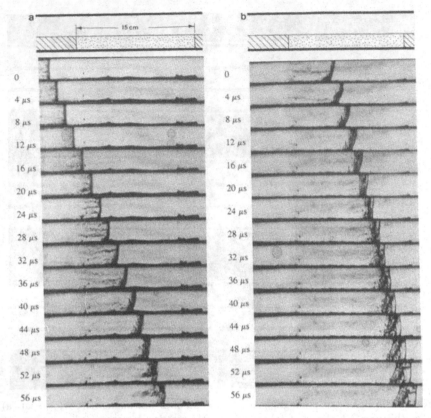

Figure 7.22. Sequence of schlieren photographs of a detonation entering from a rigid wall section into an acoustically-absorbing wall section (Teodorczyk & Lee, 1995).

The velocity deficit increases with decreasing sensitivity (increasing λ), and failure occurs at a critical sensitivity of about $w/\lambda \approx 4$, where w is the height of the two-dimensional channel section. If the zigzag mode is considered to be the lowest mode before failure in a solid wall two-dimensional channel, the detonation limit would correspond to $w/\lambda \approx 0.5$. The velocity deficit is of the order of 30% of the CJ value when failure occurs. After failure, the deflagration wave propagates at a relatively constant velocity of about 35% to 40% of the detonation velocity in the rigid tube (prior to the absorbing wall section). When the deflagration exits the absorbing wall section, it transits back to a normal detonation after a certain transitional distance.

Figure 7.22 shows a sequence of schlieren photographs as a detonation from a rigid wall section enters an acoustic-absorbing wall section. As the transverse waves of the cellular detonation front are damped, the reaction zone is seen to increase in thickness and eventually decouple from the leading shock front as the detonation fails and becomes a deflagration wave. The important role played by the transverse waves in the propagation of self-sustained detonation is evident.

The damping of shock waves incident on a porous wall is illustrated in Fig. 7.23. In Fig. 7.23a, the oblique shock incident on a solid wall results in a Mach reflection. In Fig. 7.23b, no Mach stem is observed subsequent to the reflection of the incident shock from a porous wall. Only a very weak reflected shock is obtained. Expansion

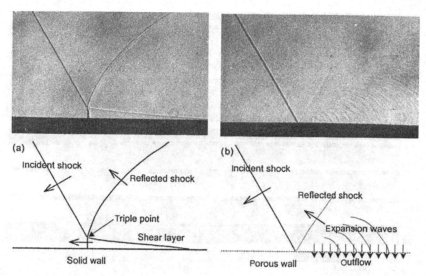

Figure 7.23. Damping of shock waves incident on a porous wall (photographs taken from Kobayashi *et al.*, 1993).

waves are generated from the mass outflow through the porous wall, which leads to the attenuation of the reflected shock.

With mass outflow from the porous wall, the flow behind the leading shock front of the detonation wave diverges in a manner similar to the boundary layer effect discussed earlier. Thus, the attenuation (i.e., velocity deficit) and failure of the detonation in the porous-wall tube may, in fact, be due to flow divergence rather than the elimination of the transverse waves. In Section 7.2, we saw that, for unstable detonation, velocity deficits do not correlate with flow divergence, because the propagation mechanism is dominated by instabilities. However, for stable detonations (e.g., C_2H_2–O_2 with a high percentage of argon dilution), the velocity deficit and failure agree well with the wave-curvature mechanisms described in Fay's theory. Thus, it would be of interest to compare the effect of the porous wall in the propagation of stable and unstable detonations. The detonations for the three mixtures shown in Fig. 7.21 are relatively unstable, as is evident from the cellular pattern shown in Fig. 7.20. One can also see clearly from Fig. 7.20b the disappearance of transverse waves in the porous wall section and the regeneration of new transverse waves due to instability. Thus, the role played by instability in the propagation of detonations in these mixtures is quite evident.

Figure 7.24 shows open-shutter photographs of detonations in stable mixtures of $C_2H_2 + 2.5\, O_2$ with a high percentage of argon dilution. Comparing Fig. 7.24 with Fig. 7.20, we can see that the cellular pattern is more regular and the transverse waves are much weaker (i.e., they are similar to Mach waves in a supersonic flow), indicating that the detonation is stable and propagates according to the classical shock ignition mechanism of the ZND model. In Fig. 7.24a, where the detonation did not quench but suffered only a slight velocity deficit, one does not observe a drastic elimination and regeneration of transverse waves. Mass outflow through the

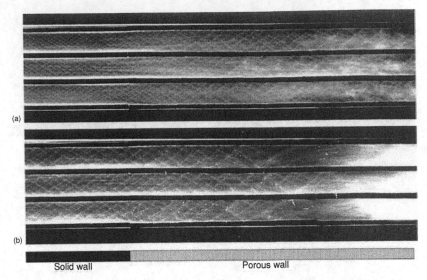

(a)

(b)

Solid wall   Porous wall

Figure 7.24. Open-shutter photographs of a detonation in stoichiometric C_2H_2–O_2–75% Ar at (a) $p_0 = 27$ kPa and (b) $p_0 = 21$ kPa entering a porous-walled section (Radulescu & Lee, 2002).

porous wall results in flow divergence and in a decrease in the detonation velocity. This is also accompanied by a slight enlargement of the cells (or transverse wave spacing). However, the transverse waves remain weak like Mach waves, and the propagation mechanism is governed by shock ignition. In Fig. 7.24b, the detonation fails after propagating a certain distance into the porous section. The failure is due to excessive curvature rather than to the inability to generate new transverse waves sufficiently fast to compensate for the damping of the transverse waves by the porous wall.

In Fig. 7.25, the velocity of the steady detonation in the porous section is shown as a function of the sensitivity for these stable detonations in mixtures of

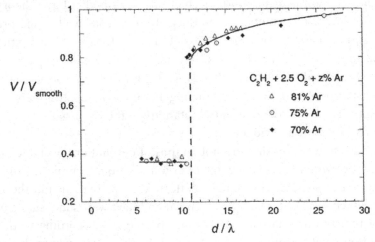

Figure 7.25. Velocity deficit in the acoustic absorbing wall section for stable detonations (Radulescu & Lee, 2002).

$C_2H_2 + 2.5\ O_2$ with high argon dilution. Note that the velocity deficit does not exceed $\approx 20\%$ before failure occurs and that this is compatible with the failure in smooth tubes due to flow divergence from boundary layer effects. After failure, the deflagration again propagates at about 40% of the normal value in the rigid-walled tube. Of particular interest is that failure now corresponds to a critical condition of $w \approx 11\lambda$ rather than the condition of $w \approx 4\lambda$ found for unstable detonations. This result is analogous to the critical-tube-diameter problem, where failure corresponds to $d \approx 13\lambda$ for unstable detonations, whereas for stable detonations (i.e., mixtures of $C_2H_2 + 2.5\ O_2$ with high argon dilution or overdriven detonations), failure occurs at $d \approx 28\lambda$ (Lee, 1995). Thus, the role played by the porous wall in the propagation of the detonation differs for stable and unstable detonations.

We have discussed the role of boundary conditions in the propagation of detonation waves in smooth-walled rigid tubes, rough or obstacle-filled tubes, and acoustically absorbing wall tubes in the previous three sections of this chapter. We see that the role played by the boundary conditions differs for stable and unstable detonations. For stable detonations, velocity deficits and failure result from curvature or flow divergence in the reaction zone, because transverse waves and instability play only a minor role if any. For unstable detonations, the boundaries exert their influence on the instability mechanism of the detonation. Of particular interest is the effect of very rough-walled tubes on the propagation of a detonation. The surface roughness and obstacles generate turbulence and transverse waves from the reflection of the leading shock off the obstacles. The interactions between transverse waves also lead to Mach stem formation and hence hot-spot formation away from the wall that facilitate ignition. This is precisely the *raison d'être* of the intrinsic instability of the detonation wave itself.

Apart from local hot spots via wave reflections and interactions, the transverse waves also provide a strong mechanism for the production of vorticity. The pressure fluctuations interact with the strong density-gradient field in the reaction zone and generate vorticity via the baroclinic torque mechanism. Autoignition and rapid combustion can result from turbulent mixing of hot products with the unburned mixture. Thus, mixing augments the role played by shock wave interactions in promoting autoignition in the detonation wave.

The relative importance of turbulence and transverse waves for the propagation of detonation waves in a rough tube is illustrated in Fig. 7.26. The top and bottom walls of the detonation channel are lined with an acoustically absorbing material to damp out transverse pressure waves. Obstacles are placed on top of the acoustically absorbing liner. As can be observed from the schlieren pictures, there is a lack of transverse waves in the reaction zone, in contrast to Fig. 7.12. The reaction front has the typical inverted V-shape because the vorticity generated near the wall promotes a higher turbulent burning rate. The overall propagation velocity of the combustion wave in this case is about an order of magnitude less than for the case without the acoustically absorbing liner beneath the obstacles (shown in Fig. 7.12). It appears that the transverse waves play a more dominant role

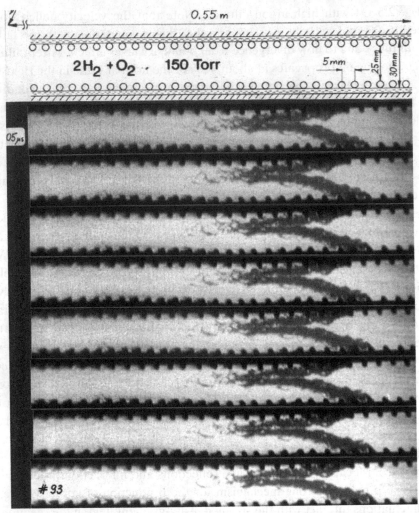

Figure 7.26. Sequence of schlieren photographs of a detonation propagating through a porous-walled tube section (Teodorczyk *et al.*, 1991).

in the propagation of the detonation wave than the turbulence generated by the obstacles.

Thus, in a rough tube, fast supersonic combustion waves that cover a wide range of propagation velocities (equal to or less than the normal CJ detonation velocity of the mixture) can be realized. When boundary conditions control the propagation mechanism, it becomes difficult to distinguish distinct propagation regimes of detonations and deflagrations. In classical theory, where the role of boundary conditions is not considered, the distinction between detonation and deflagration is based on the mechanisms of shock ignition versus heat and mass transport, respectively. The tendency for nature to involve more mechanisms to sustain the propagation of detonation via instability and the role that can be played by boundary conditions in controlling the propagation mechanism blurs the sharp distinction between the two types of combustion waves.

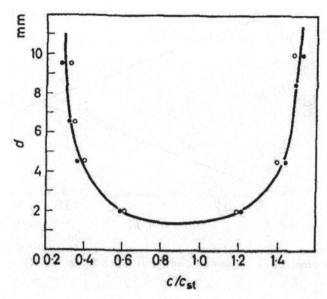

Figure 7.27. Critical tube diameter as a function of methane concentration in methane-oxygen (normalized with stoichiometric composition) (Pusch & Wagner, 1962).

7.5. DETONATION LIMITS

Detonation limits refer to the conditions outside of which the steady self-sustained propagation of a detonation wave is not possible. In the previous sections, we have discussed the important role played by boundary conditions on the propagation of the detonation wave. Thus, we would expect that the limit criteria would be different for different boundary conditions (e.g., rigid circular tubes, rigid rectangular tubes of different aspect ratios, rough tubes, acoustically absorbing wall tubes, and unconfined detonations). For each specific boundary condition, the initial condition of the mixture (i.e., its pressure, temperature, and chemical composition) should also play a role in the limit phenomena. Because most studies of the gaseous detonation phenomenon are carried out in circular rigid-wall tubes, we will concentrate on this particular boundary condition.

Given a tube diameter and initial thermodynamic state (p, T), the composition limits for a given fuel–oxidizer mixture can be determined experimentally. Figure 7.27 shows the results obtained by Pusch and Wagner (1962) for methane–oxygen mixtures where the limiting tube diameter is plotted against the fuel concentration (normalized with stoichiometric composition). The U-shaped curve indicates that the lean and rich composition limits widen with increasing tube diameter and suggest that the limits tend toward an asymptotic value as $d \to \infty$. Figure 7.28 replots the concentration limits in terms of the reciprocal of the tube diameter, which can be seen to be analogous to the dependence of the velocity deficit on $1/d$ (e.g., Fig. 7.1). This permits the extrapolation of the concentration limits as $1/d \to 0$. It seems reasonable to expect that there exists a concentration limit dictated by the chemical kinetics of the mixture only.

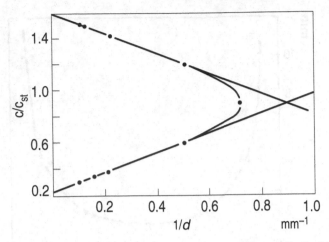

Figure 7.28. Concentration limits of methane in methane–oxygen (normalized with stoichiometric composition) as a function of inverse tube diameter (Pusch & Wagner, 1962).

For H_2–O_2 mixtures, Belles (1959) attempted to define the detonation limits on the basis of the competition between the chain-branching reaction $H_2 + O \rightarrow OH + H$ and the chain-termination reaction $H + O_2 + M \rightarrow HO_2 + M_2$. A critical temperature can be defined when the chain-termination reaction dominates. The limiting composition can then be determined from a CJ calculation when the von Neumann shock temperature falls below the critical value.

Dove and Tribbeck (1970) computed the reaction profile behind shock waves in an H_2–O_2 mixture with more detailed kinetics. They pointed out that, even though the condition may be outside the isothermal chain-branching limit, the heat release may still lead to a "thermal explosion." In an adiabatic system without heat losses, it becomes a matter of waiting long enough for eventual explosion to occur. Losses depend on boundary conditions; thus it is difficult to consider limits without defining the boundary conditions. Dove and Tribbeck also pointed out the importance of instability and strong transverse waves near the detonation limits. The three-dimensional cellular structure would invalidate one-dimensional arguments for the existence of limits.

Boundaries introduce losses of heat, momentum (friction or drag), and mass. Zeldovich (1940) had already considered heat and momentum losses in his very first formulation of the detonation structure. Fay's model treated the heat and momentum losses as boundary layer effects, which leads to a flow divergence in the reaction zone. Flow divergence can also result from the lateral expansion due to yielding boundaries and is also related to wave curvature. Thus, the various effects of boundary losses can all be considered under a general mechanism of flow divergence or wave curvature. As discussed previously in Chapter 4, the inclusion of source terms in the one-dimensional conservation equations for the reaction zone leads to an eigenvalue problem of finding the detonation velocity such that the particular solution is regular when the singular sonic condition is reached. The source terms compete to bring the flow to the sonic condition, and once it is reached, chemical equilibrium may not yet be realized. Hence, the detonation velocity does not necessarily correspond to complete energy release. A general feature of the dependence of the detonation velocity on the strength of the source terms is that the velocity decreases with increasing strength of the source term until a critical value is reached for which a steady solution can no longer be found. In other words, there corresponds a maximum velocity deficit (due to friction, curvature, etc.) beyond which no steady eigenvalue detonation solution can be obtained. This minimum velocity (or maximum velocity deficit) can be interpreted as the onset of detonation limits, which corresponds to a maximum curvature (or front divergence) or maximum frictional losses. It should also be noted that the dependence of the velocity on the strength of the term (e.g., curvature) has an S-shaped form, indicating multivalued solutions for a given source strength within the limits. However, the low-velocity detonation solutions are unstable, and it suffices to consider just the high-velocity branch.

It should be pointed out that all the theoretical consideration of limits above are based on the ZND model for the detonation structure. We have already pointed out that most detonations are unstable, and instabilities dominate the propagation mechanism. However, there are special mixtures for which the detonation is stable, and these can be described by the one-dimensional ZND model (e.g., C_2H_2–O_2 with high argon dilution).

Tsuge *et al.* (1970) carried out a detailed study of detonation limits due to flow divergence from a yielding boundary. A source term for the area divergence is introduced into the conservation of mass, and detailed kinetics of the H_2–O_2 reaction were considered. They compared their theoretical predictions with the experimental results of Dabora *et al.* (1965) and found reasonable agreement. They also claimed that the low-velocity solution corresponds to the low-velocity detonation observed by Munday *et al.* (1968), which is found to be unstable and to eventually transit to the upper high-velocity branch.

For $C_2H_2 + 2.5\,O_2$ with 75% argon dilution, the dependence of the detonation velocity on wave curvature was computed by Radulescu & Lee (2002). Figure 7.29 shows the dependence of the velocity on the front curvature for two chemical kinetic models. The broken line denotes the approximate single-step model (He & Clavin, 1994; Yao & Stewart, 1996; Klein *et al.*, 1995), whereas the solid curve is based on the full kinetic mechanism of Varatharajan and Williams (2001). A velocity deficit of about 20% is obtained, which corresponds to experimental observations shown in Fig. 7.24. For a given mixture, the limiting tube diameter has been obtained by Radulescu & Lee (2002). The mechanism that controls the propagation of the detonation (i.e., shock ignition for stable ZND detonation and cellular instability for unstable detonations), and hence also the failure, can be examined by using the appropriate length scale to characterize the detonation structure. For stable detonations, the ZND reaction length provides the length scale for the detonation structure. For unstable detonations, the cell size λ is the more representative length scale. We have observed previously (Fig. 7.21) that for the three undiluted mixtures

Table 7.1. Detonation limits in a porous-walled tube (Radulescu & Lee, 2002)

Mixture	p_0, kPa	$(d/\Delta_{ZND})_{exp}$	$(d/\Delta_{ZND})_{model}$
$C_2H_2 + 2.5\,O_2$	2.2	95.8	365
$C_3H_8 + 5\,O_2$	11.5	289	721
$CH_4 + 2\,O_2$	34.5	37	822
$C_2H_2 + 2.5\,O_2 + 70\%$ Ar	17.5	253	332
$C_2H_2 + 2.5\,O_2 + 75\%$ Ar	26.5	293	334
$C_2H_2 + 2.5\,O_2 + 81\%$ Ar	42.0	308	334

$C_2H_2 + 2.5\,O_2$, $CH_4 + 2\,O_2$, and $C_3H_8 + 5\,O_2$, the failure limit corresponds to a value of $d/\lambda \approx 4$.

For highly argon-diluted mixtures where the detonation is relatively stable (i.e., the propagation mechanism is described by the ZND model), the ZND reaction length provides the appropriate characteristic length scale. Table 7.1 gives the limiting tube diameter (referenced to the ZND length Δ_{ZND}) for stable and unstable detonations. As can be observed, the stable detonations in argon-diluted mixtures have a $d/\Delta_{ZND} \approx 300$, whereas the unstable detonations in undiluted mixtures have widely different values for the limiting value of d/Δ_{ZND}. However, if the cell size λ is used, the limiting ratio $d/\lambda \approx 4$ is found. This suggests that the failure of stable detonations is due to excessive curvature from mass leakage out of the porous wall, whereas for unstable detonations, failure is a result of the dampening of transverse waves that inhibit the development of instabilities.

The near-limit propagation of the detonation is quite complex, and the experimental determination of the detonation limits requires an operational criterion to identify the onset of detonation limits. The importance of stability near the

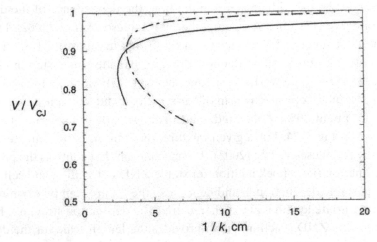

Figure 7.29. Dependence detonation velocity on frontal curvature for $C_2H_2 + 2.5\,O_2$ with 75% argon dilution with $p_0 = 26.5$ kPa (Radulescu & Lee, 2002).

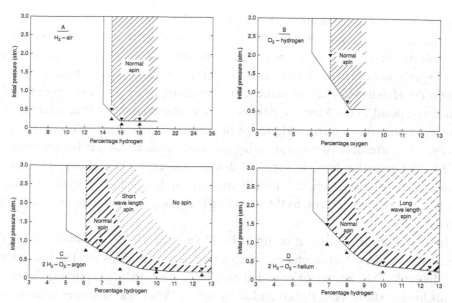

Figure 7.30. Detonation limits in different hydrogen–oxygen–diluent mixtures (Gordon *et al.*, 1959).

detonation limits was recognized in the early investigation of Campbell *et al.* (1926, 1927, 1928) in that spinning detonations were associated with near-limit conditions. Gordon *et al.* (1959) have investigated detonation limits in hydrogen–oxygen mixtures in a 20-mm-diameter tube. They used a driver section to initiate an initially overdriven detonation and then observed the subsequent decay of the overdriven wave. The strength of the initial overdriven detonation can be controlled by the pressure ratio between the driver and the test section. When the mixture in the test section is within the detonation limits, the overdriven detonation decays asymptotically to a final steady velocity. However, when the mixture is outside the limits, the overdriven detonation generally continues to decay to a deflagration with the reaction front separating from the shock. If a stronger driver were used, the initial overdriven detonation would decay first to a steady value and remain at that velocity for a certain distance of propagation. It would then decay rapidly to a deflagration if the mixture were outside the limits. For hydrogen–air mixtures, they also observed a pulsating detonation with a cyclic fluctuation of hundreds of tube diameters. The spinning mode is observed in all near-limit mixtures when the detonation is near the CJ velocity. Even in mixtures outside the limits, they observed spinning detonations during the short duration when the detonation is near the CJ velocity.

The detonation limit boundaries for four hydrogen–oxygen diluent systems are shown in Fig. 7.30, where the initial pressure of the mixture is plotted against hydrogen (or oxygen) concentration. The failure of the detonation at the limits is quite complex. The detonation limits are the regimes between the vertical solid and dotted lines. The pair of triangular points indicates the two limit conditions tested (the upper point corresponds to stable spin detonation, and the lower point indicates failure or unstable detonation). For example, the composition limit for an H_2–air mixture in the 20-mm-diameter tube is about 15% H_2. The shaded area in Fig. 7.30 represents the region where spinning detonations are observed. In argon-diluted H_2–O_2 mixtures, shorter-wavelength (or higher-frequency) multiheaded spins are observed. However, in helium-diluted H_2–O_2 mixtures, a long wavelength that does not correspond to the Manson–Fay theory was observed. Note that spinning detonation occurs over a fairly wide range of mixture composition. Thus, in Gordon's study, limits are defined when spinning detonation can no longer be obtained. The width of the vertical band in Fig. 7.30 denotes the experimental uncertainties. Beyond spinning detonation, the initially overdriven detonation decays to a deflagration or, in certain mixtures, to the pulsating (or galloping) unstable mode of propagation.

In a later study of detonation limits by Moen *et al.* (1981), the onset of spinning detonation rather than the failure of spinning detonation was used as the criterion for the limits. The argument provided by Moen *et al.* is that the spinning detonation is no longer stable past the condition of onset. When the spinning detonation is perturbed (e.g., by an obstacle such as an orifice plate or a few turns of a wire spiral), the detonation would fail and become a deflagration. However, just at the onset of spinning detonation, the perturbed detonation can reinitiate itself and form a spinning detonation again after a certain distance of propagation.

It is interesting to note that similar observations of alternate failure and reinitiation of a near-limit detonation, due to the perturbations from the confining walls of the detonation, have been reported by Murray (1984). In his experiment, a detonation in a mixture of 4.7% ethylene in air was initiated in a polyethylene tube 0.89 m in diameter. The wall thickness of 0.025 mm corresponded to the near-limit condition of the detonation. When a metal supporting hoop of the plastic tube was encountered, the perturbation generated caused the detonation to fail. However, the failed detonation reinitiated itself a short distance downstream of the hoop. The detonation failed again when perturbed by another hoop further downstream. Thus, repeated failure and reinitiation of the detonation were observed. Murray's experiment is an excellent demonstration of the effect of boundary conditions on the propagation of a detonation, particularly near the limits, where the detonation is very sensitive to perturbations. Past the first onset of spin, the detonation can no longer be considered as a bona fide detonation. The phenomenon is one of strong coupling between the chemical energy release and the fundamental transverse acoustic mode of the tube. Thus when this resonant coupling is perturbed, it cannot reestablish itself. As far as a *stable* spinning detonation is concerned, it is the first appearance of spin that results in it. There may still exist a narrow range of mixture composition near the onset of spin where the detonation is stable (i.e., can reestablish itself after suffering a large perturbation). However, this has not yet been investigated.

Since the pitch-to-diameter ratio for single-headed spin is given by

$$\frac{p}{d} = \frac{\pi D}{k_n},$$

where $k_n = 1.89$ is the first root of the derivative of the Bessel function and the ratio of the detonation velocity to the sound speed is $D/c \approx \gamma_2 + 1/\gamma_2$ (using $\gamma_2 \approx 1.2$ as the specific heat ratio of the products), we get $p/d \approx \pi$. The pitch, p, can be equated to the length of a detonation cell. Since $L_c \approx 1.6\lambda$, where λ is the cell size or transverse wave spacing, the limit criterion (based on the onset of single-headed spin) is $\lambda \approx 2d$. The cell size λ is dependent on the mixture composition and initial pressure of the mixture. Thus, for a given tube diameter d, the limiting conditions can be obtained. Moen *et al.* found reasonably good agreement with experiment for the prediction of the onset of spin in ethylene–air mixtures for tubes of different diameters.

It should be pointed out that the spin pitch is determined by the acoustic vibrations in the tube rather than by the characteristic length scale of the chemical reactions. Thus, there is no theoretical justification for equating the detonation cell length to the pitch of a single-headed spin. Lee (1984) presented a different argument and obtained a criterion $\lambda = \pi d$ instead of $\lambda = 2d$. For the single-headed spin mode, the longest characteristic length scale of the tube is the circumference, πd. Thus, the characteristic time of the transverse vibration is $\pi d/c$. In a cellular detonation, the characteristic transverse dimension is λ, and thus the characteristic time for chemical reactions in a cellular detonation is λ/c. Equating the two gives $\lambda = \pi d$, which relates the time scale for the lowest transverse mode (i.e., spin) to the time scale of chemical reactions. Kogarko and Zeldovich (1948) also proposed $\lambda = \pi d$ as the limit criterion, but gave no explanations for arriving at this choice. It should be noted that the difference between the two criteria is within the experimental variations in cell size measurements.

As instabilities set in when the limit is approached, large fluctuations in the detonation velocity associated with the instabilities will also occur. Because velocity measurements can be made with a high degree of accuracy, a criterion for the detonation limits can perhaps be formulated on the basis of the fluctuation in the detonation velocity. Manson *et al.* (1966) investigated the near-limit behavior of $C_3H_8 + 5\,O_2 + z\,N_2$ at $p_0 = 1$ atm, $T_0 = 298$ K over a range of different tube diameters (6 mm $< d <$ 52 mm). They made accurate measurements of the detonation velocity and observed the unstable detonation structure simultaneously with streak schlieren photography. They defined a parameter called the *relative deviation,* $\delta = \Delta D/\overline{D}$ (where ΔD is the maximum deviation between the local detonation, velocity D_l, and the mean detonation velocity D_m) to characterize the velocity fluctuation. The local velocity was determined by a pair of ionization probes spaced 500 mm apart, whereas the mean velocity is the average value over a distance of propagation of at least 5 m. The normalizing velocity, $\overline{D}$, is the average of D_l and D_m. Using the streak schlieren photographs of the unstable detonation front, they were able to identify the velocity fluctuation within the unstable detonation structure.

Figure 7.31 shows the stability map for $C_3H_8 + 5\,O_2 + z\,N_2$, where the inert nitrogen concentration of the mixture, z, is plotted against the inverse tube diameter for the constant-δ lines that define the different regions of stability. For detonations with

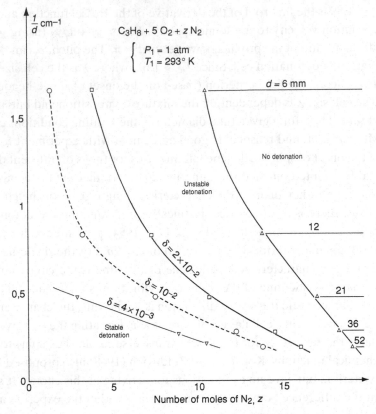

Figure 7.31. Stability map for number of moles of N_2, $C_3H_8 + 5\,O_2 + z\,N_2$ at $p_0 = 1$ atm and $T_0 = 293$ K (Manson *et al.*, 1966).

$\delta \leq 0.004$, the detonations are defined as stable detonations even though the streak schlieren photographs indicate high-frequency spin of the order of 1 to 1.5 MHz. For $0.01 \leq \delta \leq 0.02$, the detonation is considered to be unstable. For $\delta > 0.02$, the detonation corresponds to the fundamental spinning mode. For each tube diameter, there is also a concentration of nitrogen diluent, z, where no self-sustained detonation can be observed. Manson *et al.* chose a critical value of $\delta^* = 0.004$ (i.e., 0.4%) as the maximum allowable deviation from the mean for stable detonations. Thus, their criterion excludes spinning detonations even though they propagate at a remarkably constant mean velocity close to the theoretical CJ value (even though the local fluctuation of a spinning detonation is large).

It is clear that there is a certain degree of arbitrariness in defining what detonations can be considered to be within the limits. It appears that the single-headed spin should be considered as the final self-sustained mode of propagation. The galloping mode observed in certain mixtures has been considered to be within the limits also, even though for most of the galloping cycle, the shock and the reaction front are decoupled, with the complex propagating at about half the CJ velocity.

The measurements of the detonation velocity by regularly spaced ionization probes are averaged between adjacent probes. It would be of value to monitor the velocity continuously. A novel microwave Doppler interferometer using a coaxial mode was developed by Lee and Pavlasek (1994). The coaxial technique is superior to the normal microwave interferometer that uses the detonation tube as the waveguide, where many modes are excited, quite often leading to ambiguous interpretations. Lee *et al.* (1995) carried out near-limit velocity measurements in a 38.4-mm-diameter 10-m-long tube with a 0.5-m-long driver section, where a more sensitive mixture at higher initial pressures was used to initiate the insensitive near-limit mixture in the driven section of the tube. Various mixtures were used, because one mixture cannot illustrate the variety of possible near-limit behavior.

Figure 7.32 shows the velocity–time history and the corresponding velocity histograms for the different mixtures. In Fig. 7.32a, a stable mixture of $C_2H_2 + 2.5 O_2 + 75\%$ Ar was used. The velocity–time history shows that the velocity is practically constant throughout the duration of the detonation propagation in the 10-m tube. The velocity histogram indicates a very narrow band of less than 100 m/s in which the detonation velocity varies. The velocity deficit is less than 5%.

For a more unstable mixture of $C_3H_8 + 5 O_2$ closer to the limits, the velocity–time history (Fig. 7.32b) shows rapid fluctuations that are characteristic of multi- and single-headed spinning detonations. However, the mean velocity is still fairly constant throughout the duration of the 10-m-long propagation. The velocity deficit is larger, of the order of 10%, and the histogram shows a wider spectrum of velocity fluctuations.

For an unstable mixture of $C_2H_6 + 3.5 O_2$, further toward the limits (Fig. 7.32c), the detonation is observed to make periodic excursions (but at irregular intervals) to a low-velocity regime (i.e., $V \simeq 0.7 V_{CJ}$). It remains in the low-velocity regime for a short duration and then abruptly transits back to the rapid fluctuating regime of Fig. 7.32b. The velocity histogram of this behavior is shown in Fig. 7.32c, and we see a wide spectrum of velocity fluctuations in the range 1400 m/s $< V <$ 2200 m/s. This near-limit behavior was referred to as the "stuttering mode" by Lee *et al.* (1995).

For an unstable mixture of $C_3H_8 + 5 O_2$, further past the stuttering mode, we see that the duration of the low-velocity regime becomes very long (Fig. 7.32d). The low velocity has typical values of half the CJ velocity of the mixture. The histogram also illustrates that the fraction of time spent in the low-velocity regime far exceeds that in the high-velocity regime near the CJ velocity. In terms of distance of propagation, the low-velocity phase of the cycle is of the order of 100 tube diameters or more. This mode of propagation is referred to as the *galloping mode,* christened by Duff in the discussion of the paper by Manson *et al.* (1966).

During the low-velocity phase of the galloping cycle, the shock and the reaction front are decoupled even though they remain propagating at the same velocity. Theoretically, this is not possible in a one-dimensional analysis of a shock–flame complex where the flame must propagate at a lower velocity than the leading shock front (and the two fronts separate with time). However, if we consider a two-dimensional model that includes boundary layer effects, the negative displacement thickness results in a diverging flow behind the shock. Relative to the shock front, the flow is

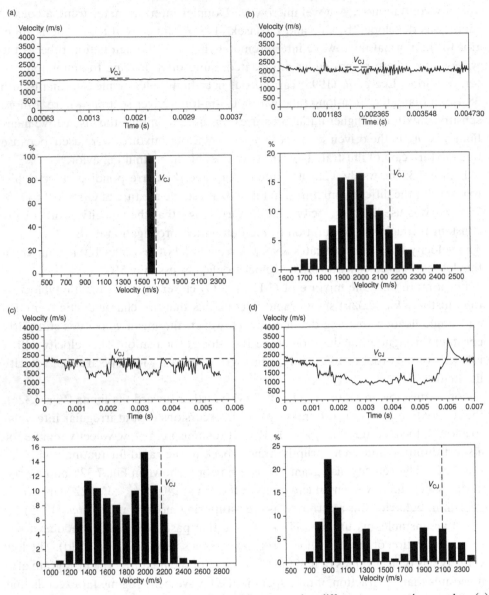

Figure 7.32. Velocity histories and velocity histograms for different propagation modes: (a) $C_2H_2 + 2.5\ O_2 + 75\%$ Ar at $P_0 = 100$ Torr, (b) $C_3H_8 + 5\ O_2$ at $p_0 = 12$ Torr, (c) $C_2H_6 + 3.5\ O_2$ at $p_0 = 10$ Torr, and (d) $C_3H_8 + 5\ O_2$ at $p_0 = 8$ Torr (Lee *et al.*, 1995).

subsonic and flow divergence leads to a decrease in the flow velocity downstream of the shock. Thus, for any given burning velocity of the reaction front, the flame will stabilize at a point where the reaction front is stationary relative to the shock, and the two fronts propagate at the same velocity. Galloping detonations are mostly observed in small-diameter tubes where the boundary layer has a larger effect on the flow behind the shock.

Beyond the galloping mode, the detonation can remain at a velocity of about half the CJ value for very long distances (hundreds of tube diameters) without reaccelerating to a high-velocity detonation. This low-velocity regime is also unstable and fails, when perturbed. The shock eventually decays to a sonic wave and the reaction front decays to a slow deflagration that propagates at a velocity of the order of tens of meters per second, depending on the level of turbulence. However, a finite perturbation may also trigger the onset of detonation from this metastable state.

It is clear that a criterion for defining the detonation limits first requires what one considers a bona fide detonation. It is reasonable to consider that the onset of single-headed spinning detonations in a rigid smooth-walled tube is the limiting condition. It has often been argued that, in terms of the damaging potential of a detonation wave, the galloping mode is equally if not more destructive than a stable detonation. However, galloping detonations are generally observed in small-diameter tubes and in so-called unstable mixtures. It is the highly unstable behavior in the reaction zone that results in the rapid growth of instabilities that lead to re-acceleration of the detonation from the metastable low-velocity regime back to the high-velocity regime. In stable mixtures, the detonation fails due to the mechanism of excessive curvature, and the velocity deficit can then be used to characterize the limiting conditions. For different boundary conditions (smooth tubes of different geometries, rough-walled tubes, flexible-walled tubes, acoustically absorbing wall tubes, etc.), the failure mechanism and thus the limiting conditions differ. For completely unconfined spherical detonations, continued steady propagation of the detonation wave requires that the growth rate of transverse waves be sufficiently rapid to cope with the rate of increase of the surface area of the detonation wave, so that the average dimension of the detonation cells can be maintained (Lee *et al.*, 1969). If the detonation is stable and propagates according to the ZND model, then there cannot be a constant-velocity spherical detonation, because the velocity depends on the curvature, which changes with increasing radius of the detonation wave. Stable constant-velocity spherical detonations in highly argon-diluted mixtures have not yet been observed.

7.6. CLOSING REMARKS

Although the CJ theory permits the detonation velocity to be determined from the conservation laws using the CJ criterion, von Neumann showed that the detailed structure must be considered to rule out weak detonations. Thus, a more rigorous detonation theory must consider the structure and use the generalized CJ criterion to determine the detonation velocity. If the real structure is to be considered, then the boundary conditions must also enter into the determination of the detonation velocity.

This chapter has discussed the influence of various boundary conditions on the propagation of detonation waves. In general, the effect of boundary conditions depends on whether the detonation is stable or unstable. For stable detonations, this means that the classical ZND model provides an adequate description of the propagation mechanism, and the boundary conditions accordingly influence the propagation of the detonation wave through their effect on the ZND structure (e.g. curvature). If the detonation is unstable, and the development of instabilities is responsible for the propagation of the detonation, then the influence of boundary conditions on the growth of instabilities controls the propagation of the detonation. For example, porous-walled tubes that absorb the transverse waves will inhibit and suppress instabilities, thus causing the detonation to attenuate and fail. Detonation limits are a direct consequence of the influence of boundary conditions, and the limit criterion will be specific to the boundary conditions imposed by a given geometry. For unstable detonations, the operational definition of the detonation limit is difficult to specify, due to the diversity of unstable phenomena near the limits (e.g., spinning and galloping detonations).

Bibliography

Belles, F. 1959. In *7th Int. Symp. on Combustion* 745–751.

Brochet C. 1966. Contribution a l'étude des detonations instable dans les mélanges gazeux. Doctoral thesis, University of Poitiers, France.

Campbell, C., and A.C. Finch. 1928. *J. Chem. Soc.* 131:2094.

Campbell, C., and D.W. Woodhead. 1926. *J. Chem. Soc.* 129:3010.

Campbell, C., and D.W. Woodhead. 1927. *J. Chem. Soc.* 130:1572.

Chapman, W.R., and R.N.V. Wheeler. 1926. *J. Chem. Soc. Lond.* 129:2139–2147.

Dabora, E.K., J.A. Nicholls, and R.B. Morrison. 1965. *Tenth Int. Symp. on Combustion*, 817.

Dove, J., and T.D. Tribbeck. 1970. *Astronaut Acta* 15:387–397.

Dupré G., O. Pcraldi, J.II.S. Lee, and R. Knystautas. 1988. *Prog. Astronaut and Aeronaut.* 114:248–263.

Evans, M., F. Given, and W. Picheson. 1955. *Found. Appli. Phys.* 26(9):1111–1113.

Eyring H., R.E. Powell, G.H. Duffrey, and R.B. Paolin. 1949. *Chem. Rev.* 45:69.

Fay, J. 1959. *Phys. Fluids* 2(3).

Gooderum, P.B. 1958. *NACA Tech. Note* 4243.

Gordon, W.E., A.J. Mooradian, and S.A. Harper. 1959. In *7th Int. Symp. on Combustion*, 752–759.

Guénoche, H. 1949. *Rev. Inst. Français Pétrole* 4:15–36, 48–69.

He, L., and P. Clavin. 1994. *J. Fluid Mech.* 277:227.

Jones, H. 1947. *Proc. R. Soc. Lond A.* 189:415.

Kistiakowsky, G.B., and P.H. Kydd. 1956. *J. Chem. Phys.* 25:824.

Kistiakowsky, G., and W.G. Zinman. 1955. *J. Chem. Phys.* 23:1889.

Kistiakowsky, G., H. Kwight, and M. Malin. 1952. *J. Chem. Phys.* 20:594.

Klein, R., J.C. Krok, and J.E. Shepherd. 1995. Explosion Dynamics Laboratory Report FM95-04, California Institute of Technology.

Knystautas, R., J.H.S. Lee, I.O. Moen, and H. Gg. Wagner. 1979. In *17th Int. Symp. on Combustion*, 1235–1245.

Kobayashi, S., T. Adachi, T. Suzuki. 1993. In *Shock waves at Marseille IV: Shock structure and kinematics, blast waves and detonations*, ed. R. Brun and L.Z. Dumitrescu, 175–180. Springer.

Kogarko, S.M., and Ya. B. Zeldovich. 1948. *Dokl. Akad. Nauk SSSR* 63:553.

Laberge, S., M. Atanasov, R. Knystautas, and J.H.S. Lee. 1993. *Prog. Astronaut and Aeronaut.* 153:381–396.

Lafite, P. 1923. *C. R. Acad. Sci.* 176:1392–1394.

Lee, J.H.S. 1984. *Am. Rev. Fluid Mech.* 16:311–336.

Lee, J.H.S. 1986. In *Advances in chemical reaction dynamics*, ed. P.M. Reutzejois and C. Capellos, 345–378. Reidel.

Lee, J.H.S. 1995. In *Dynamics of exothermicity*, ed. J.R. Bowen *et al.*, 321–335. Gordon and Breach.

Lee, J.H.S., R.I. Soloukhin, and A.K. Oppenheim. 1969. *Acta Astronaut.* 14:565–584.

Lee, J.J., G. Dupré, R. Knystautas, and J.H.S. Lee. 1995. *Shock Waves* 5:175–181.

Lee, J.J., and T. Pavlasek. 1994. In *Non-intrusive combustion diagnostics*, ed. K.K. Kuo and T.P. Parr, 285–293. Begell House, New York.

Lyamin, G.A., V.V. Mitrofanov, A.V. Pinaev, and V.A. Subbotin. 1991. In *Dynamic structure of detonation in gaseous and dispersed media*, ed. A.H. Borisov, 51–75. Kluwer Academic.

Makris, A. 1993. The propagation of gaseous detonation in porous media. Ph.D. thesis, McGill University, Montreal, Canada.

Makris, A., H. Shafique, J.H.S. Lee, and R. Knystautas. 1995. *Shock Waves* 5:89–95.

Manson, N., Ch. Brochet, J. Brossard, Y. Pujol. 1962. In *9th Int. Symp. on Combustion*, 461–469. See also 1966. Ch. Brochet. Contribution a l'étude de détonations instables dans les mélanges gazeux. Thèse du Docteur, Université de Poitiers, France.

Manson, N., and H. Guénoche. 1956. In *Sixth Int. Symp. on Combustion*. 631–691.

Manzhalei, V.I. 1992. *Fiz. Goreniya Vzyva* 28(3):93–99.

Mitrofanov, V.V. 1997. In *Advances in combustion sciences in honor of Ya. B. Zeldovich*, 327–340.

Moen, I., M. Donato, R. Knystautas, and J.H.S. Lee. 1981. In *18th Int. Symp. on Combustion*, 1615.

Moen, I.O., A. Sulmistras, G.O. Thomas, D. Bjerketvedt, and P. Thibault. 1985. In *Dynamics of explosion*, ed. J.R. Bower, J.C. Reyer, and R.I. Soloukhin.

Munday, G., A.R. Ubbelohde, and I.E. Wood. 1968. *Proc. R. Soc. A* 306:159–170.

Murray, S.B. 1984. The influence of initial and boundary conditions on gaseous detonation waves. Ph.D. thesis, McGill University.

Murray, S.B., and J.H.S. Lee. 1986. *Prog. Astronaut. and Aeronaut.* 106:329–355.

Peraldi, O., R. Knystautas, and J.H.S. Lee. 1986. Criteria for transition to detonation in tubes. *Proc. Combust. Inst.* 21:1629.

Pusch, W., and H. Gg. Wagner. 1962. *Combust Flame.* 6:157–162.

Radulescu, R., and J.H.S. Lee. 2002. *Combust. Flame* 131:29–46.

Renault, G. 1972. Propagation des détonations dans les mélanges gazeux contenus dans des tubes de section circulaire et de section rectangulaire: Influence de l'état de la surface interne des tubes. Ph.D. thesis, Université de Poitiers, France.

Shchelkin, K.I. 1940. *Zh. Eksp. Teor. Fiz.* 10:823–827.

Sommers, W.P., and R.B. Morrison. 1962. *Phys. Fluids* 5:241.

Teodorczyk, H., and J.H.S. Lee. 1995. *Shock Waves* 4:225–236.

Teodorczyk, A., J.H.S. Lee, and R. Knystautas. 1988. In *22nd Int. Symp. on Combustion* 1723–1731.

Teodorczyk, A., J.H.S. Lee, and R. Knystautas. 1991. *Prog. Astronaut. and Aeronaut.* 133:223–240.

Tsuge, S., H. Furukawa, M. Matsukawa, and T. Noyakawa. 1970. *Astronaut Acta* 15:377–386.

Varatharajan, B., and F. Williams. 2001. *Combust. Flame* 125:624.

Wood, W., and J.G. Kirkwood. 1954. *J. Chem. Phys.* 22:1920.

Yao, J., and D.S. Stewart. 1996. *J. Fluid Mech.* 309:225.

Zeldovich, Ya. B. 1940. *Zho. Eksp. Teor. Fiz.* 10:542. Translated in NACA Tech. Memo. 1261 1950.

Zeidovich, Ya. B., AA. Borisov, B.E. Gelfand, S.M. Frolov, and A.E. Mailkov. 1988. Nonideal detonations waves in rough tubes. *Prog. Astronaut. Aeronaut.* 144:211–231.

Zeldovich, Ya. B., and A.S. Kompaneets. 1960. *Theory of detonations*. 185–191. Academic Press.

8 Deflagration-to-Detonation Transition

8.1. INTRODUCTION

The two combustion modes of deflagration (flame) and detonation can be generally distinguished from each other in a number of ways: by their propagation speed, the expansion (deflagration) versus compression (detonation) nature of the wave, subsonic (deflagration) versus supersonic (detonation) speed relative to the mixture ahead of the wave, and the difference in the propagation mechanism. A deflagration wave propagates via the diffusion of heat and mass from the flame zone to effect ignition in the reactants ahead. The propagation speed is governed by heat and mass diffusivity, and the diffusion flux is also dependent on the reaction rate that maintains the steep gradient across the flame. On the other hand, a detonation wave is a supersonic compression shock wave that ignites the mixture by adiabatic heating across the leading shock front. The shock is in turn maintained by the backward expansion of the reacting gases and products relative to the front, thus providing the forward thrust needed to drive the shock. A propagating flame generally has a precursor shock ahead of it, and the flame can therefore propagate at supersonic speeds with respect to a stationary coordinate system. In theory, there is still a pressure drop across the flame itself, but there may still be a net pressure increase in the products across the shock–flame complex with respect to the initial pressure of the reactants. Therefore, there may not be a clear distinction between a deflagration and a detonation with respect to the stationary frame of the laboratory coordinates, in that both can be supersonic with a net pressure increase in the products. However, in the case of a deflagration, the reaction front still propagates via diffusive transport, in contrast to autoignition across the leading shock front for a detonation.

Deflagrations typically require fractions of a millijoule of energy for ignition, whereas direct initiation of a detonation requires joules (or kilojoules) even for sensitive fuel–oxygen mixtures. Therefore, a deflagration is generally the more probable mode of combustion, due to the ease of its initiation. However, a self-propagating deflagration is generally unstable and tends to accelerate subsequent to its ignition. Under appropriate boundary conditions, a flame can continuously accelerate and undergo an abrupt transition to a detonation wave. The onset of detonation occurs locally in the flame zone when the appropriate critical conditions are met, and the history of how these critical conditions are achieved is irrelevant to the process for the onset of detonation. In general, there is no critical maximum flame speed that has to be attained before the onset of detonation can occur provided that the necessary critical conditions are achieved. However, if we consider the classical experiment of transition from deflagration to detonation in a long smooth-walled tube, then a deflagration will generally accelerate to some maximum velocity of the order of about half the CJ detonation speed when spontaneous onset of detonation takes place. We shall restrict ourselves to a discussion of the transition from deflagration to detonation in a smooth-walled tube to bring out all the salient features of the processes involved when an initial laminar flame kernel accelerates and eventually transits to a detonation. The *direct* generation of the required critical conditions for the onset of detonation (e.g., turbulent mixing of products and reactants in a jet) will be discussed later as a direct initiation process.

From gasdynamic considerations, deflagration solutions are represented on the lower branch of the Hugoniot curve, and the maximum deflagration speed (for a given initial state ahead of the flame) corresponds to the point where the Rayleigh line and the Hugoniot curve are tangent. For detonations, the solution is represented on the upper branch of the Hugoniot curve. The minimum velocity detonation solution corresponds to the tangency of the Rayleigh line to the Hugoniot curve on the upper branch. Transition from deflagration to detonation can thus be thought of as a jump from the lower branch to the upper branch of the Hugoniot curve. However, for a propagating deflagration wave, a precursor shock is usually generated ahead of the flame. The precursor shock changes the initial state ahead of the flame and, accordingly, the Hugoniot curve, because the Hugoniot curve depends on the initial state ahead of the flame. For different precursor shock strengths, we will have different Hugoniot curves that represent different deflagration solutions. On the other hand, when a detonation is formed, the initial state is taken to be the original undisturbed state of the mixture, because the detonation wave is supersonic. If we consider the quasi-steady shock–flame complex during the acceleration phase of the deflagration, it may be possible to go from this deflagration branch to the detonation branch of the Hugoniot curve (based on the initial undisturbed state) without crossing the forbidden quadrant of the Hugoniot curve where pressure increases while density decreases, which would require an imaginary velocity.

The transition from deflagration to detonation in a long, smooth tube can generally be divided into a flame acceleration phase and the detonation onset phase. The initial flame acceleration phase can involve the entire spectrum of flame acceleration mechanisms (instability, turbulence, acoustic interaction, etc.). Which flame acceleration mechanisms dominate (or are absent) depends strongly on the initial and boundary conditions. Hence, it is not possible to formulate a general theory for this phase of the transition phenomenon. However, for the final phase of the onset of detonation, it appears possible to describe at least qualitatively the critical conditions required for the spontaneous onset of detonation. In a smooth tube, experiments indicate that the flame generally accelerates to a velocity of the order of the CJ deflagration speed of the mixture. This state is, however, unstable and eventually results in the formation of local explosion centers. Overdriven detonations are then formed from these hot spots, which grow and then decay to a CJ detonation.

In order to show the rich diversity of combustion phenomena associated with the transition process, we shall discuss the standard transition processes in a long smooth-walled tube where there is an acceleration phase followed by the onset of detonation and the subsequent decay of the overdriven detonation to its CJ velocity.

8.2. GASDYNAMICS OF DEFLAGRATION WAVES

The propagation of a deflagration wave is described by the lower branch of the Hugoniot curve, where the pressure and density behind the reaction front (i.e., the flame) are lower than in the initial state (p_0, ρ_0). Furthermore, a deflagration wave propagates at a subsonic speed relative to the unburned reactants ahead of it. Therefore, the boundary conditions downstream of the deflagration can influence the state ahead of the flame in the reactants. For example, if the deflagration propagates from the closed end of a tube where the particle velocity must vanish, then the increase in specific volume (or decrease in density) across the flame front results in a displacement of the reactants ahead of it. The flame acts like a leaky piston that displaces the reactants, causing them to move in the same direction as the flame. The rate of increase in the specific volume of the mixture across the flame also causes compression waves to be generated, which radiate ahead of the flame front. The compression waves eventually catch up with one another, forming a precursor shock wave. Eventually, the flow field corresponding to the propagation of a deflagration wave from the closed end of a tube will consist of a precursor shock wave (which sets the reactants behind it into motion) and a flame front that follows the precursor shock. The flame (or reaction front) now propagates into the shock-compressed reactants instead of the mixture at its initial state. Behind the reaction front, the particle velocity is zero to accommodate the closed-end boundary condition of the tube. Because the pressure behind the flame is higher than the initial pressure ahead of the precursor shock wave, the momentum of the gas between the shock and the flame must increase with time. As a result, the separation distance between the shock and flame increases (i.e., the flame and shock speeds are different).

The initial state ahead of the deflagration front changes for different deflagration speeds because the precursor shock strength is different. The possible final states behind the accelerating flame will therefore be represented on different Hugoniot curves. The states for a propagating deflagration can be readily determined from the conservation laws across the shock and the flame. We shall consider the propagation of the deflagration from a closed-end tube.

Referring to the sketch in Fig. 8.1, the states behind the two fronts are denoted by the subscripts 1 and 2. For convenience, we shall define u_1 as the particle velocity

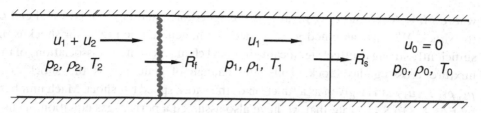

Figure 8.1. Sketch of a propagating flame following a precursor shock wave in a smooth tube.

behind the precursor shock relative to the gas ahead (i.e., $u_0 = 0$), and u_2 as the particle velocity behind the flame relative to the mixture ahead that moves at velocity u_1. Thus, the particle velocity behind the flame relative to the fixed laboratory coordinates will be $u_1 + u_2$. We can write the conservation laws across the precursor shock as

$$\rho_0 \dot{R}_s = \rho_1(\dot{R}_s - u_1), \tag{8.1}$$

$$p_0 + \rho_0 \dot{R}_s^2 = p_1 + \rho_1(\dot{R}_s - u_1)^2, \tag{8.2}$$

$$h_0 + \frac{\dot{R}_s}{2} = h_1 + \frac{(\dot{R}_s - u_1)^2}{2}, \tag{8.3}$$

$$h = \frac{\gamma}{\gamma - 1} \frac{P}{\rho}, \tag{8.4}$$

where $\dot{R}_s$ is the shock velocity with respect to the fixed laboratory coordinates. Across the flame, the conservation laws are

$$\rho_0(\dot{R}_f - u_1) = \rho_1(\dot{R}_f - u_1 - u_2), \tag{8.5}$$

$$p_1 + \rho_1(\dot{R}_f - u_1)^2 = p_2 + \rho_2(\dot{R}_f - u_1 - u_2)^2, \tag{8.6}$$

$$h_1 + Q + \frac{(\dot{R}_f - u_1)^2}{2} = h_2 + \frac{(\dot{R}_f - u_1 - u_2)^2}{2}, \tag{8.7}$$

where Q is the chemical energy release per unit mass across the flame front, and $\dot{R}_f$ is the velocity of the flame with respect to the fixed laboratory coordinates. Solving for the density, pressure, particle velocity, and temperature ratios across the normal shock gives

$$\frac{\rho_1}{\rho_0} = \frac{\gamma_0 + 1}{(\gamma_0 - 1) + \frac{2}{M_s^2}}, \tag{8.8}$$

$$\frac{p_1}{p_0} = \frac{2\gamma_0}{\gamma_0 + 1} M_s^2 - \frac{\gamma_0 - 1}{\gamma_0 + 1}, \tag{8.9}$$

$$\frac{u_1}{c_0} = \frac{2(M_s^2 - 1)}{(\gamma_0 + 1)M_s}, \tag{8.10}$$

$$\frac{T_1}{T_0} = \left(\frac{p_1}{p_0}\right)\left(\frac{\rho_0}{\rho_1}\right), \tag{8.11}$$

where $M_s = \dot{R}_s/c_0$ and $c_0^2 = \gamma_0 p_0/\rho_0$ is the speed of sound in the mixture ahead of the shock. We have assumed $\gamma_0 = \gamma_1$, because in general the precursor shock is insufficiently strong to effect significant chemical changes (namely, dissociation) of the mixture crossing the shock. Thus, the change of state across the shock (i.e., p_1, ρ_1, T_1, u_1) can be given as a function of the shock speed (or shock Mach number) by Eqs. 8.8 to 8.11. Note that we have also assumed a perfect-gas equation of state $(p = \rho RT)$.

Across the flame front, the conservation of mass and momentum (i.e., Eqs. 8.5 and 8.6) gives

$$\rho_1 S^2 = \frac{p_2 - p_1}{1 - \frac{\rho_1}{\rho_2}},$$

where $S = \dot{R}_f - u_1$ is the burning velocity of the flame relative to the mixture ahead of it. Defining $y = p_2/p_1$ and $x = \rho_1/\rho_2$, we write

$$\frac{S^2}{c_0^2} = \frac{T_1(y-1)}{\gamma_0 T_0(1-x)}, \tag{8.12}$$

where we have used the equation of state $p = \rho RT$ and assumed $\gamma_1 = \gamma_0$. Equation 8.5 can be written as

$$\rho_1 S = \rho_2(S - u_2),$$

which gives

$$\frac{u_2}{S} = 1 - x. \tag{8.13}$$

Solving the energy equation (Eq. 8.7) together with the conservation of mass and momentum gives the Hugoniot relationship across the flame, which can be conveniently expressed as

$$(x - \alpha)(y + \alpha) = \beta, \tag{8.14}$$

where

$$x = \frac{\rho_1}{\rho_2}, \qquad y = \frac{p_2}{p_1}, \qquad \alpha = \frac{\gamma_2 - 1}{\gamma_2 + 1},$$

and

$$\beta = \alpha \left\{ \frac{\gamma_0 + 1}{\gamma_0 - 1} + \frac{2\gamma_0 Q}{c_1^2} - \alpha \right\},$$

because we have assumed $\gamma_1 = \gamma_0$ and the sound speed ahead of the flame to be $c_1^2 = \gamma_0 p_1/\rho_1$. The boundary conditions behind the flame also need to be specified. For a closed-end tube, we write

$$u_1 + u_2 = 0. \tag{8.15}$$

We have eight equations (Eqs. 8.8 to 8.15) for the quantities

$$M_s, \quad \frac{p_1}{p_0}, \quad \frac{\rho_1}{\rho_0}, \quad \frac{u_1}{c_0}, \quad \frac{T_1}{T_0}, \quad \frac{S}{c_0}, \quad \frac{u_2}{S}, \quad x = \frac{\rho_1}{\rho_0}, \quad \text{and } y = \frac{p_2}{p_1}.$$

If the burning velocity and the properties of the reactants ahead of the flame are known, then we can complete the solution of the problem. However, the burning velocity depends not only on the thermodynamic state of the reactants ahead of it, but also on the fluid mechanical state (i.e., turbulence parameters). This is, in general, not known. Nonetheless, we can solve the inverse problem by finding the burning velocity that satisfies the boundary conditions for a given shock velocity M_s. We first choose a precursor shock strength, and then compute the state behind it from Eqs. 8.8 to 8.11. We then iterate for the burning velocity, which provides a solution that satisfies the boundary condition at the back (viz., $u_1 + u_2 = 0$).

Since the sound speed in the hot product gases (c_2) is relatively high, the back boundary condition can be communicated to the flame front rapidly. However, when the flame speed $\dot{R}_f \geq c_2$, the back boundary condition can no longer influence the state behind the flame. This is referred to as the *first critical flame speed* by Taylor and Tankin (1958). If the flame speed is greater than this critical value, a rarefaction fan (similar to that for a detonation) will follow behind the flame front. The head of the rarefaction fan propagates at $u_1 + u_2 + c_2$. Therefore, for a flame speed equal to or greater than the first critical speed, it is possible to replace the boundary condition $u_1 + u_2 = 0$ (Eq. 8.15) by

$$\dot{R}_f = u_1 + u_2 + c_2. \tag{8.16}$$

To determine the first critical speed, $\dot{R}_f = c_2$, we need to compare $\dot{R}_f/c_0$ with c_2/c_0 for various values of an assumed shock strength M_s. The sound speed behind the flame can be written as

$$\frac{c_2}{c_0} = \frac{\gamma_2 p_2}{\rho_2} \frac{\rho_0}{\gamma_0 p_0} = \frac{\gamma_2}{\gamma_0} \left(\frac{p_2}{p_1}\right)\left(\frac{p_1}{p_0}\right)\left(\frac{\rho_1}{\rho_2}\right)\left(\frac{\rho_0}{\rho_1}\right) = xy\frac{\gamma_2}{\gamma_0}\left(\frac{p_1}{p_0}\right)\left(\frac{\rho_0}{\rho_1}\right). \tag{8.17}$$

For any given shock Mach number, p_1/p_0 and ρ_0/ρ_1 can be found from the Rankine–Hugoniot equations (Eqs. 8.8 and 8.9). The desired value of the burning velocity $S = \dot{R}_f - u_1$ is determined by iteration for a solution that satisfies the boundary conditions $u_1 + u_2 = 0$ (i.e., Eq. 8.15). The first critical flame speed will be obtained when

$$\frac{\dot{R}_f}{c_0} = \frac{c_2}{c_0}.$$

For flame speeds greater than this first critical flame speed, a rarefaction fan will be formed in the product gases, which essentially isolates the back boundary condition from the flame. To determine the solution, we proceed as before, except that the iteration for the burning velocity (or flame speed) for a given shock strength is now based on the condition given by Eq. 8.16 instead of by $\dot{R}_f = c_2$.

A flame speed of $\dot{R}_f = u_1 + u_2 + c_2$ corresponds to the CJ deflagration speed where the flow is sonic behind the flame (similar to a CJ detonation). Accordingly, the Rayleigh line from the initial state (ahead of the flame) will be tangent to the lower branch of the Hugoniot curve. Note that the state ahead of the flame corresponds to the state behind the precursor shock and not the initial undisturbed state. Unlike a CJ detonation, which propagates into the undisturbed mixture (p_0, ρ_0), a CJ deflagration propagates into the perturbed mixture behind the precursor shock. For different, precursor shock strengths, the shocked states (p_1, ρ_1) are different, and hence the Hugoniot curve, which depends on the initial state, also varies with the precursor shock strength. For a given mixture with given initial conditions (p_0, ρ_0), there correspond multiple possible CJ deflagration speeds, whereas there is only one CJ detonation speed.

We note from the above discussions that there is a first critical flame speed when the back boundary condition can no longer influence flame propagation. There is also a second critical flame speed that occurs when the flame speed $\dot{R}_f$ and the precursor shock speed $\dot{R}_s$ are equal. When this happens, the separation between the shock and the flame becomes constant. Note that the boundary condition given by Eq. 8.16 is still satisfied (i.e., one has sonic conditions behind the flame). For the second critical speed,

$$\dot{R}_s = \dot{R}_f = u_1 + u_2 + c_2,$$

and since the separation between the two fronts is now constant, we can write the conservation laws across the initial state (p_0, ρ_0) and the product state (p_2, ρ_2) behind the flame. The second critical deflagration speed is also the CJ detonation speed. The CJ detonation solution corresponds to the tangency of the Rayleigh line from the initial state (p_0, ρ_0) to the Hugoniot curve (based on the initial state). However, it also corresponds to the CJ deflagration solution, where the Rayleigh line from the state ahead of the flame but behind the precursor shock, (p_1, ρ_1), is tangent to the Hugoniot curve based on the shocked state (p_1, ρ_1).

Although it does not do so in reality, the transition from deflagration to detonation might proceed via the following path: Starting from the closed end of a tube, the flame accelerates until the first critical speed $\dot{R}_f = c_2$ is reached. This first critical speed corresponds to a CJ deflagration with $u_1 + u_2 = 0$. Further increases in the flame speed will still correspond to a CJ deflagration with $\dot{R}_f = u_1 + u_2 + c_2$ until the second critical flame speed is reached when $\dot{R}_f = \dot{R}_s = u_1 + u_2 + c_2$, and a CJ detonation is obtained.

Prior to reaching the second critical speed of a CJ detonation, the precursor shock would have already attained a strength capable of igniting the mixture via adiabatic shock compression. For most of the hydrocarbon gas mixtures with oxygen, the autoignition temperature is of the order of 1200 K, corresponding to a shock strength $M_s \approx 3.5$, which is about half the second critical speed. When autoignition occurs, the reaction front (flame) is no longer independent of the precursor shock. In fact, the reaction front trails behind the shock by an induction distance determined by the shock temperature. If a flame originally accelerates the precursor to its autoignition limit, a second reaction front may appear ahead of the flame. The subsequent burnout (explosion) of the mixture in between these two reaction fronts may trigger the onset of detonation when the shock wave from this explosion catches up with the precursor shock itself.

To illustrate the magnitudes of these various critical flame speeds, we consider the same mixture of $C_2H_2 + O_2$ that Taylor and Tankin used. The Hugoniot equation for this mixture can be closely approximated by

$$(x - 0.08)(y + 0.08) = 0.474 + 22.326 \frac{288}{T_1}, \tag{8.18}$$

where $x = \rho_1/\rho_2$ and $y = p_2/p_1$. The value of $\gamma_2 = 1.174$ is obtained from these Hugoniot equations. Equation 8.18 is valid for different initial temperatures, and up to $T_1 = 700$ K, Taylor and Tankin showed that it agrees closely with the computation by Manson (1949). Using Eq. 8.18, the precursor shock strength, M_s, for different values of the burning velocity S/c_0 is shown in Fig. 8.2. Also plotted is the temperature ratio behind the precursor shock front. The three critical flame speeds are indicated by the vertical dashed lines.

The first critical flame speed, where $\dot{R}_f = c_2$ (and $u_1 + u_2 = 0$), was found to correspond to a value of $S/c_0 = 0.53$. With a value of $c_0 = 340$ m/s, the burning velocity is 180.2 m/s. The laminar burning velocity for C_2H_2–O_2 mixtures is 11.4 m/s. Thus, an equivalent turbulent burning velocity of about 16 times the laminar burning velocity is required to reach the first critical flame speed.

For the second critical flame speed where $\dot{R}_f = \dot{R}_s$, a value $S/c_0 = 1.54$ is obtained. This requires a burning velocity of 523.6 m/s. The shock and flame speed is found to be $\dot{R}_f/c_0 = \dot{R}_s/c_0 = 8.69$ (or 2954.6 m/s), which is very close to the CJ detonation velocity of 2951 m/s calculated from Eq. 8.18 based on the initial state (p_0, ρ_0). Note that this close agreement is coincidental. For other mixtures, the difference between the second critical speed and the CJ detonation speed may be greater, but with the Hugoniot curve given by Eq. 8.18 for C_2H_2–O_2 mixtures, it appears that the upper tangency point of the Rayleigh line to the Hugoniot curve, based on the initial state (p_0, ρ_0), corresponds to the lower tangency point of the Rayleigh line to the Hugoniot curve based on the shocked state (p_1, ρ_1). For the case of a perfect gas with a constant value of γ and heat release Q (no dissociation), Courant and Friedrich (1948) have demonstrated the equivalence of a CJ detonation in the initial state (p_0, ρ_0) and a CJ deflagration based on the shocked state (p_1, ρ_1).

Assuming an autoignition temperature of 1200 K, the third critical flame speed for C_2H_2–O_2 mixtures was found to be $S/c_0 = 0.3$, which gives a shock strength of $M_s \approx 3.5$. The induction time corresponding to a precursor strength of $M_s = 3.5$ is of the order of 10 μs. Thus, for the particular case of C_2H_2–O_2, we note that $(S/c_0)_3 < (S/c_0)_1 \leq (S/c_0)_2$. However, for other explosive mixtures, this may not necessarily be the case.

According to the classical concept of the transition from deflagration to detonation, it was thought that the deflagration must accelerate to a critical velocity so that the precursor shock strength is such that autoignition occurs in the shocked mixture. Thus, the third critical flame speed sets the criterion for the onset of detonation. However, experiments indicate that the onset of detonation usually occurs in

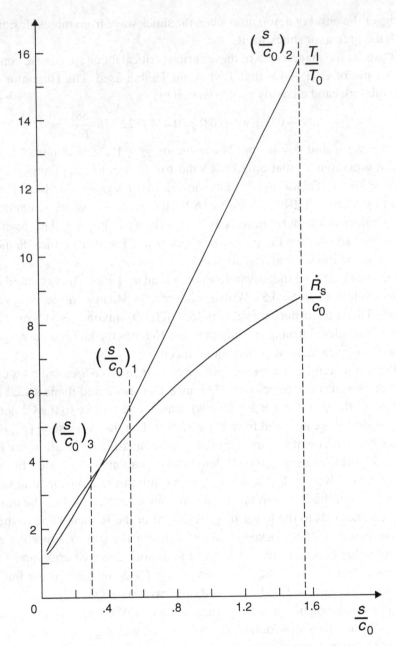

Figure 8.2. Precursor shock strength M_s for different values of burning velocity (Lee & Moen, 1980).

the turbulent burning zone of the deflagration independent of the precursor shock strength. The critical condition for the onset of detonation thus remains unresolved.

8.3. SALIENT FEATURES OF THE TRANSITION PHENOMENON

The transition from deflagration to detonation can occur under a multitude of conditions, depending on the particular way in which the onset of detonation is being

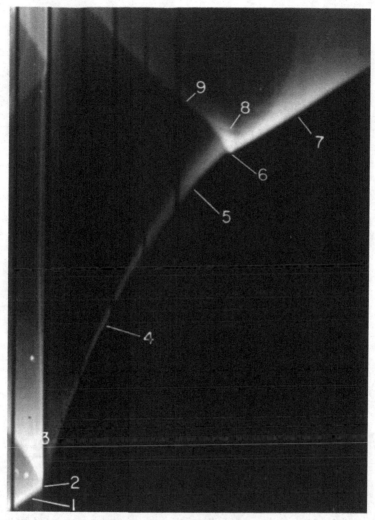

Figure 8.3. Typical streak photograph of the abrupt transition to detonation via a jet of hot combustion products (Lee *et al.*, 1966).

triggered. However, we shall consider first the classical transition phenomenon in a long smooth-walled tube in which the deflagration accelerates continuously subsequent to ignition until an abrupt transition to detonation occurs. Figure 8.3 shows a typical streak photograph of the transition process in a C_2H_2–O_2 mixture. Ignition is via a small jet of hot products issuing from a tiny orifice (2) subsequent to the reflection of a detonation wave (1) upstream of the orifice plate. The orifice plate (3) essentially serves as the closed-end of the tube where ignition takes place. The deflagration propagates initially as a laminar deflagration and eventually becomes turbulent (5), as indicated by the wider and more intense turbulent reaction zone. Abrupt transition to detonation occurs at (6), whereby an overdriven detonation (7) is first formed that then decays to the CJ velocity of the mixture. A shock wave known as the retonation wave (9), formed at the same time as the detonation, propagates back into the combustion products. Transverse shock waves (8) are also formed at

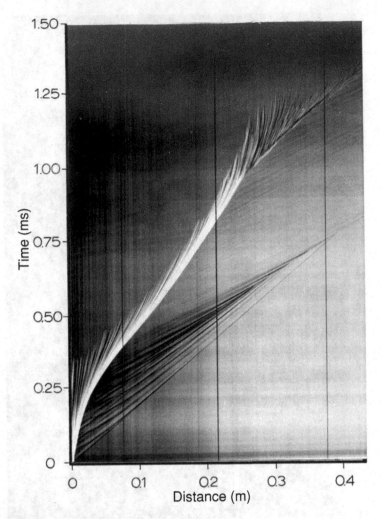

Figure 8.4. Streak schlieren photograph illustrating pressure waves generated by the increase in burning rate of the flame (Urtiew & Tarver, 2005).

the onset of detonation, which then reflect off the wall of the tube, forming the periodic bands of luminosity as the product gases are being compressed. The distance from the ignition (2) to the onset of detonation (6) is referred to as the "induction distance" in the older literature. However, it is now commonly called the *run-up distance* to avoid any confusion with the induction-zone thickness of the ZND detonation structure. The run-up distance depends not only on the properties of the explosive mixture, but also on the initial and boundary conditions (type, strength, and location of the ignition source, tube dimensions and geometry, wall roughness, open or closed ends of the tube, etc.). Therefore, attempts to correlate the run-up distance with the chemical composition of the mixture have not been successful or meaningful, in general.

An important feature of an accelerating deflagration is the generation of pressure waves by the flame front. Figure 8.4 is a streak schlieren photograph illustrating the pressure waves generated as a result of the increase in the burning rate of the flame subsequent to ignition. The train of compression waves generated at the beginning is due to the increase in the burning rate of the developing spherical flame kernel after spark ignition. The deceleration at around 0.1 m (or 0.50 ms) is due to the decrease in flame area when the initial spherical flame kernel touches the wall and the flame flattens out, decreasing the flame area. The second acceleration at around 1 ms is due to the flame becoming turbulent, forming a so-called tulip-shaped flame that spreads faster near the wall than near the axis of the tube. The large increase in flame area when the tulip-shaped flame is formed causes the second flame acceleration in the streak photograph shown in Fig. 8.4. The increase in burning rate due to the change in the flame morphology gives rise to an increase in the effective burning velocity of the flame, which increases not only the displacement flow velocity ahead of the flame but also the generation of pressure waves associated with the increase in burning rate.

A better illustration of the onset of detonation can be obtained via high-speed schlieren movies. Figure 8.5 shows the initial stages of the flame acceleration phase subsequent to ignition. The flame front has a cellular structure due to instabilities, and a train of weak compression waves can be observed in front of the propagating flame. Figure 8.6 shows the later stages of the turbulent flame acceleration process. The cellular structure of the flame is now of a finer scale. Also, the compression waves coalesce, and a train of strong compression waves can be seen ahead of the flame. This train of compression waves will eventually coalesce to form a precursor shock wave ahead of the flame.

The onset of detonation is illustrated in Fig. 8.7. In the third frame, the formation of two local explosion centers can be observed at the bottom wall of the channel in the turbulent flame zone. These explosion centers grow with time, but no detonation is generated from them. In the fifth frame from the top, we note the formation of a third explosion center in between the previous two near the bottom wall of the channel. A detonation bubble is already formed in the fifth frame. The hemispherical detonation bubble from this third explosion center propagates forward in the compressed reactants as an overdriven detonation. The shock wave that propagates back into the reaction products is referred to as a retonation wave. The transverse shock wave that reflects from the top and bottom wall of the channel is associated with the hemispherical wave that originates from the local explosion. The multiple reflections of this transverse wave from the top and bottom walls of the channel correspond to the periodic waves in between the detonation and retonation wave trajectories of Figs. 8.3 and 8.4.

Figures 8.3 to 8.7 describe the typical transition phenomenon in a long smooth tube. However, the transition processes are not unique, and a detonation can be formed from a deflagration in many different ways. It is for this reason that a general theory cannot be developed for the transition phenomenon, even on a qualitative basis. Thus, we shall discuss all the relevant flame acceleration mechanisms and the various ways in which the onset of detonation can occur. Under different initial and

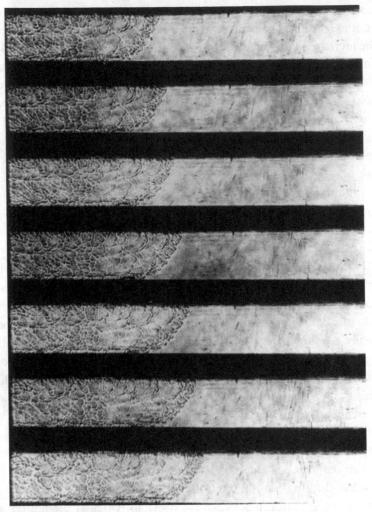

Figure 8.5. Sequence of schlieren photographs illustrating the initial flame acceleration phase subsequent to ignition in stoichiometric H_2–O_2 at $p_0 = 83.7$ Torr; $\Delta t = 5$ μs between frames (courtesy of A.K. Oppenheim).

boundary conditions, different combinations of some of these mechanisms may be dominant in effecting the transition to detonation.

8.4. FLAME ACCELERATION MECHANISMS

From the streak photographs shown earlier (Figs. 8.2 and 8.3), we see that the flame front velocity increases continuously until transition to detonation occurs. We will now discuss the mechanisms that are responsible for this continuous increase in flame speed. For a one-dimensional planar flame front propagating in a tube, the "flame speed" refers to the rate at which the planar flame surface propagates with respect to a fixed coordinate system. However, the flame front in the predetonation

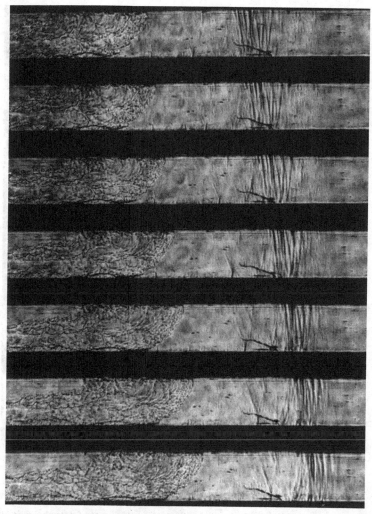

Figure 8.6. Sequence of schlieren photographs illustrating the later stages of flame acceleration in stoichiometric H_2–O_2 at $p_0 = 83.7$ Torr; $\Delta t = 5\,\mu$s between frames (courtesy of A.K. Oppenheim).

phase of its propagation is three-dimensional and transient. Therefore, it is important to first define what is meant by the flame speed (in the direction of propagation) of this three-dimensional fluctuating burning surface. Locally, the burning velocity (relative to the unburned mixture ahead of the flame) can vary considerably on the turbulent flame surface, and parts of the flame surface may even be quenched due to excessive curvature and stretch. However, one may still be able to draw two control surfaces (planes) normal to the direction of propagation: one ahead of the leading edge of the turbulent flame in the unburned reactants, and one in the wake of the flame zone where the gases are combustion products. The progress of this finite turbulent burning zone between the two planes with time can be measured, and we can then define a flame speed (or a burning velocity) within the context of a one-dimensional propagation along the axis of the tube. The burning

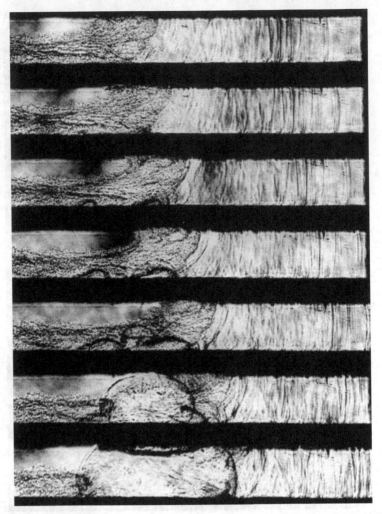

Figure 8.7. Sequence of schlieren photographs illustrating the onset of detonation in stoichiometric H$_2$–O$_2$ at $p_0 = 83.7$ Torr; $\Delta t = 5$ μs between frames (courtesy of A.K. Oppenheim).

velocity, so defined, is essentially an effective burning velocity that, when multiplied by the cross-sectional area of the tube, gives the volumetric burning rate of the three-dimensional transient "turbulent" flame. The word "turbulent" is in quotation marks because there are numerous mechanisms that can give rise to the three-dimensional burning surface other than turbulent fluctuations in the flow of the unburned mixture ahead of the flame. The volumetric burning rate in the "turbulent" flame zone depends on the burning surface area and the local burning velocity of the flame surface. Any mechanism that results in an increase in the surface area of the flame will increase the volumetric burning rate. The local burning velocity can be enhanced or quenched by turbulent transport, flame curvature, and stretch. Thus, flame acceleration refers to the rate of increase of the *averaged* burning velocity in the three-dimensional "turbulent" flame zone.

Propagating laminar flames are intrinsically unstable, and an initially planar flame front can develop into a three-dimensional cellular flame surface from gasdynamic effects and from a *thermodiffusive* instability that results from the competition between the diffusion of heat and mass. A flame surface is also a strong density interface with a typical density ratio of about 7 across the flame. The influence of acceleration and of acoustic excitation on the steep density gradients can also result in instability and hence a subsequent increase in the flame area as the instability develops. The convection of the flame surface into a velocity gradient and turbulent flow field will also result in an increase in the burning surface area from flame folding and wrinkling. Thus, there are numerous mechanisms to render a propagating flame unstable, all resulting in an increase in the burning rate (thus the effective burning velocity).

All propagating flames are gasdynamically unstable due to the diffuser (or nozzle) effect of diverging (or converging) streamlines ahead of a perturbed flame front. The decrease (or increase) in the approaching flow velocity then results in the growth of flame perturbations. This is known as the Landau–Darrieus instability. Different diffusivities of the fuel and oxidizer molecules tend to enrich or deplete the deficient species in a mixture along a perturbed flame front, leading to a variation in the local burning velocity. If there is a variation in the diffusion of heat along the perturbed flame surface, some combinations of mass and heat diffusion can result in the growth of the flame perturbations. This is referred to as the thermodiffusion instability. Flame instability is thoroughly summarized by Markstein (1964). More recent review articles on flame front instabilities are given by Matalon and Matkowsky (1982), Sivashinsky (1983), and Clavin (1985).

Figure 8.8 shows the morphology of an unstable cellular flame in rich mixtures of heavy hydrocarbon fuels in air. Similar cellular structures can be observed in lean hydrogen– and methane–air flames. There exists a natural tendency for many laminar flames to become unstable, taking on a cellular structure that results in a larger burning surface area.

Deflagrations are subsonic, which allows flow perturbations generated by the flame to be propagated ahead into the unburned mixture and hence, can feed back to influence the subsequent flame propagation. The two important effects generated by a propagating flame in the mixture ahead of it are a displacement flow and pressure waves. The displacement flow is a consequence of the increase in the specific volume of the gas as it crosses the flame front. A typical increase in specific volume (or decrease in the density) across a stoichiometric hydrocarbon–air flame is about 7. From the conservation of mass, the velocity of the unburned mixture ahead of the flame will acquire a velocity $u_1 = (\Delta v / v_1)S$, where S is the burning velocity and $\Delta v = v_2 - v_1$ is the increase in specific volume across the flame. The relationship given above is based on the propagation of the flame from a closed-end tube where $u_2 = 0$. For a typical laminar burning velocity of about 0.5 m/s, the displacement flow velocity is $u_1 \approx 3$ m/s.

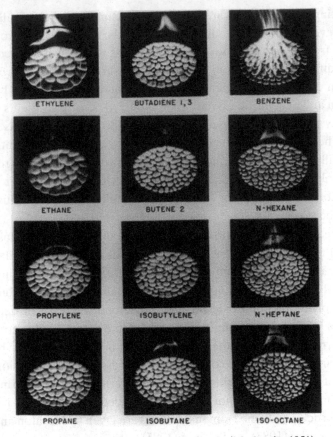

Figure 8.8. Morphology of cellular flames (Markstein, 1951).

The displacement flow alone does not result in an increase in the burning rate, because the flame is simply convected along with the flow. However, in the presence of boundary surfaces (e.g., tube wall and obstacles) in the path of the flame, a velocity-gradient field will be developed. When the flame is convected into a velocity-gradient field, the flame surface will be deformed, because different portions of the flame surface are convected along with different velocities in the gradient field. Also, when the Reynolds number of the displacement flow becomes sufficiently large, the velocity-gradient field results in the formation of turbulent eddies and the flame becomes wrinkled to comply with the fluctuating velocity field of the turbulent flow.

Figure 8.9 shows an excellent example of the convection of a laminar flame into a turbulent flow field. A hot wire generates an initially buoyant plume. At some distance downstream, the plume becomes turbulent, similar to the smoke trail arising from a smoldering cigarette. The convection of the flame subsequent to ignition and the breakup of the laminar flame surface as it propagates into the turbulent portion of the plume are clearly demonstrated.

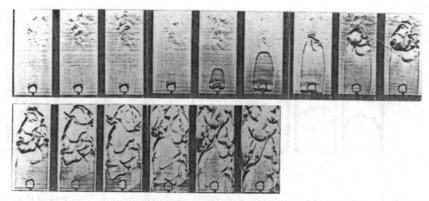

Figure 8.9. Convection of a laminar flame into a turbulent flow field (courtesy of R. Strehlow).

A more dramatic effect of the influence of the nonuniform flow field in the reactants ahead of the flame due to the presence of obstacles is illustrated in Fig. 8.10. The flame propagates from the closed end of a square channel with periodic baffles on the lower wall, acting as flow obstructions. The displacement flow over the first baffle is rather slow, and a laminar vortex is generated downstream of the baffle. As the flame propagates into the vortex, it is rolled up into the vortex, significantly increasing the surface area of the flame. The increase in burning rate increases the flow velocity ahead of the flame. In the second window of the channel, the wake flow now consists of a shear layer rather than a large laminar vortex. The flame gets

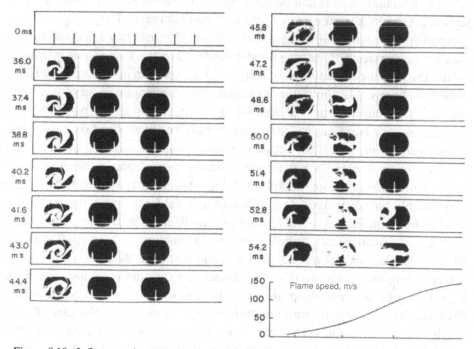

Figure 8.10. Influence of a nonuniform flow field on flame propagation (Chan *et al.*, 1983).

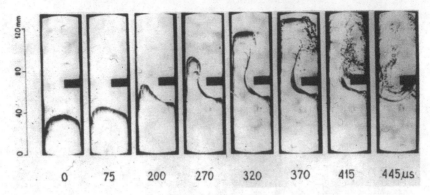

Figure 8.11. Propagation of a flame past an obstacle; onset of detonation (courtesy of P. Wolanski).

convected rapidly downstream with the flow above the shear layer and then propagates into the recirculating region behind the baffle through the shear layer. We
see that obstacles can greatly deform the flow field ahead of the flame, leading to
a significant enhancement of the burning rate past each obstacle. For a sufficiently
sensitive mixture, the rapid burning rate in the turbulent shear layer downstream of
the obstacle can even give rise to the onset of detonation.

Figure 8.11 shows the propagation of the flame past an obstacle in a C_2H_2–O_2
mixture (perhaps the most sensitive of all detonable gas mixtures) at an initial pressure of 0.27 atm. The onset of detonation occurs as the turbulent flame is entrained
into the recirculation zone immediately behind the baffle. A retonation wave can be
seen to propagate back upstream in the last frame of Fig. 8.11.

Apart from the nonuniform gradient field and the turbulence that can be generated by the obstacles, the strong acceleration field as the flame front is being convected through the nonuniform velocity-gradient field can also induce interface instabilities. Figure 8.12 shows the propagation of the flame past an obstacle in the
channel. The reduction in flow area due to the presence of the baffle causes the
streamlines to converge as in a converging nozzle. The flame front, as it is convected
in this accelerating flow field, becomes unstable, and one can first observe the appearance of small periodic perturbations on the flame front in the second frame.
The perturbations then grow explosively in subsequent frames to turn the flame
into a highly turbulent front. In a turbulent flow field, the convection of the flame
front through the eddies would also subject the flame to an acceleration field. The
breakup into smaller-scale turbulent structures could result from the Taylor instability mechanism acting on the density gradient across the flame.

Depending on the geometry of the obstacles, a very dramatic increase in burning rate can be achieved downstream of the obstacle. Figure 8.13 shows schlieren
photographs of a flame transmitting from a hemispherical flow blockage (Kumagai
& Kimura, 1952). In Fig. 8.13a, there is no flow blockage in the path of the flame.
The hemispherical flame is initially laminar and becomes turbulent from acoustic
vibration-induced instabilities inside the combustion chamber. The total burnout

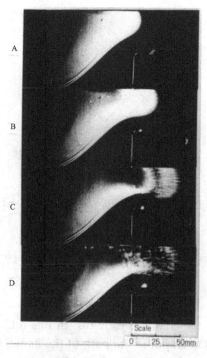

Figure 8.12. Propagation of a flame past an obstacle in a channel; formation of interface instabilities at the flame front (courtesy of T. Hirano).

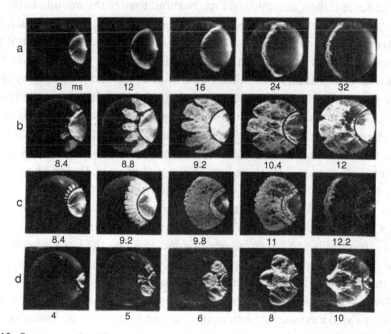

Figure 8.13. Sequence of schlieren photographs illustrating the effect of obstacle geometry on flame propagation: (a) no obstacle; hemispherical blockages with (b) five, (c) seventy, and (d) three holes (Kumagai & Kimura, 1952).

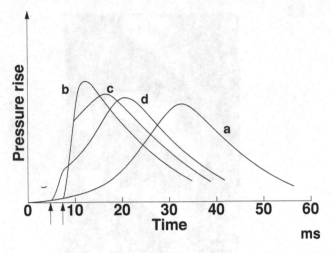

Figure 8.14. Pressure traces corresponding to schlieren photographs shown in Fig. 8.13 (Kumagai & Kimura, 1952).

time of the mixture in the chamber is 32 ms. In Fig. 8.13b, a hemispherical block-age with five holes has been put in the path of the flame. Prior to the arrival of the flame, five jets of unburned gas are first formed downstream. As the flame transmits through the blockage, we can initially see five burning jets of flame tongues that merge to form a folded surface. The edges of the folded flame surface burn rapidly with finer-scale turbulence, and the total burnout time of the mixture in the chamber is 12 ms, about a third of the time without the obstacle. Figure 8.13c and 8.13d show two extreme cases of seventy holes and three holes in the hemispherical blockage plate, respectively. The burnout time is roughly the same. In the seventy-hole case, there is a large initial increase in the flame area due to flame folding by large-scale velocity gradients. However, the smaller flame folds rapidly burn themselves out, resulting in a finer-scale flame front for the burning of the second half of the mixture in the chamber, which is similar to the case of Fig. 8.13a without any obstacle. For Fig. 8.13d, the initial increase in flame area due to only three jets is smaller, but the larger-scale flame folds take a longer time to burn out. Hence, the flame retains a larger burning surface area for a longer period of time.

Figure 8.14 shows the pressure traces corresponding to the four cases illustrated in Fig. 8.13. Without any obstacle, the pressure shows a gradual rise to a peak at about 32 ms, corresponding to the flame position indicated by the last frame in Fig. 8.13a. For the five-hole obstacle, the rapid pressure rise corresponds to the significant increase in burning rate downstream of the obstacle at about 9 ms (third frame in Fig. 8.13b). For the seven- and three-hole obstacles shown in Fig. 8.13c and 8.13d, one notes a rapid increase in the burning rate and a pressure rise immediately downstream of the obstacle.

Apart from generating a displacement flow, a propagating flame also generates pressure waves ahead of it. A sudden change in the energy release rate (due to a

change in the flame's surface area or the local burning rate) will result in the generation of pressure waves due to the change in the rate of the specific volume increase
across the flame. Theoretical analysis of the generation of pressure waves at a flame
front due to a change in flame speed, heat release, specific heat ratio, the entropy,
pressure, and velocity of the unburned mixture ahead of the flame has been considered by Chu (1952). As is the case with displacement flow, the pressure waves
generated can result in a strong positive feedback loop that enhances the burning
rate of the flame. A flame is essentially a heat source with an energy release rate
very sensitive to flow perturbations. Resonant conditions for the coupling of the
acoustic vibrations in a system with the flame is given by Rayleigh's criterion, which
says that the increase in heat release must correspond to the compression phase of
the vibration cycle. The increase in heat release rate can easily be accomplished by a
change in the surface area of the flame. Therefore, if the surface area fluctuations of
the flame can be tuned to the acoustic vibrations with the proper phase relationship,
a self-sustained resonant vibration can be realized.

Figure 8.15 shows the shape of a flame in resonant coupling with the acoustic
oscillation of the system. The system is a vertical tube with an upward flow of a
combustible gas (methane–air) from the bottom that stabilizes the downward propagation of a flat flame at the top. For mixture compositions away from stoichiometry, small acoustic vibrations involving fluctuations of that flat flame about a mean
position are observed. However, when the mixture composition is varied toward
stoichiometry, the amplitude of the flame oscillations increases rapidly, followed by
an increase in the flame area due to the acoustic excitation.

Figure 8.15a shows the progressive increase in the flame area as the amplitude
of the oscillations increases. Each frame in Fig. 8.15a corresponds to one cycle of
oscillation at the high-pressure phase of the cycle. Thus, eight cycles are shown in
which the amplitude of the acoustic vibration increases until the flame flashes back
down the flame tube. In Fig. 8.15b, the flame shape for one cycle is shown, and we
can see the alternate increase and decrease in the flame area corresponding to the
high- and low-pressure phases of the cycle. The burning rate can be coupled to the
acoustic vibrations in the system in accord with Rayleigh's criterion.

In an oscillating flow field, it is found that a periodic cellular structure is induced
in the flame front (Markstein & Somers, 1952). The cell size was found to be primarily dependent on the amplitude and frequency of the oscillation, rather than on
the mixture composition as in thermodiffusive instability. Figure 8.16 illustrates the
vibration-induced cellular structure in a flame propagating in a closed vessel.

In Fig. 8.16a, the schlieren movie obtained by Leyer (1970) in a tube 78.5 cm
long and with a 3.3×3.3-cm square cross-section is shown. The mixture is stoichiometric propane–air at an initial pressure of 1 atm. The portion of the curved
flame front that is normal to the tube axis is subjected to the oscillatory flow associated with the acoustic vibration, and thus it has a smaller cell size than the
curved flank of the flame that is parallel to the flow oscillations. The cell size in the
curved flank of the flame front is larger and corresponds to the intrinsic instability

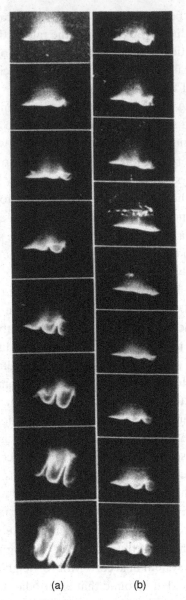

(a)   (b)

Figure 8.15. Flame shape due to resonant coupling with acoustic oscillation in a vertical tube (Kaskan, 1952).

of the flame rather than the vibration-induced cells in the central portion of the flame.

Figure 8.16b is a single frame of a schlieren movie obtained by Markstein (1952) for a downward-propagating rich butane–air flame. Again, the spontaneous cell structure characteristic of the intrinsic instability of the mixture on the curved flank of the flame is much coarser than the vibration-induced cells in the portion of the flame that is normal to the axis and subjected to the oscillatory flow of the acoustic

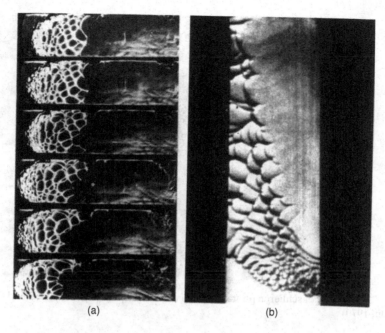

(a)   (b)

Figure 8.16. Cellular structure of a propagating flame induced by vibrations ((a) Leyer, 1970; (b) Markstein, 1952).

vibrations. It should be noted that vibration-induced cells are periodic, appearing and disappearing according to the cyclic excitation by the flow field.

When the amplitude of the vibration increases, the flame is observed to shed *bubbles* of unburned mixture that get embedded into the flame zone. The flame takes on a *foamy* structure that resembles the structure of a violent turbulent *flame brush*. Figure 8.17 shows frames from a schlieren movie of a flame propagating in a 10-cm-long 4 × 2.5-cm closed-end tube at an initial pressure of 1 atm in a mixture of $C_3H_8 + 4.17 O_2 + 6 N_2$ (Leyer & Manson, 1970). The transition from the initial laminar structure to the final turbulent foamy structure of the flame front due to acoustic excitation is illustrated by these schlieren photographs. The burning rate is significantly increased when the flame takes on the foamy structure. The pressure waves that radiate from the burnout of the bubbles of unburned mixture embedded in the flame zone further enhance the combustion wave when they reflect off the side walls of the tube to traverse the flame zone again. The transverse pressure fluctuations interact with the coaxial density gradient and produce vorticity via the baroclinic mechanism. Even though there is no flow turbulence from velocity-gradient fields, the flame zone turbulence can arise from the baroclinic vorticity production mechanism of interacting pressure- and density-gradient fields.

From the schlieren photographs shown earlier in Fig. 8.4, we can see that the pressure waves generated ahead of the flame steepen to form shock waves. These shock waves can reflect from the end of the tube and return to interact with the flame. The instability of a density interface subsequent to the interaction with a shock

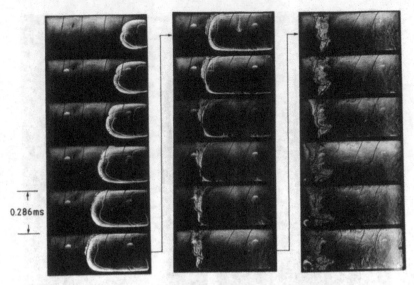

Figure 8.17. Sequence of schlieren photographs of a flame propagating in a closed-end tube (Leyer & Manson, 1970).

wave has been studied by Markstein (1957), Rudinger (1958), Richtmyer (1960), and Meshkov (1970). Figure 8.18 shows a sequence of schlieren photographs of an initially laminar flame interacting head-on with a shock front. The first six frames at the top of Fig. 8.18 show the development of a central indentation of the curved flame front following the passage of the incident shock through the flame. From the theoretical analysis of Markstein (1964), this is the stabilizing phase when the flame

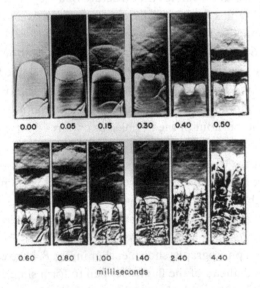

Figure 8.18. Sequence of schlieren photographs of the head-on collision of a shock wave with an initially laminar flame (Markstein, 1964).

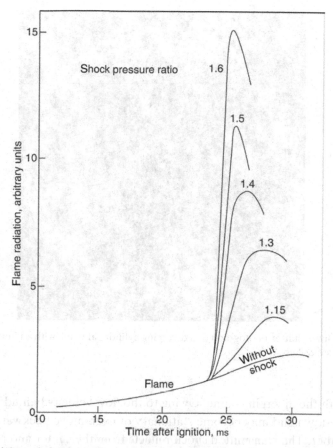

Figure 8.19. Light emission intensity resulting from shock–flame interactions (Markstein, 1957).

attempts to reverse its initial curvature upward. Upon reflection from the bottom of the tube, the transmitted shock then propagates upward to interact with the tulip-shaped flame again. This is the destabilizing phase, and the perturbations on the flame front grow rapidly, as indicated in the last few frames of Fig. 8.18.

As an indication of the increase in combustion intensity resulting from this shock–flame interaction, Markstein recorded the light emission from the flame during the interaction process. Figure 8.19 shows the intensity of the light emission subsequent to the interaction of the flame with shock waves of different strengths. We can see that even with fairly weak shock waves, the combustion intensity can be increased by an order of magnitude as a result of this Markstein–Richmeyer–Meshkov instability.

In a closed chamber, the shock can reflect back and forth from the boundaries of the chamber and interact with the flame many times as it propagates. With each interaction, the perturbation grows, and accordingly the burning rate increases. Figure 8.20 shows the radial propagation of a cylindrical flame from the center of a cylindrical chamber where ignition is effected. As the flame expands, a cylindrical precursor shock is generated. Upon reflection from the wall, a converging shock

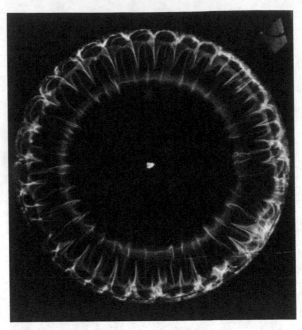

Figure 8.20. Open-shutter photograph of a diverging cylindrical flame with an inward-propagating shock wave (Lee & Lee, 1966).

interacts with the diverging flame. Owing to the flow velocity behind the shock, the flame slows down and may become stationary or even move backward with respect to the chamber. The transmitted shock reflects from the center and propagates out to interact with the flame again. These two interactions are similar to those shown previously in Fig. 8.18. When the shock reflects off the wall, it returns to interact with the flame again. During their interaction, the flame decelerates (relative to the fixed laboratory coordinates) and remains stationary, or it is even pushed back slightly by the inward flow associated with the converging wave.

Because the flame remains nearly stationary with respect to the chamber during each of the interactions with the emerging shock, it is possible to obtain an open-shutter photograph that shows the flame shape during each of these interactions with an inward-propagating shock. The innermost bright ring corresponds to the flame shape during the first interaction with a converging shock. We can see that the flame is still fairly cylindrical. In the second bright ring, the folded cellular structure of the cylindrical flame can easily be observed. It becomes difficult to distinguish the cylindrical flame clearly for subsequent head-on interactions with an inward-propagating shock, because the flame is nearing the end of its travel. Owing to repeated shock-flame interactions, the burning rate increases rapidly, and in many cases the onset of detonation is observed as the cylindrical flame nears the end of its propagation.

We have briefly surveyed the various mechanisms that can result in an increase in the burning rate of a propagating flame. Many of the mechanisms have little to do with the actual turbulence in the mixture ahead of the flame itself. Regarding the influence of turbulence in the unburned mixture on the burning rate of the propagating flame, the combustion mechanism depends strongly on the scale and intensity of the turbulence as compared to the characteristic scales of the combustion process itself (e.g., the laminar flame thickness).

The various regimes of turbulent combustion can be classified comprehensively by the Borghi diagram (Borghi, 1985), which is essentially a map with coordinates defined by a pair of parameters: $\sqrt{k}/S_l$ (the ratio of the square root of the turbulent kinetic energy to the laminar burning velocity) and l/δ (the ratio of the turbulent integral scale to the laminar flame thickness). In the transition from deflagration to detonation, the regime of interest corresponds to high turbulent intensities and small turbulent scales. The reaction zone is a thick extended region where bubbles, or pockets, of unburned mixture are embedded in a sea of hot products and reacting gases. The embedded bubbles often autoignite, generating pressure waves. A strong and often dominating mechanism of vorticity production is now due to the interaction of the pressure- and density-gradient fields. Shock–shock interactions, shock-density interfaces, and shock–vortex interactions are important mechanisms for high-speed deflagrations prior to the onset of detonation. Formulation of a quantitative description of the flame acceleration phase of the transition from deflagration to detonation presents a formidable task.

8.5. ONSET OF DETONATION

The onset of detonation occurs when the appropriate local conditions are attained in the turbulent flame zone. The previous history of the flame acceleration process to achieve the required local condition is irrelevant to the process of the formation of the detonation wave. For example, it has been shown that the onset of detonation can occur in the mixing zone of a turbulent jet of hot combustion products with the reactants (Knystautas et al., 1979). In this case, there is no flame acceleration phase, because the jet can be produced in a variety of ways. However, if we consider the classical experiment of the transition to detonation in a long smooth-walled tube, then the appropriate condition in the turbulent flame zone is achieved when the deflagration has accelerated to some critical velocity.

Figure 8.21 shows a typical streak schlieren photograph illustrating the final phase of the flame acceleration process and the onset of a detonation wave. A simultaneous single flash photograph is also shown, taken at the particular instant (about 130 μs), indicated by the white dash-dot horizontal line. A piezoelectric pressure transducer is located at about 0.25 m, and its time axis is indicated by the white vertical dashed line. The pressure–time history from this pressure transducer is shown in the insert of Fig. 8.21, but replotted on the same time scale as the streak photograph (shown as the white trace in the figure).

Prior to the onset of detonation, the turbulent reaction zone extends over a length of about 0.05 m, corresponding to about two characteristic dimensions of the tube

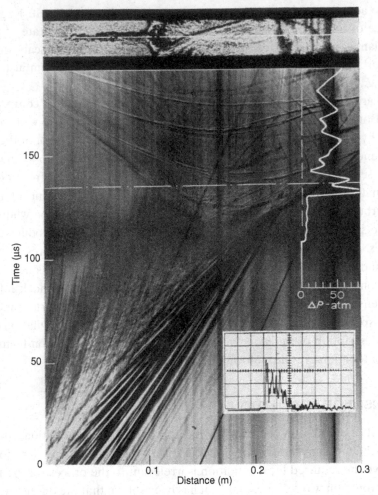

Figure 8.21. Streak schlieren photograph and single-flash photograph illustrating the final phase of detonation onset (Urtiew & Oppenheim, 1966).

cross section. The shock velocity is about 1480 m/s, whereas the reaction-zone ve-locity appears to be just slightly slower. The CJ velocity for this mixture is 2837 m/s, and thus the precursor-shock and reaction-zone speeds are of the order of half the CJ detonation speed (corresponding also to about the CJ deflagration velocity of the mixture). The onset of detonation is seen to occur locally in the turbulent flame brush near the bottom wall, forming a hemispherical detonation and a retonation wave that propagates in the opposite direction. From the pressure trace, the precur-sor shock followed by the overdriven detonation is clearly indicated. The pressure history corresponds to the wave diagram shown by the streak schlieren record. The periodic reflections of the transverse wave from the top and bottom walls of the channel are also illustrated in both the streak schlieren photograph and the pres-sure record. The velocity of the transverse wave and the retonation wave is of the order of 1400 m/s, which corresponds to the sound speed of the detonation products.

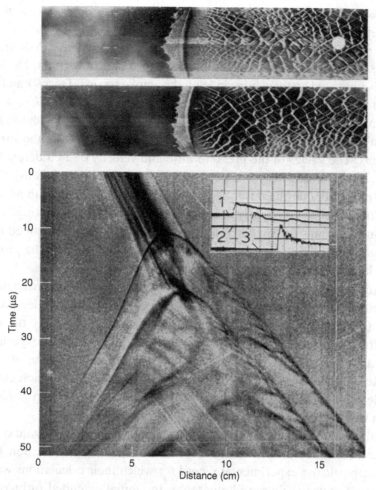

Figure 8.22. Smoked foil and streak schlieren photograph of the transition process (Urtiew & Oppenheim, 1966).

Figure 8.21 gives a complete description of the typical wave processes at the onset of detonation in a smooth tube.

Definitive proof that the wave formed at the explosion center is indeed a detonation wave can be obtained from the soot imprints that were obtained simultaneously with the streak schlieren photograph. Figure 8.22 shows the soot records from the top and bottom walls of the detonation channel, as well as the accompanying streak schlieren photograph of the transition process. The streak record is a superposition of the streaks from two slits: one located near the top and one near the bottom wall of the detonation channel. The abrupt appearance of the typical fish-scale pattern of a detonation on the smoked foil is definitive proof that a detonation is indeed initiated at the explosion center of the turbulent flame brush. Also, the smaller-scale cellular structure shown in the soot imprint at the initial abrupt appearance of the cellular pattern indicates that the detonation is highly overdriven when first formed.

The scale of the cellular pattern increases as the detonation decays to its CJ value subsequent to initiation.

Three pressure traces are also shown in the inset of Fig. 8.22. Pressure gauges 1 and 2 are further upstream to the left of the edge of the photograph. The location of gauge 3 is indicated by the white circle on the smoke record. Traces 1 and 2 indicate a pressure ratio of about 15, corresponding to a precursor shock $M_s \approx 3.6$, which is in agreement with the velocity from the streak record. Gauge 3 shows a pressure ratio of about 24.5, and the velocity of the detonation wave from the streak record is 2500 m/s. The values of the pressure ratio and velocity of a CJ detonation in the mixture are 17.3 and 2240 m/s, respectively. This indicates that the wave is still over-driven. It appears that the onset of a detonation always begins with an overdriven detonation, which subsequently decays to the CJ value of the mixture. The violent explosion that forms the overdriven detonation is always accompanied by the formation of a retonation wave that propagates back upstream into the product gases in accordance with the conservation of momentum.

According to classical theory, a deflagration is distinguished from a detonation by the difference in the ignition mechanisms: diffusion (for a deflagration) versus auto-ignition (for a detonation). Thus, transition should be defined by the switchover from diffusion to autoignition via shock compression. Indeed, the experimental observations presented previously indicate that an abrupt explosion always occurs when the detonation is formed. The cause of this sudden explosion could be the adiabatic compression of the reactants by the precursor shock or by precursor compression waves ahead of the flame zone.

Meyer *et al.* (1970) carried out a detailed analysis of the adiabatic compression processes by the precursor waves prior to the abrupt onset of detonation. Figure 8.23 shows the particular experimental record for which their calculations were carried out. In the framing schlieren photographs, the initial extended turbulent burning zone and the train of intense compression pulses ahead of it are clearly indicated. An explosion first occurs in the turbulent flame zone, as shown in the third frame from the top. However, the volume of unburned reactants in between the two flame tongues appears to be insufficient to result in the formation of a detonation from the shock wave generated by this explosion. In the frame indicated by the time 730 μs, another explosion occurs near the upper channel wall. This explosion leads to the formation of the detonation.

A detailed analysis of the temperature history of a reactant particle as it is processed by the train of compression pulses ahead of the flame zone was carried out. From the temperature–time history, the induction time was also computed. It was concluded that the gasdynamic processes of the compression waves ahead of the flame are entirely insufficient to bring about auto-explosion and the onset of detonation. Meyer *et al.* suggested that heat and mass transfer from the flame could be responsible for bringing about the explosion that led to the detonation. It is also important to note that there may be more than one explosion site (or center) in the

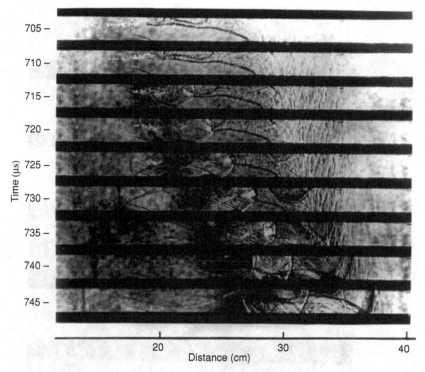

Figure 8.23. Sequence of schlieren photographs illustrating the onset of detonation (Meyer *et al.*, 1970).

turbulent flame brush. However, the condition of the mixture in the neighborhood of the explosion center must also be conducive to the amplification of the shock wave from the explosion center in order to result in the formation of the overdriven detonation wave.

Turbulent flame acceleration and the generation of the pressure wave processes are highly nonlinear and irreproducible, being dependent on initial and boundary conditions and thus the phenomenon of the onset of detonation cannot be unique. Figure 8.24 shows the onset of detonation occurring at the leading edge of the turbulent flame front behind a number of precursor shock waves generated earlier in the flame acceleration process. The detonation evolves from the flame front between the second (55 μs) and the third (60 μs) frame. However, an explosion center is also observed at the top wall in the boundary layer. The backward-propagating retonation wave appears to originate from this explosion center.

In Fig. 8.25, the detonation occurs at the foot of the precursor shock when the turbulent flame that spreads along the boundary layer reaches the shock front. The detonation that is formed then spreads downward along the shock front, and a retonation wave is also formed that propagates back into the products. The formation of the transverse wave that subsequently reverberates between the upper and lower walls of the detonation channel is clearly illustrated.

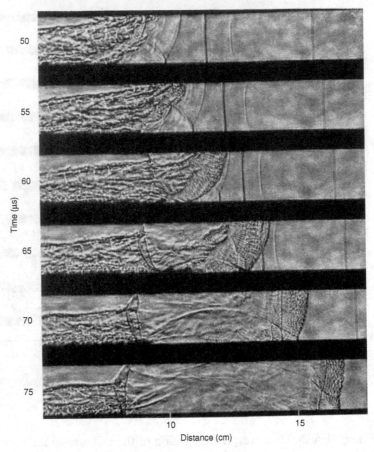

Figure 8.24. Sequence of schlieren photographs showing the onset of detonation, originating in the turbulent flame brush (Urtiew & Oppenheim, 1966).

It should be noted that the detonation does not necessarily have to originate in the turbulent flame brush. In Fig. 8.26, the onset of detonation occurs at the contact surface subsequent to the merging of two shock waves. Two precursor shocks merge, and the first frame (40 μs) shows the shock, the contact surface, and the reflected shock. These waves are ahead of the turbulent flame front, which trails behind. The contact surface has a higher local temperature than elsewhere in the flow field, due to the merging of the two shocks when they catch up with one another. After an induction period, autoignition of the contact surface occurs, and then the reaction front from this explosion catches up with the leading shock ahead of it, initiating a detonation. An accompanying retonation is observed to be formed, which propagates back into the combustion products. It is interesting to note that, subsequent to the formation of the detonation, another explosion occurs in the turbulent flame zone behind the detonation.

Figure 8.27 shows the subsequent frames of the schlieren movie from the same experiment. In the first frame (80 μs), the cellular detonation that is formed can

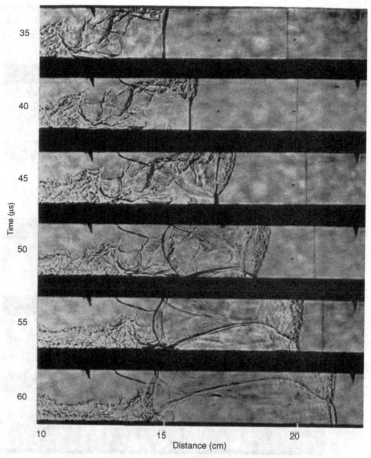

Figure 8.25. Sequence of schlieren photographs showing the onset of detonation, originating at the foot of the precursor shock wave (Urtiew & Oppenheim, 1966).

be seen propagating to the right, and the retonation wave to the left. In the fourth frame, the spherical shock from a large explosion in the turbulent flame brush can be observed. The forward-propagating shock from this second explosion merges with the first retonation, while the backward-propagating shock now forms the second retonation wave. Two transverse shocks are formed from this spherical explosion, which occurs near the center of the detonation channel.

We see that, as a result of the different kinds of flow fields that are generated during the highly irreproducible flame acceleration phase, the onset of detonation is not a unique phenomenon. Hence, it is not possible to have a unique theory for the onset of detonation. However, in a long smooth-walled tube, it appears that the flame must first achieve some critical velocity before these various modes for the onset of detonation can be spontaneously triggered. Experiments indicate that the critical flame speed is of the order of the sound speed of the products, or about half the CJ detonation velocity. This also corresponds to the CJ deflagration speed of this mixture. It seems reasonable to expect that, in the transition process, the flame

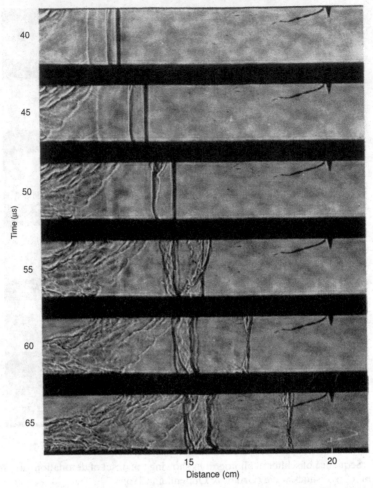

Figure 8.26. Sequence of schlieren photographs illustrating the onset of detonation at the contact surface subsequent to the merging of two precursor shock waves (Urtiew & Oppenheim, 1966).

must first accelerate to the maximum possible deflagration speed (i.e., for CJ deflagration), and then instabilities will develop and trigger the onset of detonation. Of course, with artificially induced large perturbations, the onset of detonation can be effected for any flame speed. However, the spontaneous onset of detonation appears to occur when the flame has reached a velocity of about the CJ deflagration speed.

Although the details may vary, it appears that the onset of detonation always seems to be prompted by an *explosion within the explosion* (a term coined by A.K. Oppenheim to describe the local explosion within the turbulent flame zone). A spherical detonation appears to be formed instantaneously from this local explosion center and propagates outward to form the combined detonation–retonation–transverse-wave pattern. When the detonation is first formed from the hot spot, it is highly overdriven, which is analogous to the direct initiation of a spherical

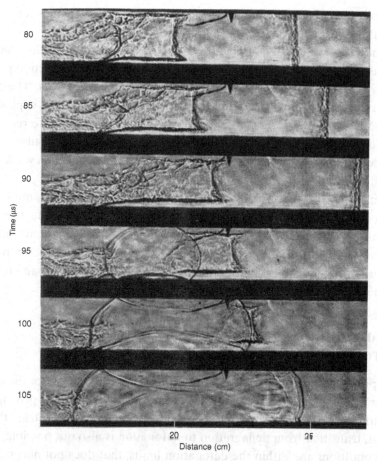

Figure 8.27. Continuation of Fig. 8.26: sequence of schlieren photographs illustrating a local explosion behind the newly formed detonation (Urtiew & Oppenheim, 1966).

detonation by a powerful energy source. If one considers the instantaneous constant volume explosion of an explosion center, the strength of the shock wave formed in a constant volume explosion is only of the order of $M_s \approx 2$. Since we see the instantaneous emergence of a highly overdriven detonation from the explosion center and not a weak shock, it appears that there must exist a very effective amplification mechanism that amplifies the shock wave from a constant volume explosion to an overdriven detonation state in a very short distance (of the order of a centimeter or less).

The SWACER mechanism was proposed by Lee *et al.* (1978) to account for the rapid formation of an overdriven detonation from the relatively weak shock of a volumetric explosion at a local hot spot. The SWACER mechanism is based on a proper synchronization of the chemical energy release in the mixture with the traveling compression pulse (or shock) as it propagates through it. SWACER is analogous to the resonant coupling of the acoustic oscillation of a system with a heat source. Rayleigh's criterion requires that the heat added must be in phase with the compression phase of the cycle. Thus, the SWACER mechanism is, in essence, Rayleigh's criterion for a traveling compression pulse.

To achieve the resonant coupling, Lee *et al.* proposed that an induction time gradient must exist in the medium. Each particle undergoes the induction process and is on the verge of exploding when the compression pulse arrives. The additional temperature rise associated with the compression then triggers the explosion of the particle. With the appropriate energy release function, the pressure rise due to the energy release can be fed in phase to enhance the compression pulse as it propagates through the gradient field. With the SWACER mechanism, a weak shock can be amplified rapidly to a detonation wave.

It should be noted that Zeldovich *et al.* (1970, 1980) carried out a numerical computation of the development of an explosion in a nonuniform, preheated mixture in an attempt to elucidate the phenomenon of knock in internal combustion engines. Their computations also indicated that, with the appropriate temperature gradient, a very large pressure spike is developed suggesting that the shock and the reaction zone are coupled as in a detonation.

8.6. CRITERION FOR TRANSITION FROM DEFLAGRATION TO DETONATION

For the propagation of a detonation wave, there exists a criterion for the detonation limit (for a given boundary condition) beyond which self-sustained propagation of a detonation wave is not possible. It is reasonable to assume that, outside the detonation limit, transition from deflagration to detonation is also not possible. However, even if conditions are within the detonation limits, that does not necessarily imply that transition can be effected. Near-limit detonations are usually initiated by a powerful ignition source, which generates an initially overdriven detonation that decays eventually to the CJ value. The mechanisms for flame acceleration are totally different from those for the propagation of a detonation. Thus, conditions that permit the self-sustained propagation of a detonation can be quite different from those for flame acceleration. It is neither meaningful nor possible to define the criterion for flame acceleration, because a whole spectrum of different mechanisms is involved. However, it might be possible to define a criterion for the onset of detonation. In other words, assuming a deflagration has already accelerated to a certain maximum velocity compatible with the given boundary conditions, there are perhaps certain requirements that still have to be met before the spontaneous onset of detonation can occur. Attempts have been made by Lee and co-workers (Lee *et al.*, 1984; Knystautas *et al.*, 1986; Peraldi *et al.*, 1986; Lee, 1986) to establish a transition criterion in tubes. Boundary conditions play an important role, so it is only meaningful to discuss transition for specific boundary conditions (smooth-walled circular tubes, rough-walled tubes, unconfined spherical geometry, etc.).

Consider first the special case of a smooth-walled circular tube. Assuming flame acceleration mechanisms have brought the flame to the maximum deflagration state,

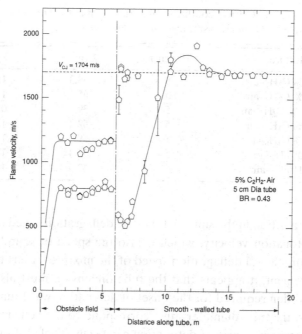

Figure 8.28. Flame velocity as a function of distance through an obstacle section followed by a smooth tube (Knystautas *et al.*, 1986).

we then ask: "What criterion must be met before the onset of detonation can occur?" In the studies carried out by Lee and co-workers, a section of rough-walled tube (viz., a tube with periodic obstacles) was used to first rapidly accelerate the flame to a certain velocity. The flame then emerged into a smooth section of the tube. Transition was then determined by varying the flame velocity and mixture sensitivity (through the initial pressure, composition, and concentration of inert diluents).

Figure 8.28 shows a typical result for the transition in the smooth tube for two different flame velocities obtained in the obstacle section of the tube. For a mixture composition of 5% C_2H_2 in air and an initial flame velocity of about 800 m/s (about half the CJ velocity of the mixture, 1704 m/s), the flame velocity is observed to drop rapidly as it emerges into the smooth tube. However, it reaccelerates rapidly and undergoes transition to detonation further downstream. For a higher flame velocity of about 1200 m/s in the rough section, the flame spontaneously transits to a detonation within a couple of tube diameters upon exiting into the smooth tube.

In Fig. 8.29, the mixture composition is 4% C_2H_2 in air, and we observe the onset of detonation subsequent to the emergence of the flame into the smooth section. In certain experiments, the flame accelerates after an initial drop in velocity (upon emerging into the smooth section), and transition to detonation is observed. However, in some cases the flame tries to re-accelerate initially, but fails to continue the acceleration and then decays. Figure 8.29 shows two representative cases of successful and unsuccessful transition. The initial flame velocity is about 700 m/s, which again corresponds to about half the CJ detonation velocity of 1595 m/s. It appears

Table 8.1. Critical conditions for transition in the smooth-walled section (Knystautas *et al.*, 1986)

Mixture	D (cm)	λ (mm)	λ/D
4% C_2H_2–air	5	58.3	1.18
5% C_2H_4–air	5	65.1	1.32
10% C_2H_4–air	5	39.1	0.80
4% C_3H_8–air	5	52.2	1.06
5% C_3H_8–air	5	59.0	1.19
20% H_2–air	5	55.4	1.12
51% H_2–air	5	52.5	1.06

that, to effect transition in the smooth tube, the deflagration speed required is about half the CJ detonation velocity, which is also the speed of sound in the combustion products and the CJ deflagration speed of the mixture. Apart from a minimum velocity requirement, it appears that the tube dimension must also be sufficiently large. The condition required for the onset of detonation was found to be $\lambda/d \approx 1$. In other words, the tube diameter must correspond to the cell size of the mixture before the spontaneous onset of detonation can occur. Table 8.1 lists a range of fuels and mixture compositions, and the ratio λ/d where transition to detonation is successful. We note that λ/d for the mixtures shown are all of the order of unity. Note that, for the detonation limit, the ratio $\lambda/d \approx \pi$. Thus, for transition, a more

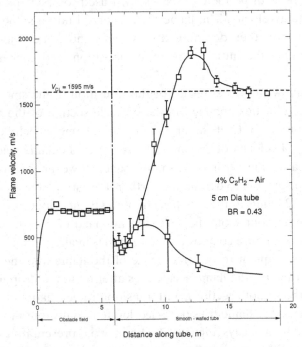

Figure 8.29. Successful and unsuccessful transition to detonation in a smooth tube (Knystautas *et al.*, 1986).

Table 8.2. Conditions for transition in a rough-walled tube
(Peraldi *et al.*, 1986).

D (cm)	d (cm)	Transition Mixture	λ (cm)	λ/d
5	3.74	22% H_2–air	3.07	0.82
		47.5% H_2–air	4.12	1.10
		4.75% C_2H_2–air	1.98	0.51
		6% C_2H_4–air	3.78	1.01
		9% C_2H_4–air	3.01	0.81
15	11.4	18% H_2–air	10.7	0.94
		57% H_2–air	11.7	1.03
		4% C_2H_2–air	5.8	0.51
		4.5% C_2H_4–air	8.7	0.76
		13.5% C_2H_4–air	11.5	1.01
		3.25% C_3H_8–air	11.2	0.98
		5.5% C_3H_8–air	11.6	1.02
30	22.86	18% H_2–air	21.0	0.92
		57% H_2–air	18.5	0.81
		4% C_2H_2–air	10.6	0.46
		4.5% C_2H_4–air	18.0	0.79
		13.5% C_2H_4–air	20.0	0.87
		3.25% C_3H_8–air	21.0	0.92
		5.5% C_3H_8–air	9.2	0.40

D (cm)	d (cm)	No transition Mixture	λ_{min} (cm)	λ_{min}/d
5	3.74	CH_4–air	30.0	8.02
		C_3H_8–air	5.2	1.40
15	11.4	CH_4–air	30.0	2.63
30	22.86	CH_4–air	30.0	1.31

sensitive mixture is required than that for detonation propagation in the same tube diameter.

In a rough tube (filled with obstacles in the form of orifice plates), it is found that a flame will accelerate to a flame speed also of the order of the sound speed in the combustion products (or the CJ deflagration speed) prior to transition to a detonation (or quasi-detonation). For transition to occur in the rough tube, it is found that $\lambda/d \approx 1$ is also required. However, in the rough tube, d refers to the orifice diameter and not the tube diameter, because the orifice diameter is the characteristic dimension of the unobstructed tube in the direction of propagation (i.e., along the tube axis). Table 8.2 lists the values of λ/d when transition is observed in the rough tube. We can see that, with the exception of acetylene, the fuels give the same value of $\lambda/d \approx 1$ as in the smooth tube. Acetylene is a particularly sensitive fuel, and cell size measurements often reveal that the cell size can fluctuate by a factor of two. No satisfactory explanation has been advanced as yet, though it has been speculated

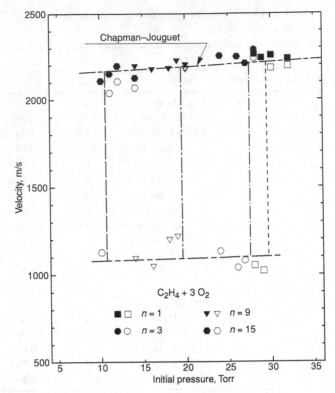

Figure 8.30. Wave velocity upstream (black symbols) and downstream (open symbols) of acoustically absorbing wall section with n layers of acoustically absorbing material (Dupré *et al.*, 1988).

that diacetylene polymers can often form, which are more sensitive than acetylene.
However, no definitive investigations have been made to date to resolve the acetylene puzzle.

Examining older evidence of transition in smooth tubes, it may be concluded that
the onset of detonation usually occurs when the deflagration speed is of the order of
half the CJ detonation speed (or the sound speed of the products). Recent experiments by Eder and Brehm (2001) also indicate that the maximum deflagration speed
at the end of an obstacle section required for transition to detonation corresponds
closely to the sound speed of the products.

Of particular significance in the understanding of the conditions required for the
transition to detonation is the observation that, when a detonation fails as the transverse waves are dampened out (thus destroying the cellular structure), the velocity
drops to the CJ deflagration speed. Figure 8.30 shows the results of Dupré *et al.*
(1988), in which a detonation fails after the transverse waves have been dampened
out by acoustically absorbing lining on the tube wall. As can be observed, the velocity drops from the CJ detonation value to about half that value.

The importance of this observation is that, when the mechanism necessary for
self-sustained propagation of the detonation is removed, the detonation becomes a
CJ deflagration. Therefore, we may conclude that, for transition to occur, the deflagration must first accelerate to its maximum CJ deflagration velocity (which is

about $\frac{1}{2}V_{CJ}$, the sound speed of the products). Then, the onset of detonation requires the generation of transverse waves to form the regular cellular structure of self-sustained detonations. The onset of detonation can be looked at as the process of generating transverse waves that couple with the shock front in order to form the cellular structure of a self-sustained detonation. As discussed in the previous section, this process requires the formation of a blast wave from a local explosion center, the amplification of this blast wave to an overdriven detonation, and the subsequent formation of transverse waves from the spontaneous development of instabilities in the decaying overdriven detonation front. The overdriven condition is required for small-amplitude, high-frequency perturbations to grow, and the eventual transverse perturbation takes on the frequency and strength corresponding to the mixture as the CJ condition is approached. The spontaneous appearance of cells is always observed as an overdriven detonation decays to its CJ condition.

Under confined conditions (i.e., a rigid-walled tube), it is also possible to develop the transverse waves of the cellular detonation by repeated amplification as the transverse perturbations reflect back and forth from the tube wall and traverse the reaction zone. When the Rayleigh criterion is satisfied, the transverse waves can be progressively amplified, which eventually leads to the onset of detonation. Figures 8.28 and 8.29 illustrate that the formation of detonation is more progressive in nature, taking many tube diameters of propagation before the eventual onset occurs. This progressive mode contrasts to the abrupt onset from an explosion center that forms a highly overdriven detonation via the SWACER mechanism. In an unconfined geometry, this slow mode for the onset of detonation would not be possible, for the geometry does not permit the repeated reflections of the transverse waves to be progressively amplified upon traversing the reaction zone.

The two types of onset of detonation are illustrated in the streak schlieren records in Fig. 8.31. In Fig. 8.31a, the detonation is formed by the progressive amplification

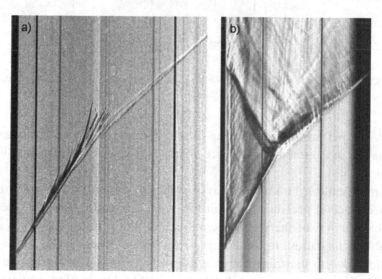

Figure 8.31. Streak schlieren photographs illustrating the two different modes of detonation onset (courtesy of J. Chao).

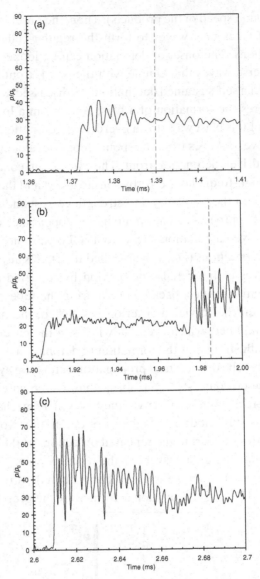

Figure 8.32. Pressure profiles illustrating the progressive amplification of transverse waves leading to the onset of detonation (courtesy of J. Chao).

of transverse perturbations reflecting from the tube wall. No abrupt transition and no formation of a retonation wave are observed in this progressive amplification to detonation.

The amplification of transverse waves leading to the onset of detonation is shown in the sequence of pressure traces in Fig. 8.32. The vertical dashed lines indicate the time of arrival of the reaction zone behind the precursor shock wave of the metastable deflagration. In the first pressure profile, shown in Fig. 8.32a, pressure fluctuations due to the transverse waves can be seen behind the precursor shock front within the reaction zone. As the metastable deflagration propagates down the tube, the transverse waves amplify as they reflect from the tube walls and as they traverse the reaction zone. The amplified transverse waves can been seen in Fig. 8.32b, where the pressure fluctuations in the reaction zone (indicated by the dashed vertical line) are now much greater in amplitude than those in Fig. 8.32a. Due to this progressive amplification of the pressure fluctuations through the reaction zone, the onset of detonation occurs, and a fully developed CJ detonation is observed in Fig. 8.32c.

The progressive amplification leading to the onset of detonation is in contrast to the mechanism shown in the streak schlieren photograph of Fig. 8.31b, where the onset of detonation is due to local explosions in the turbulent flame zone. Amplification to a highly overdriven detonation occurs via the SWACER mechanism, and the formation of the accompanying retonation wave, as required by the conservation of momentum, can be seen clearly in the figure. In geometries that do not permit the repeated reflections of transverse waves, only the abrupt onset from local explosion centers is possible.

We may conclude that the transition from deflagration to detonation is not a unique phenomenon. The formation of a detonation results from the development of transverse perturbations at the shock front to form a self-resonating system where the gasdynamic oscillations are supported by the energy release by chemical reactions. The requirement for transition is to create the condition for instability to develop and grow, leading to the formation of the cellular structure of a detonation front.

8.7. CLOSING REMARKS

The transition from deflagration to detonation is an ill-defined phenomenon in that it is difficult to distinguish the two types of combustion waves unambiguously. From the point of view of the propagation velocity of the two types of combustion waves, they can overlap one another. A detonation propagates at a supersonic speed relative to the reactants ahead of it, which are at rest. Thus, the detonation propagates at the same velocity relative to the fixed laboratory coordinates. For a propagating deflagration wave, the precursor shock and the trailing reaction front can also propagate at supersonic velocities relative to the reactants ahead of the precursor shock. Thus, with respect to the fixed laboratory coordinates, the deflagration also propagates at supersonic speeds even though the reaction front itself propagates at a subsonic speed relative to the reactants ahead of it. The reactants are set into motion by the precursor shock, and the reaction front is convected by this flow behind the precursor shock.

In terms of the nature of the two types of combustion waves, detonations are generally considered as compression waves, whereas deflagrations are expansion waves with a pressure drop across them. However, if one considers a precursor shock–reaction front as a complex propagating at supersonic speeds relative to the undisturbed reactants (or the laboratory frame), then the pressure of the product gases can be higher than the initial pressure of the reactants, and the overall effect of the shock–flame complex is that due to a compression wave.

In terms of the propagation mechanism, classical theory defines detonations as propagating via shock-induced autoignition. However, in an unstable cellular detonation, the mechanisms for ignition and the subsequent combustion of the reactants crossing the detonation are no longer clear. The local velocity of the leading shock front of an unstable cellular detonation can vary considerably, from $1.8\,V_{CJ}$ to $0.5\,V_{CJ}$, and upon crossing the leading front, the shocked particle immediately undergoes rapid expansion, because the local shock velocity decays like a blast wave. The particle may further be compressed by transverse shocks as well as undergo turbulent mixing in the complex flow field of the detonation reaction zone. Often large unreacted pockets of reactants are swept downstream into the products when shock compression and turbulent diffusion do not provide the conditions for ignition and for rapid burning of the reactants that cross the leading front. Thus, in an unstable cellular detonation, the combustion mechanism is no longer a straightforward autoignition mechanism via adiabatic shock heating as described by classical CJ theory.

For a deflagration, classical theory describes it as a diffusion-controlled wave (i.e., heat and radical species diffuse from the reaction zone to effect chemical reactions in the reactants ahead of it). However, for a high-speed propagating deflagration wave, the reaction front is highly turbulent with strong pressure fluctuations in the reaction zone. We have shown earlier that when pressure fluctuations in a high-speed deflagration are damped out by acoustically absorbing walls, the deflagration speed drops significantly, indicating that turbulence alone cannot maintain the reaction rates of these high-speed deflagrations. Thus, the combustion mechanism of a high-speed deflagration is not unlike those of an unstable cellular detonation.

Without a sharp distinction between deflagration and detonation in terms of the propagation velocity, the nature of the wave (i.e., compression or expansion), and the combustion mechanism, it is difficult to discern an abrupt transition from one type of wave to the other.

Bibliography

Borghi, R. 1985. On the structure and morphology of turbulent premixed flames. In *Recent advances in aerospace sciences*, ed. C. Casci, 117–138. Plenum Press.

Chan, C., I.O. Moen, and J.H.S. Lee. 1983. Influence of confinement on flame acceleration due to repeated obstacles. *Combust. Flame* 49:27–39.

Chu, B.T. 1952. In *4th Int. Symp. on Combustion*, 603.

Clavin, P. 1985. Dynamical behavior of premixed flame fronts on laminar and turbulent flows. *Prog. Energy Combust. Sci.* 11:1.

Courant, R., and K.O. Friedrich. 1948. *Supersonic flow and shock waves*, 226. New York: Interscience.

Dupré, G., O. Peraldi, J.H. Lee, and R. Knystautas. 1988. *Prog. Astronaut. Aeronaut.* 114:248–263.

Eder, A., and N. Brehm. 2001. *Heat Mass Transfer* 37:543–548.

Kaskan, W. 1952. In *4th Int. Symp. on Combustion*, 575.

Knystautas, R., J.H. Lee, I. Moen, and H. Gg. Wagner. 1979. Direct initiation of spherical detonation by a hot turbulent gas jet. *Proc. Combust. Inst.* 17:1235–1245.

Knystautas, R., J.H.S. Lee, O. Peraldi, and C.K. Chan. 1986. *Prog. Astronaut. Aeronaut.* 106:37–52.

Kumagai, S., and I. Kimura. 1952. In *4th Int. Symp. on Combustion*, 667.

Lee, B.H.K., J.H.S. Lee, and R. Knystautas. 1966. *AIAA J. Tech. Note* 4:365.

Lee, J.H.S. 1986. *Prog. Astronaut. Aeronaut.* 106:3–18.

Lee, J.H.S., R. Knystautas, and A. Freiman. 1984. *Combust. Flame* 56:227–239.

Lee, J.H.S., R. Knystautas, and N. Yoshikawa. 1978. *Acta Astronaut.* 5:971–982.

Lee, J.H.S., and B.H.K. Lee, 1996. *AIAA J.* 4:736.

Lee, J.H.S., and I. Moen. 1980. *Prog. Energy Combust. Sci.* 6:359.

Leyer, J.C. *Rev. Gén. Thermique* 1970. 98:121–138.

Leyer, J.C., and N. Manson. 1970. In *13th Int. Symp. on Combustion*.

Manson, N. 1949. Propagation des détonations et des déflagrations dans les mélanges gaseux. L'Office National d'Etudes et de Recherches Aéronautiques et L'Institut Francais de Pétrole.

Markstein, G.H. 1951. *J. Aeronaut. Sci.* 18:428.

Markstein, G. 1952. In *4th Int. Symp. on Combustion*, 44–59.

Markstein, G. 1957. In *6th Int. Symp. on Combustion*, 387–389.

Markstein, G. 1964. *Non-steady flame propagation*. Pergamon Press.

Markstein, G., and L.M. Somers. 1952. In *4th Int. Symp. on Combustion*, 527.

Matalon, M., and M.J. Matkowsky. 1982. *Fluid Mechanics*. 124:239–259.

Meshkov, Y.Y. 1970. Instability of a shock wave accelerated interface between two gases. NASA TT F-13 074.

Meyer, J.W., P.A. Urtiew, and A.K. Oppenheim. 1970. *Combust. Flame* 14:13–20.

Peraldi, O., R. Knystautas, and J.H.S. Lee. 1986. In *21st Int. Symp. on Combustion*, 1629–1637.

Richtmyer, R.D. 1960. Taylor instability in shock acceleration of compressible flows. *Comm. Pure Appl. Math.* 23:297–319.

Rudinger, G. 1958. Shock wave and flame interactions. In *Combustion and Propulsion 3rd AGARD Colloq.*, 153–182. London: Pergamon Press.

Sivaskinsky, G.I. 1983. *Ann. Rev. Fluid Mech.* 15:179–199.

Taylor, G.I., and R.S. Tankin. 1958. In *fundamentals of gasdynamics*, ed. H.W. Emmons, 622–686. Princeton Univ. Press.

Urtiew, P., and A.K. Oppenheim. 1966. *Proc. R. Soc. Lond. A* 295:13.

Urtiew, P., and C. Tarver. 2005. Shock initiation of energetic materials at different initial temperatures. *Combust. Explos. Shock Waves* 41:766.

Zeldovich, Ya. B. 1980. Regime classification of an exothermic reaction with nonuniform initial conditions. *Combust. Flame* 39:211–214.

Zeldovich, Ya. B., A. A. Borisov, B. E. Gelfand, S. M. Frolov, and A. E. Mailkov. 1988. Nonideal detonation waves in rough tubes. *Prog. Astronaut. Aeronaut.* 144:211–231.

Zeldovich, Ya. B., V.B. Librovich, G.M. Makhviladze, and G.I. Sivashinsky. 1970. Astronaut. Acta 15:313–321.

9 Direct Initiation of Detonations

9.1. INTRODUCTION

Direct initiation refers to the instantaneous formation of the detonation without going through the predetonation stage of flame acceleration. By "instantaneous," it is meant that the conditions required for the onset of detonation are generated directly by the ignition source rather than by flame acceleration, as in the transition from deflagration to detonation.

Direct initiation was first used to generate spherical detonations because, in an unconfined geometry, the various flame acceleration mechanisms are ineffective (or absent), and hence the transition from a spherical deflagration to a spherical detonation cannot generally be realized. In 1923, Lafitte (1925) used a powerful igniter consisting of 1 g of mercury fulminate to directly initiate a spherical detonation in a mixture of $CS_2 + 3 O_2$. He also used a planar detonation emerging from a 7-mm-diameter tube into the center of a spherical flask containing the same mixture of $CS_2 + 3 O_2$. With this method, Lafitte was not successful in initiating a spherical detonation. Using the same powerful igniter of mercury fulminate, spherical detonations were also initiated in $2 H_2 + O_2$ mixtures. From the streak photographs taken by Lafitte, the detonation is observed to form instantaneously by the igniter, without a noticeable predetonation period. The detonation of a condensed explosive charge generates a very strong blast wave that decays rapidly as it expands to form the CJ detonation (hence, this mode of initiation is also referred to as blast initiation). In the immediate vicinity of the igniter, the initial blast decay is governed by the ignition energy as in a non-reacting blast. As the blast wave expands, the chemical energy released at the shock front begins to influence its propagation. The blast wave then becomes an overdriven detonation wave. Eventually, far from the igniter, the chemical energy release completely controls the wave motion, and a self-sustained CJ detonation is obtained. Note that the decay of an initially overdriven detonation to a CJ wave in direct initiation is not unlike the onset of detonation in the transition from deflagration to detonation, where also an overdriven detonation is first formed from a local explosion center in the turbulent flame zone and then decays to a CJ wave.

Because a detonation is essentially a strong shock supported by the chemical energy release in its wake, it is reasonable to expect that direct initiation requires a strong shock of sufficient duration to be generated by the igniter. Direct initiation of detonations using a shock tube was demonstrated by Shepherd (1949), Berets *et al.* (1950), Mooradian and Gordon (1951), and Fay (1953). Ignition of a deflagration by a shock wave can also occur at lower shock strengths (as compared to a shock at the CJ detonation velocity). Once a deflagration is ignited, the subsequent onset of detonation would require further flame acceleration as in a normal transition process. Hence, only when a very strong shock is used that immediately forms an overdriven detonation can direct initiation be obtained without the pre-detonation deflagration phase. Instead of a shock tube, a detonation wave from one mixture (the driver mixture) can also generate a strong shock in another mixture that is initially separated from the driver mixture by a thin diaphragm. Such a detonation driver was used by Berets *et al.* (1950) as well as by Gordon *et al.* (1959) to achieve direct initiation in a tube. Particularly for insensitive near-limit mixtures, a detonation can only be obtained via direct initiation using a strong detonation driver section. Direct initiation by a detonation is similar to blast initiation where the initial blast wave forms an overdriven detonation that decays to a CJ detonation in the test mixture downstream of the driver section. The rate of decay depends on the steepness of the expansion wave behind the initiating detonation (i.e., on the length of the driver section).

Apart from strong blast waves and detonation waves, direct initiation can be achieved by a variety of other means. For example, both a turbulent jet of hot combustion products and a jet carrying a highly reactive chemical (e.g., fluorine, chlorine trifluoride) have been shown to initiate detonations directly in the mixing zone of the jet with the reactants (Knystautas *et al.*, 1978; Murray *et al.*, 1991). This process is similar to the onset of detonation that occurs in the turbulent mixing zone of the deflagration at the end of the flame acceleration phase. Direct initiation by flash photolysis has also been demonstrated by Wadsworth (1961) and Lee *et al.* (1978). In a flash photolysis initiation, a free radical gradient is generated in the absorption path of an ultraviolet pulse in the reactants. This free radical gradient gives rise to an induction-time gradient. The subsequent explosion and the coupling of the explosion wave with the chemical energy release in the gradient field give rise to a rapid amplification to an overdriven detonation in the gradient field. The overdriven detonation subsequently decays to a CJ detonation, as in all detonation formation processes.

In essence, direct initiation refers to the generation of the critical conditions for the onset of detonation directly by the initiating source, bypassing the flame acceleration phase. Because different initiation sources create a variety of conditions for the onset of detonation, it is of interest to describe the various means by which direct initiation can be achieved. The onset of detonation itself is found to be common to all the different sources, and is similar to the transition from deflagration to detonation. In the transition process, the critical conditions for the onset of detonation are produced via flame acceleration, in contrast to direct initiation, where the initiation source generates the critical conditions directly.

9.2. BLAST INITIATION (EXPERIMENTAL OBSERVATIONS)

Although Lafitte was the first to successfully initiate spherical detonations in a mixture of $CS_2 + 3O_2$ using a powerful igniter (a charge of 1 g of mercury fulminate condensed explosive), he did not investigate the critical conditions for direct initiation. In a later study, Zeldovich *et al.* (1957) studied the direct initiation process in more detail and found that a minimum spark energy is required for direct initiation in tubes of various diameters for mixtures of $C_2H_2 + 2.5O_2$ and $2H_2 + O_2$. In a finite-diameter tube, the reflection of the blast wave from the side wall can lead to the initiation of detonation. Thus, the spark energy required is less in smaller-diameter tubes. Only in the limit of a sufficiently large-diameter tube does the spark energy approach the value required for an unconfined spherical detonation. As recognized by Zeldovich *et al.*, the direct initiation process involves the generation of a strong shock wave of sufficiently long duration by the igniter. Initially, the blast wave generated by the igniter is generally much stronger than the CJ detonation of the mixture. As the blast decays, the energy released by the chemical reaction that is initiated behind the shock begins to influence the shock motion, and the strong blast becomes an overdriven detonation wave. If the igniter energy is sufficiently large, the overdriven detonation will decay to a CJ detonation. The decay characteristics of the blast wave also depend on the energy–time profile of the igniter (i.e., the power density). In the limit of infinite power density where the source volume is vanishingly small and the energy release duration approaches zero, the igniter can be considered as an ideal instantaneous point (line, or planar) energy source. For these ideal sources, the initial strong blast decay is characterized by only the source energy. For a condensed explosive charge, the power density is sufficiently high that the blast decay can be closely approximated by the blast from an ideal instantaneous point source. Therefore, direct initiation of spherical and cylindrical detonations by ideal point and line sources can be realized experimentally using condensed explosive charges. However, the instantaneous initiation of a planar sheet explosive is much more difficult to achieve experimentally, and thus we shall only describe the experimental results for the direct initiation of spherical and cylindrical detonations by point and line sources.

The experimental study of the direct initiation of spherical detonations by Bach *et al.* (1969) indicates that there exist three regimes corresponding to whether the initiation energy is below, above, or near the critical value. To achieve an instantaneous ideal point source, they used a laser-induced breakdown spark from a Q-switched ruby laser. The spark volume was of the order of fractions of a cubic millimeter, and the duration of the spark energy release was 10 ns. Thus, the laser spark approached an ideal instantaneous point energy source. Figure 9.1 shows the schlieren photographs of the spherical blast when the blast energy is below

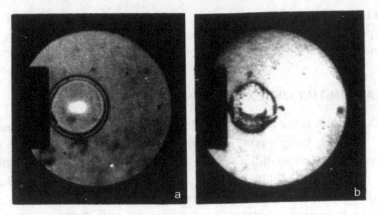

Figure 9.1. Schlieren photographs of (a) subcritical and (b) supercritical blast initiation of a spherical detonation by a laer spark.

(Fig. 9.1a) and above (Fig. 9.1b) the critical value. For the subcritical energy regime, the reaction front progressively decouples from the leading shock as the overdriven detonation decays. Subsequently, the shock decays asymptotically to an acoustic wave, while the reaction front propagates as a laminar deflagration. When the blast energy is above the critical value (the supercritical regime), the reaction front is always coupled to the shock front of the decaying overdriven detonation. Instabilities set in as the detonation approaches its CJ velocity, and the detonation front takes on its characteristic turbulent cellular structure as shown in Fig. 9.1b. The corresponding streak schlieren photographs of the subcritical and the supercritical regimes are shown in Fig. 9.2.

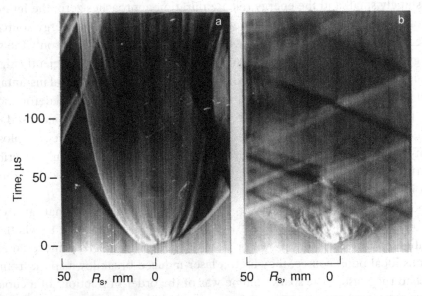

Figure 9.2. Streak schlieren photographs of (a) subcritical and (b) supercritical blast initiation of a spherical detonation (Bach *et al.*, 1969).

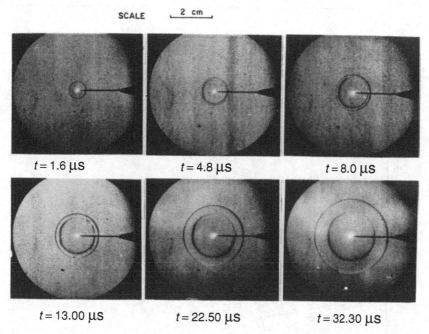

Figure 9.3. Sequence of schlieren photographs of the subcritical regime (Bach *et al.*, 1969).

Figure 9.3 shows a more detailed time sequence of schlieren photographs of the subcritical regime. At early times ($t = 1.6\ \mu$s), the shock and the reaction zone are still coupled as an overdriven detonation wave. At later times ($t > 4.8\ \mu$s), progressive decoupling of the reaction front from the shock can be observed. Eventually, the decoupled shock decays to an acoustic wave, and the reaction front now propagates as a laminar deflagration. The blast and reaction front trajectories for the subcritical energy regime are shown in Fig. 9.4. The shock- and reaction-front trajectories for two limiting cases are also plotted for comparison.

Assuming that the blast wave decays as a non-reacting blast (i.e., the chemical energy release does not influence the shock motion), the shock trajectory and the thermodynamic histories of the particle paths behind the decaying blast wave can be obtained from blast wave theory. For each particle crossing the shock, the induction time can then be computed, and the reaction front behind the decaying blast wave can then be obtained. However, because the blast is assumed to be non-reacting, the theoretical trajectory does not correspond to the reacting blast for the subcritical case. If the reaction zone is assumed to be coupled to the shock, as a reacting shock front, the blast wave will decay eventually to a CJ detonation. The trajectory can then be determined from reacting blast wave theory (Lee, 1965; Korobeinikov, 1969). This trajectory is also plotted in Fig. 9.4 for comparison. We note that, for early times when the reaction front is coupled to the shock, the blast trajectory for the subcritical regime coincides with that for the reacting blast wave solution. However, when the reaction zone starts to decouple from the shock, the blast trajectory deviates from the reacting blast solution.

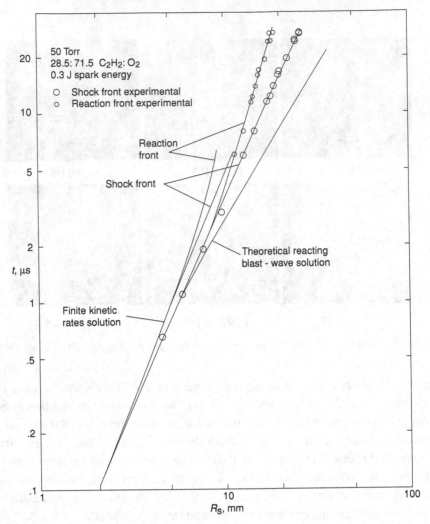

Figure 9.4. Shock- and reaction-front trajectories for the subcritical energy regime; stoichiometric acetylene–oxygen at $p_0 = 50$ Torr (Bach *et al.*, 1969).

A time sequence of schlieren photographs for the supercritical regime is shown in Fig. 9.5. The initiation source is an electrical spark from a high-voltage, low-inductance capacitor discharge. We note now that the reaction zone is always coupled to the shock, and, as the overdriven detonation decays rapidly toward a CJ detonation, instabilities set in, and the characteristic cellular detonation can be observed. A typical trajectory, showing the decay of the blast to a CJ detonation in the supercritical energy regime, is shown in Fig. 9.6. The theoretical reacting blast wave solution that describes the decay of the blast to its CJ velocity is plotted together with the experimental results. Good agreement is observed, but we note that the final steady-state velocity observed experimentally is slightly below the CJ value for the mixture. This could be a consequence of the effect of curvature, which causes a slight reduction in the detonation velocity.

2 cm

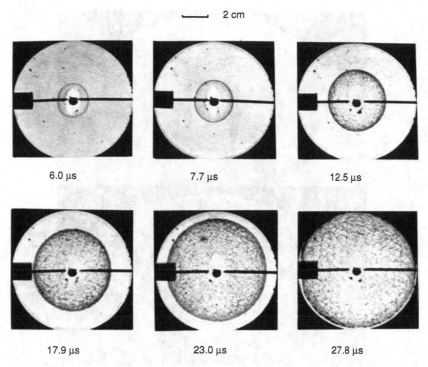

6.0 μs   7.7 μs   12.5 μs

17.9 μs   23.0 μs   27.8 μs

Figure 9.5. Sequence of schlieren photographs of the supercritical regime (Lee *et al.*, 1972).

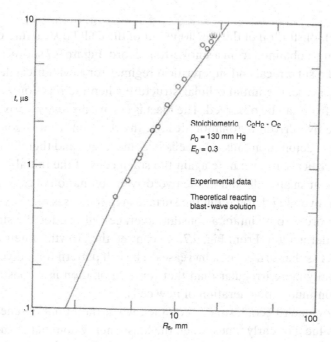

Stoichiometric C_2H_2 - O_2
$p_0 = 130$ mm Hg
$E_0 = 0.3$

Experimental data

Theoretical reacting
blast-wave solution

Figure 9.6. Blast wave trajectory in the supercritical regime; stoichiometric acetylene–oxygen at $p_0 = 130$ Torr (Bach *et al.*, 1969).

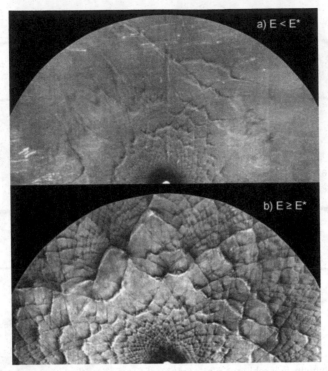

Figure 9.7. Smoked foils of (a) subcritical and (b) supercritical spherical blast initiation (courtesy of A.A. Vasiliev).

A good demonstration of the development of the cellular structure of the detonation wave can be obtained from a smoke-foil record. Figure 9.7 shows the smoke-foil records for the subcritical and supercritical regimes for a cylindrical detonation. For the subcritical case, the initial cellular structure when the reaction zone is coupled to the shock front can be observed. The blast is essentially an overdriven detonation wave. As the overdriven detonation decays, the reaction front decouples from the shock, and the detonation fails. The cells become larger and then finally disappear. In the supercritical case, we note again the small cells of the initial overdriven detonation wave at small radii, but as the overdriven detonation decays, the cells grow larger. For a diverging detonation, the surface area increases and new cells must be generated in order to maintain a constant average cell size for the stable propagation of a CJ detonation. From Fig. 9.7, we can see the growth of new cells from the enlarged cells as the surface area increases. The cell pattern for a diverging detonation is generally more irregular than that for a detonation in a constant-area tube, due to the continuous regeneration of new cells.

The phenomenon becomes more complex when the initiation energy is around the critical value. For early times, when the blast energy dominates, the blast decays progressively to an overdriven detonation. Its further decay toward the CJ velocity is similar to the subcritical regime, where the shock continues to decay past the CJ velocity as the reaction front decouples from it. However, the shock wave stops

SCALE   2   cm

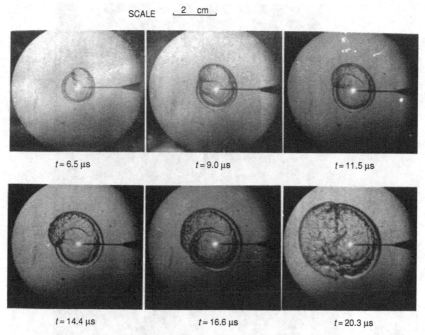

Figure 9.8. Sequence of schlieren photographs of the critical energy regime (Bach *et al.*, 1969).

decaying after reaching some sub-CJ velocity and then continues to propagate at a relatively constant velocity as a coupled metastable shock–reaction-front complex. After a certain duration of propagation at this quasi-steady metastable velocity, local explosion centers are observed to develop in the reaction front. Overdriven detonation bubbles are formed from the explosion centers, which then grow and engulf the leading shock surface to form an asymmetrical detonation wave. A time sequence of schlieren photographs illustrating the critical energy regime is shown in Fig. 9.8. In the first frame at $t = 6.5$ μs, the detonation bubble from a local explosion center can be observed. The detonation bubble grows to engulf the shock–reaction-front complex in the later frames. Eventually, an asymmetric cellular detonation is formed.

In Fig. 9.9, a streak schlieren photograph of the critical regime illustrates the formation of the detonation on the right-hand side, whereas on the left-hand side the shock and the reaction front are observed to decouple from each other after the quasi-steady regime, as in the subcritical case.

However, the detonation bubble that forms on part of the shock surface will eventually engulf the entire leading shock front to form a highly asymmetrical detonation. Thus, in the critical regime, the formation of the detonation appears to be a local phenomenon. The initial spherical blast only serves to generate the critical conditions in the vicinity of the reaction front for the onset of detonation to occur. Once explosions occur locally, the subsequent growth of the detonation bubble is independent of the initial spherical blast wave flow field. The detonation bubble from

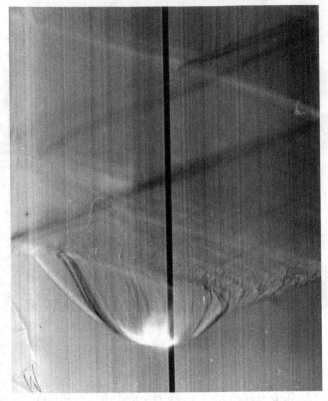

Figure 9.9. Streak schlieren photograph of the critical energy regime (Bach *et al.*, 1969).

an explosion center can grow and sweep over the surface of the spherical detonation in one direction or both. Figure 9.10a illustrates the growth of a detonation bubble that spirals in the clockwise direction, whereas in Fig. 9.10b, the bubble grows in both clockwise and counterclockwise directions. Eventually, in both cases, the detonation bubbles engulf the entire shock surface to form an asymmetrical cellular detonation.

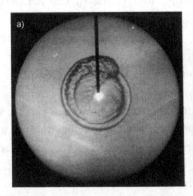
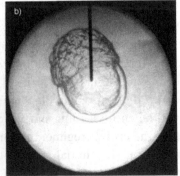

Figure 9.10. Schlieren photographs of the growth of a detonation bubble; (a) growth in clockwise direction and (b) growth in both clockwise and counterclockwise directions (Bach *et al.*, 1969).

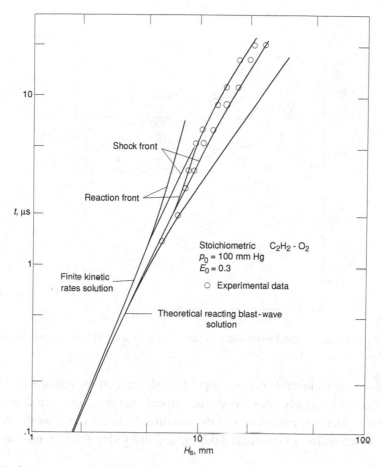

Figure 9.11. Shock- and reaction-front trajectories in the critical energy regime; stoichiometric acetylene–oxygen at $p_0 = 100$ Torr (Bach *et al.*, 1969).

The shock- and reaction-front trajectories corresponding to the critical energy regime are shown in Fig. 9.11. The blast and overdriven detonation trajectories initially agree well with the reacting blast wave solution. As the blast decays past the CJ velocity and the reaction front decouples from the leading shock, the propagation can no longer be described by the reacting blast solution that assumes a coupled detonation front at all times. The constant separation between the two waves during the quasi-steady metastable regime is also shown in Fig. 9.11. Subsequent to the formation of the detonation, the highly asymmetrical detonation can no longer be represented on the one-dimensional $R(t)$ diagram of Fig. 9.11.

The detonation velocity versus time (or equivalently radius) for the different initiation energy regimes is shown in Fig. 9.12. In the subcritical regime, the blast decays to an acoustic wave with time. However, for the critical and supercritical regimes, the decaying overdriven detonation levels off at some velocity below the CJ value for the mixture. For the supercritical case, the experimental results indicate that the final steady detonation velocity is less than the CJ value, which could be an effect of the curvature of the spherical detonation. For the critical case, the velocity of

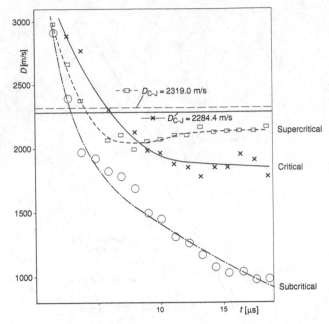

Figure 9.12. Velocities of the blast wave as a function of time for spherical blast initiation (Bach *et al.*, 1969).

the quasi-steady metastable regime appears to depend on the mixture and initiation energy. The quasi-steady velocity for the critical energy regime can drop to as low as half the CJ detonation velocity of the mixture. Figure 9.13 shows the detonation velocity versus radius for cylindrical detonation in a $C_2H_2 + 2.5\,O_2$ mixture. For the

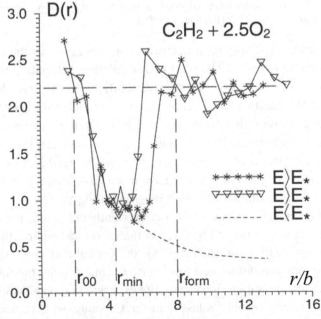

Figure 9.13. Velocity of the blast wave for cylindrical blast initiation; E^* is the critical energy (courtesy of A.A. Vasiliev).

subcritical regime, the decay of the blast to an acoustic wave in the asymptotic limit is observed. For the supercritical cases, the blast is observed to decay to slightly less than half the CJ detonation velocity in the metastable regime, prior to reacceleration back to a CJ detonation. A brief excursion to sub-CJ velocities appears to be characteristic of the direct initiation phenomenon, even when the initiation energy is above the critical value.

For the direct initiation of cylindrical detonations, an electric spark or exploding wire can provide the necessary short-duration line source if its length is small (less than a couple of centimeters). The cylindrical detonation will be confined between two plates, and boundary layer effects as well as wave reflections from the walls may influence the initiation process. It is difficult to obtain a long instantaneous line source for an unconfined cylindrical detonation. However, an instantaneous line source can be approximated as a point source traveling at infinite speed along the x-axis. If the source travels at finite speeds, an axisymmetrical conical blast wave will be formed with an angle at any section that is given by $\theta = \sin^{-1}(V_s/V_{DC})$ where V_s is the shock speed and V_{DC} is the velocity of the source along the x-axis. For $\theta \ll 1$, the shock and its flow field at any section distance x from the source can be considered to be that of a cylindrical blast wave at an instant of time $t = x/V_{DC}$. This, in essence, is the hypersonic blast wave analogy: the hypersonic flow past an axisymmetric blunt body is equivalent to that of a cylindrical blast wave (Lin, 1954). The blast energy corresponds to the drag of the body per unit length in hypersonic flow. Figure 9.14a illustrates the bow shock of a sphere traveling at hypersonic speed. The equivalent cylindrical blast wave flow field is shown in Fig. 9.14c.

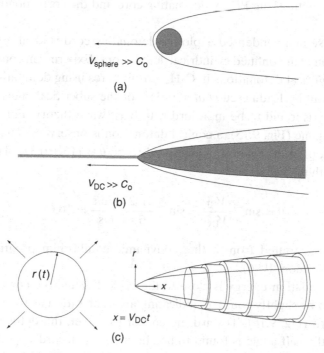

Figure 9.14. The cylindrical blast wave analogy (Radulescu *et al.*, 2003).

Table 9.1. Strength of PETN detonating cord and the corresponding
explosion length

PETN content (gm^{-1})	Equivalent line energy E_s (kJ/m)	Cylindrical explosion length $R_0 = (E_s/p_0)^{0.5}$ (m)
4.9	31	0.58
10.6	67	0.85
14.8	94	1.01
21.2	134	1.21
42.4	269	1.71
84.8	537	2.42

For the blast wave analogy to be valid, the shock radius must be large compared to the diameter of the sphere. Furthermore, the dimension of the source must be small compared to the blast radius in order for the line source approximation to be valid. To obtain a powerful line source, a high explosive detonating cord can be used instead of a hypersonic blunt body projectile. For a pentaerythritol tetranitrate (PETN) detonating cord, the detonation velocity along the cord is of the order of 6.5 km/s, and the analogous speed of a hypersonic projectile would be of the order of $M \approx 18.5$. Different strengths of detonating cord can be used, giving a characteristic explosion length for the cylindrical blast of the order of a meter. Thus, the dimension of the cord and the detonation products of the condensed explosive of the cord will be small compared to the explosion length, and the flow field can be closely approximated by ideal blast wave theory (Sakurai, 1953, 1954). Table 9.1 gives values for the strength of the PETN detonating cord and the corresponding explosion lengths.

Thus, the use of a condensed explosive detonating cord is ideal for the study of direct initiation of unconfined cylindrical detonations. Experiments on the direct initiation of cylindrical detonations in C_2H_4–air mixtures using detonating cords have been carried out by Radulescu et al. (2003). For the subcritical energy regime, the air blast decay is found to be in accord with blast wave theory. For the supercritical energy regime (Fig. 9.15), a conical detonation is observed. The half angle of the oblique conical detonation is measured to be $\theta \approx (16 \pm 0.3)°$. This is in good agreement with the value

$$\theta = \sin^{-1} \frac{V_{CJ}}{V_{DC}} = \sin^{-1} \frac{1.825 \text{ km/s}}{6.4 \text{ km/s}} = 16.6°,$$

where V_{CJ} is computed from a thermodynamic equilibrium program like CEA (McBride & Gordon, 1996).

When the initiation energy is close to the critical value, the conical detonation is observed to be slightly concave, indicating an acceleration as the detonation expands outward (Fig. 9.16). Toward the end of its travel, the detonation becomes conical, and the half angle is found to be about 16°, in accord with the theoretical

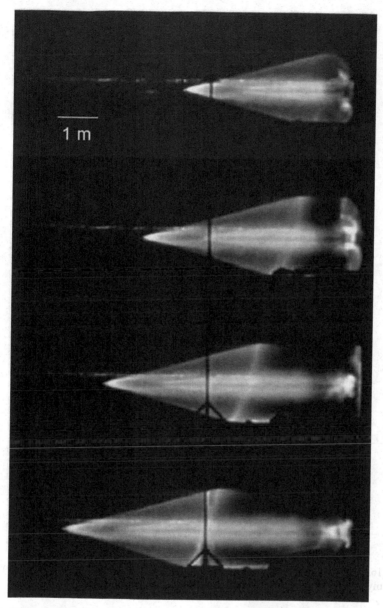

Figure 9.15. High-speed self-luminous photographs of supercritical initiation of a cylindrical deto-nation; ethylene–air at $\phi = 1.07$, $E = 67$ kJ/m (Radulescu *et al.*, 2003).

value. Thus, near the critical energy, some unsteady and acceleration processes are observed in the early stages of the propagation of the detonation wave.

At the critical value of initiation energy (shown in Fig. 9.17), the initiation phe-nomenon is much more complex. The initial strong blast (near the source) does not decay smoothly to an overdriven detonation that eventually asymptotes to a conical CJ detonation, as in the supercritical case. Instead, we note the appearance of local explosion centers around the circumference of the decaying blast wave. The deto-nation bubbles from these explosion centers merge and eventually form a conical

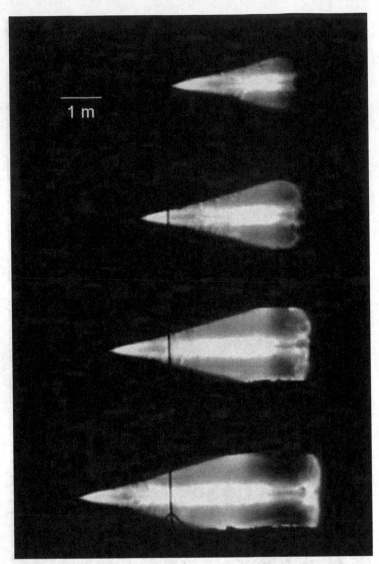

Figure 9.16. High-speed self-luminous photographs of near-critical initiation of a cylindrical detonation; ethylene–air at $\phi = 1.03$, $E = 67$ kJ/m (Radulescu *et al.*, 2003).

detonation around the axis. However, the surface of the detonation front is now
highly irregular from the merging of all the detonation bubbles. In the third frame
of Fig. 9.17, the bow shock trajectory, as measured by contact gauges, is superimposed on the framing photograph. We can observe the slight acceleration of the
shock front during the onset of detonation when the appearance of the detonation
bubbles is observed.

An instantaneous planar energy source for the generation of an ideal planar blast
wave is difficult to realize experimentally. Thus, there are few experimental results
on planar blast initiation. Direct initiation of planar detonations can be achieved
using a planar shock or a detonation wave. However, the decay rate of the initiating

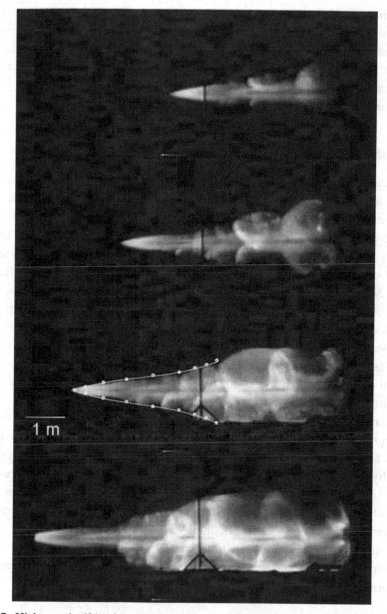

Figure 9.17.   High-speed self-luminous photographs of critical initiation of a cylindrical detonation; ethylene–air at $\phi = 1.0$, $E = 67$ kJ/m (Radulescu *et al.*, 2003).

planar blast wave is peculiar to the technique used in generating the planar shock wave and can influence the critical energy.

From the experimental observations of direct initiation of spherical and cylindrical detonations by point and line sources, respectively, we note that the phenomenon is almost identical. A strong blast decays to an overdriven detonation and then asymptotes to a CJ detonation for the supercritical energy case, or it continues to decay to an acoustic wave for the subcritical case. For the critical case, there exists a metastable quasi-steady sub-CJ period of propagation of the shock wave. This is followed by the formation of local explosion centers in the reaction zone from which detonation bubbles are formed. These detonation bubbles eventually engulf the blast wave front to form an asymmetrical cellular detonation. The formation of the detonation in the critical regime is similar to the onset of detonation process in the transition from deflagration to detonation.

9.3. NUMERICAL SIMULATION OF BLAST INITIATION

Numerical simulations of blast initiation can be readily carried out, and much more detailed information on the transient flow structure behind the blast can be obtained than from experimental studies. There exist numerous numerical schemes for the accurate integration of the reactive Euler equations. Chemical reaction models of different degrees of sophistication (from a single-step Arrhenius rate law to the full detailed chemistry of the reactions) can be incorporated into the numerical integration. Most of the early numerical studies used simple single-step Arrhenius rate models. Although a general qualitative description can be obtained with a single-step model, there are serious drawbacks. For example, it is not possible to define a unique value of the critical energy for direct initiation, because the single-step Arrhenius law gives a finite reaction rate at a finite temperature. Thus, without losses, chemical reactions will always go to completion, and initiation can be achieved with any arbitrary finite-strength shock wave if one waits long enough.

Although it is possible to use a detailed chemistry model involving the simultaneous integration of all kinetic rate equations for the elementary reactions, it is seldom justified to do so, for there are still unresolved difficulties in the numerical modeling of the physical processes at the onset of detonation. We shall choose a three-reaction model (first described by Short & Quirk, 1997, and later used by Ng & Lee, 2003, in their study of the blast initiation phenomenon) to provide a cutoff temperature and thus permit a critical energy to be defined.

The three-reaction chain-branching model of Short and Quirk consists of two temperature-dependent radical-producing reactions and one temperature-independent chain termination reaction:

1. Chain initiation: $\mathrm{F} \rightarrow \mathrm{Y}$, $k_\mathrm{I} = \exp\left[E_\mathrm{I} \left(\dfrac{1}{T_\mathrm{I}} - \dfrac{1}{T} \right) \right]$

2. Chain branching: $\mathrm{F} + \mathrm{Y} \rightarrow 2\,\mathrm{Y}$, $k_\mathrm{B} = \exp\left[E_\mathrm{B} \left(\dfrac{1}{T_\mathrm{B}} - \dfrac{1}{T} \right) \right]$

3. Chain termination: $\mathrm{Y} \rightarrow \mathrm{P}$, $k_\mathrm{c} = 1$

where F, Y, and P denote fuel, radical, and product, and E_I and E_B are the activation energies of the initiation and the branching reactions. The characteristic temperatures T_I and T_B are referred to as the crossover temperatures for the chain-initiation and -branching reactions when the initiation or branching rate becomes of the same

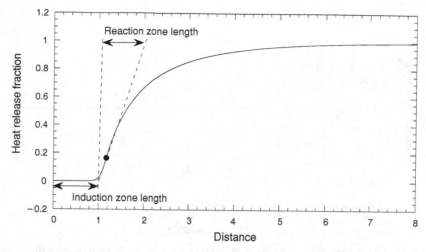

Figure 9.18. Heat-release profile using a three-step kinetic model (Ng & Lee, 2003).

order as the chain-termination rate. To represent typical chain-branching reactions, the choice of the activation energies and characteristic temperatures should be such that

$$E_I \gg E_B, \qquad T_I > T_{shock}, \quad \text{and} \quad T_B < T_{shock},$$

where T_{shock} is the von Neumann temperature of the detonation. The typical heat-release (or temperature) profile of a ZND detonation using this three-step reaction model is given in Fig. 9.18. There exists a definite thermally neutral induction zone and a reaction zone that match the heat-release profile obtained when detailed chemistry is used. For a single-step Arrhenius rate law, the heat-release profile is continuous with no distinct induction and reaction zones that can be identified. For a single-step Arrhenius rate law, the parameter that governs the stability of the detonation is the activation energy. For the three-reaction chain-branching model, stability studies of ZND detonations by Ng (2005) indicate that stability is controlled by the ratio of the induction zone to the reaction zone length:

$$\delta = \frac{\Delta_{induction}}{\Delta_{reaction}}.$$

Stable detonations can be realized for values of $\delta < 1$, and when $\delta > 1$, the ZND detonation becomes unstable.

To simulate numerically the blast initiation phenomenon, the classical similarity solution of Taylor (1950), Sedov (1946), and von Neumann (1941) for ideal planar, cylindrical, or spherical blast waves can be used to provide the starting conditions for the numerical integration. Choosing a value of $\delta = 0.604$ for the stability parameter (where the detonation is stable), the plot of shock pressure versus distance for planar blast initiation is shown in Fig. 9.19. For an initiation energy below a certain critical value (corresponding to the parameters chosen in Fig. 9.19) the blast decays continuously, with the reaction zone receding from the shock front as the

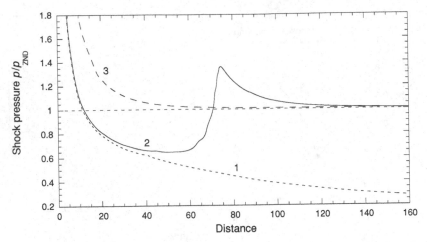

Figure 9.19. Shock pressure as a function of distance for a stable detonation with different initiation energies: $E_{s_1} = 350$, $E_{s_2} = 362$, $E_{s_3} = 746$ (Ng & Lee, 2003).

shock strength decreases until the blast becomes an acoustic wave, as in the case of a non-reacting blast (curve 1). For an initiation energy in excess of the critical value, the blast wave is found to decay asymptotically to a CJ detonation wave (curve 3), and the reaction zone is always coupled to the leading shock front in this case. For blast energies in a narrow range near the critical value, the blast first decays past the CJ value with the reaction zone progressively decoupling from the shock front. However, when the decaying blast reaches some critical value (a shock pressure of $p/p_{ZND} \approx 0.65$ for the particular values of the parameters used), the shock appears to stop decaying but continues to propagate at the critical shock strength. A quasi-steady metastable regime of propagation is obtained, and at the end of this quasi-steady regime, the shock is observed to accelerate rapidly to an overdriven detonation, which subsequently decays asymptotically to a CJ detonation (curve 2). It is of interest to note that, in the critical energy regime, the blast decays first to some sub-CJ state and remains at this metastable state for a short duration before accelerating rapidly to an overdriven detonation. This phenomenon is analogous to the onset of detonation phase in the transition from deflagration to detonation.

The temperature profiles behind the blast wave for the three regimes of subcritical, supercritical, and critical energy are illustrated in Figs. 9.20, 9.21, and 9.22, respectively. In the subcritical energy regime shown in Fig. 9.20, we note that the reaction zone progressively recedes from the leading shock and decouples from it as the blast decays. In the supercritical case shown in Fig. 9.21, the reaction zone initially recedes from the leading shock as the blast decays. Then, the reaction zone stops receding from the shock and remains at some fixed induction distance as the wave approaches the CJ state. For the critical energy regime shown in Fig. 9.22, the reaction zone progressively recedes from the shock as it decays past the CJ state to the metastable state. At the end of the quasi-steady regime, the reaction zone is seen to accelerate toward the shock again to form an overdriven detonation. Then,

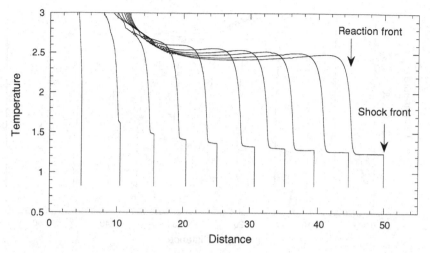

Figure 9.20. Temperature profiles at different times for subcritical planar detonation initiation (Ng & Lee, 2003).

as the overdriven detonation decays, the reaction zone again recedes slightly from the shock front to take its final position corresponding to the CJ detonation state of the mixture.

The pressure profiles for the critical energy regime are shown in Fig. 9.23. The initial pressure distribution is indicated in the first frame of Fig. 9.23. It is of interest to note that, in the reacceleration phase after the quasi-steady period, it is the pressure buildup in the reaction zone that changes the pressure gradient behind the shock from zero (when the shock is propagating at constant velocity during the quasi-steady period) to a negative gradient (i.e., pressure increases away from the shock) that results in the acceleration of the shock to an overdriven detonation. The

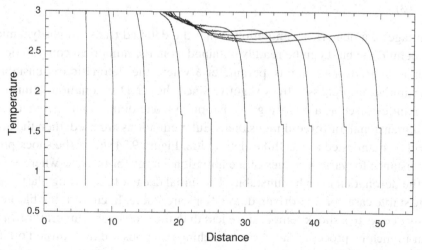

Figure 9.21. Temperature profiles at different times for supercritical planar detonation initiation (Ng & Lee, 2003).

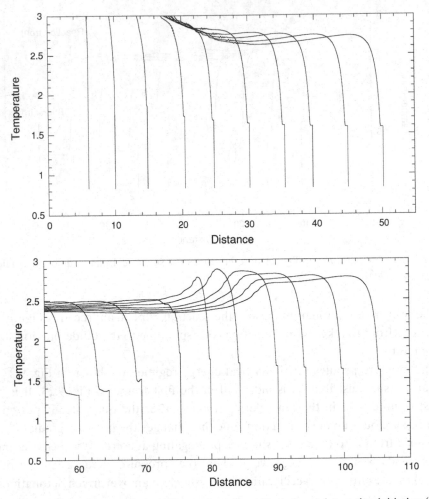

Figure 9.22. Temperature profiles at different times for critical planar detonation initiation (Ng & Lee, 2003).

advantage of a numerical simulation is that the detailed transient gasdynamic flow field behind the blast can be readily obtained from the numerical computation.

For a value of the stability parameter δ where the detonation is unstable, the direct initiation process differs slightly. Since the ZND detonation is unstable, it will manifest itself as a pulsating detonation in a one-dimensional geometry. If the two-dimensional (or three-dimensional) Euler equations are used, then the instability will be manifested as a cellular detonation. Figure 9.24 shows the shock pressure with distance for various values of the initiation energy for the case where $\delta = 1.429$ and the detonation is highly unstable. The initial decay of the strong blast is similar to the stable case, for overdriven detonations are stable. In curve 1, the blast continuously decays to a shock temperature less than the crossover temperature T_B of the chain-branching process. The chain-branching reactions are then turned off, and as the chain termination reaction takes over, the blast decays further to a sonic wave. For a higher value of the initiation energy (curve 2), the blast decays first to some

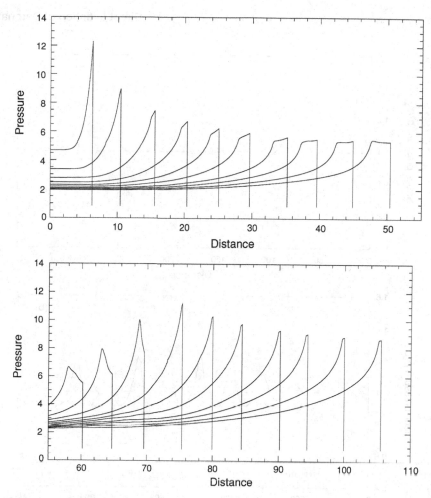

Figure 9.23. Pressure profiles at different times for critical planar detonation initiation (Ng & Lee, 2003).

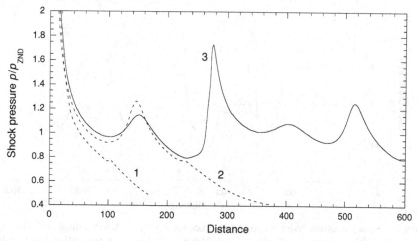

Figure 9.24. Shock pressure as a function of distance for a highly unstable detonation with different initiation energies: $E_{s_1} = 1195$, $E_{s_2} = 1371$, $E_{s_3} = 1445$ (Ng & Lee, 2003).

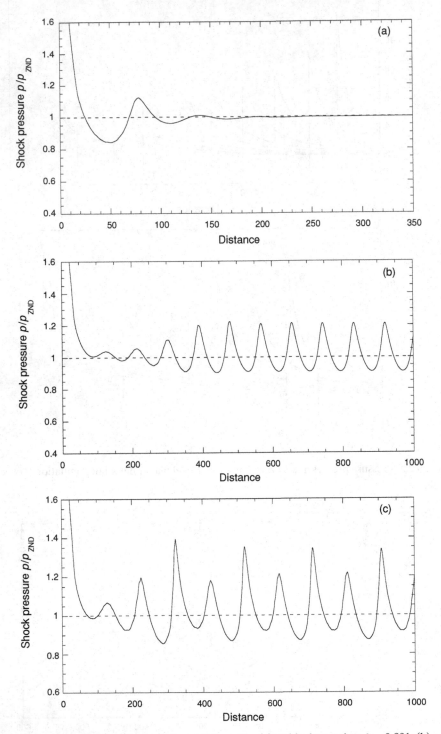

Figure 9.25. Shock pressure as a function of distance for (a) stable detonation, $\delta = 0.891$; (b) unstable detonation (single-mode oscillation), $\delta = 1.240$; and (c) unstable detonation (period-doubling bifurcation), $\delta = 1.328$ (Ng & Lee, 2003).

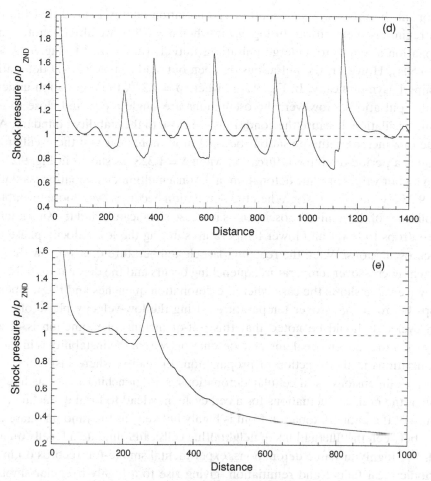

Figure 9.25. (*continued*). Shock pressure as a function of distance for (d) unstable detonation (multimode oscillation), $\delta = 1.429$, and (e) quenching (detonation limit), $\delta = 1.468$ (Ng & Lee, 2003).

strength slightly less than the CJ state. The blast reaccelerates and tries to start the pulsating cycle of oscillations due to instabilities. However, the temperature of a reacting particle in the low-velocity phase of the first pulsating cycle drops below the crossover chain-branching temperature, and the chain-termination reaction takes over, causing the detonation to fail. For higher initiation energy (curve 3), the front survives the low-velocity phase of the first pulsation cycle because, with a higher initiation energy, the expansion gradient behind the blast wave is less steep and the reacting particle temperature does not drop below the crossover chain-branching temperature. Thus, the detonation can reaccelerate and continues the pulsations. For unstable detonations, the final asymptotic state is a pulsating detonation instead of a steady ZND detonation, as shown in Fig. 9.19 for a lower value of $\delta = 0.604$.

It is of interest to look at the asymptotic behavior of the detonation after initiation for various values of the stability parameter from stable to highly unstable detonations. Figure 9.25 shows a sequence of the shock-pressure–distance plots for a range of the stability parameter $0.891 \leq \delta < 1.468$.

For all these cases, a sufficiently large initiation energy is used that the initiation regime is supercritical. In Fig. 9.25a when $\delta = 0.891$, we already note that the detonation attempts to undergo pulsations initially (in contrast to Fig. 9.19, where $\delta = 0.604$). However, the pulsations dampen out, and a steady ZND detonation is obtained asymptotically. In Fig. 9.25b, where $\delta = 1.240$, the detonation undergoes regular pulsations. However, the oscillations are single mode and close to a harmonic oscillation, because the conditions are close to the stability boundary. As the value of δ increases, the pulsations become more nonlinear and the oscillation undergoes a period-doubling bifurcation when $\delta = 1.328$, as shown in Fig. 9.25c. For even higher values of δ, the detonation takes on a multimode oscillation, as shown in Fig. 9.25d, where $\delta = 1.429$. When the detonation becomes even more unstable, the amplitudes of the nonlinear oscillations increase significantly and the shock temperature drops to lower and lower temperatures during the low-velocity phase of the shock oscillations. When the reacting particle temperature drops below the chain-branching crossover temperature, quenching occurs and the detonation fails.

Figure 9.25e shows the case when the detonation quenches and the temperature drops below the crossover temperature during the low-velocity phase of a pulsation cycle. It should be noted that this self-quenching phenomenon is observed only for a one-dimensional unstable detonation, where the instability is in the form of pulsations in the direction of propagation. In reality, where instabilities occur in three dimensions as in cellular detonations, self-quenching is not as likely. Although the cyclical fluctuations for a cell cycle may lead to local quenching, global failure of the entire detonation front is highly unlikely, as the random phase difference between the fluctuations of neighboring cells can reinitiate a locally quenched cell. For highly unstable detonations, experimental smoke-foil records do indicate periodic local failure and reinitiation, giving rise to a highly irregular smoke-foil pattern.

Thus far, our discussions have been concentrated on the planar geometry. For cylindrical and spherical geometries, the detonation has curvature. Thus, there is adiabatic expansion of the reacting gas behind the curved shock in addition to the expansion associated with the decaying blast wave. For a stable value of $\delta = 0.604$, the initiation regimes for the cylindrical and spherical geometries are illustrated in Fig. 9.26. Comparing them with the corresponding planar geometry shown in Fig. 9.19, we note that, for all the three initiation regimes, the blast always decays to a sub-CJ state. Even for values of δ well below the stability limit, the critical regimes for both cylindrical and spherical cases indicate a periodic damped oscillation as the detonation asymptotically approaches its final steady state. This is perhaps a consequence of the effect of curvature, which introduces additional lateral expansion of the particles behind the shock. The additional expansion due to curvature results in a lowering of the temperature and thus an increase in the induction zone length and an increase in the effective value of the stability parameter δ, making the detonation more unstable. However, the curvature decreases with a larger radius as the detonation expands, and the wave becomes more and more stable.

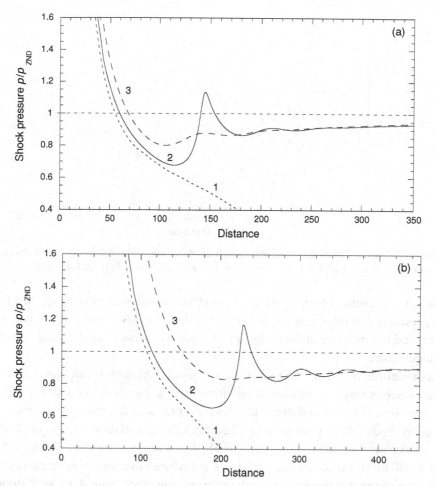

Figure 9.26. Three regimes of direct initiation for (a) cylindrical and (b) spherical geometries (Ng & Lee, 2003).

It is also of interest to note that the approach to an asymptotic state occurs extremely slowly. For the cylindrical and spherical geometries, the detonation formed is below the normal CJ state of the mixture, and it persists at the sub-CJ state for a long time. A possible explanation for this is that, with curvature, the eigenvalue detonation velocity is below the normal CJ value. However, the presence of a sonic surface effectively isolates the front from the flow field in the rear. Thus, any adjustment must rely on the local conditions between the shock and the reaction zone.

For a slightly higher value $\delta = 0.891$ where a stable ZND detonation is obtained in the case of the planar geometry, we note from Fig. 9.27 that, in the cylindrical case, a pulsating detonation is now obtained instead. However, the pulsations are almost harmonic, and the damping is very slow as the wave expands outward after initiation. Thus, for diverging detonations where curvature provides an additional lateral expansion behind the shock, the induction zone length and the stability parameter δ are effectively increased. Although curvature effects diminish with increasing

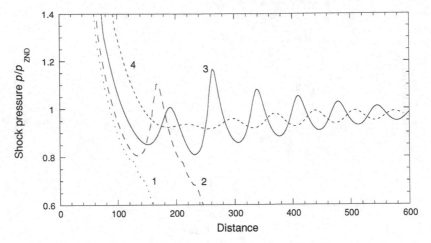

Figure 9.27. Direct initiation of unstable cylindrical detonation with different nondimensional initiation energies: $E_{s_1} = 2.74 \times 10^5$, $E_{s_2} = 3.01 \times 10^5$, $E_{s_3} = 3.74 \times 10^5$ (Ng & Lee, 2003).

radius, the presence of a sonic surface effectively isolates the nonsteady flow field in the products from the front. This leads to a slower adjustment of conditions at the front, and the memory of the early-time initiation process at small radius persists for a longer time.

As mentioned earlier, the use of the single-step Arrhenius rate law fails to provide a distinct value of the critical initiation energy. Figure 9.28 shows the initiation process for a range of initiation energies for a planar detonation using a single-step reaction model. As can be observed, a decrease in the initiation energy simply delays the reacceleration process (curves 1 to 6). When one waits longer (for case 7), the shock will always reaccelerate back to an overdriven detonation as in curves 1 to 6. This is in contrast to the three-step chain-branching reaction model (Fig. 9.19) where

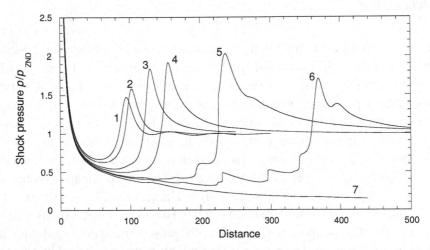

Figure 9.28. Shock pressure as a function of distance using a one-step kinetic rate law with different nondimensional initiation energies: $E_{s_1} = 3243$, $E_{s_2} = 3285$, $E_{s_3} = 3302$, $E_{s_4} = 3361$, $E_{s_5} = 3420$, $E_{s_6} = 3601$, $E_{s_7} = 3724$ (Ng & Lee, 2003).

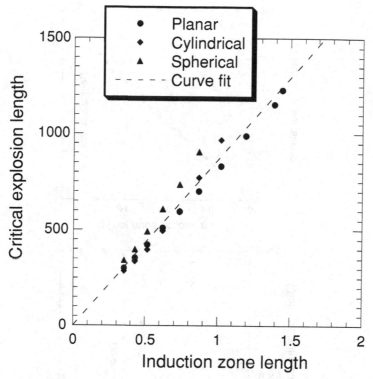

Figure 9.29. Explosion-length invariance for blast initiation in different geometries (Ng & Lee, 2003).

a sharp cutoff is observed for an initiation energy that permits the temperature to drop below the crossover chain-branching temperature. Thus, a more realistic chemical reaction model must be used in the numerical simulation of blast initiation.

The results of the numerical simulation for different geometries can also be used to demonstrate the concept of explosion-length invariance for blast initiation. For different geometries, the units of the blast energy are different. For example, the blast energy has units of energy, energy per unit length, and energy per unit area for spherical, cylindrical, and planar geometries, respectively. Thus, it is not possible to directly compare the critical energy for direct initiation between the different geometries. Noting that the decay of a strong blast wave scales according to the explosion length from the classical similarity solution of Taylor, Sedov, and von Neumann, Lee (1977) proposed that the critical explosion length for direct initiation should be the same for the different geometries:

$$R_0^* = \left(\frac{E_{\text{spherical}}^*}{p_0}\right)^{1/3} = \left(\frac{E_{\text{cylindrical}}^*}{p_0}\right)^{1/2} = \left(\frac{E_{\text{planar}}^*}{p_0}\right),$$

where E^* denotes the critical energy and has the appropriate units for the different geometries. From the numerical simulation by Ng and Lee (2003), the critical explosion length for the different geometries is plotted in Fig. 9.29 against the

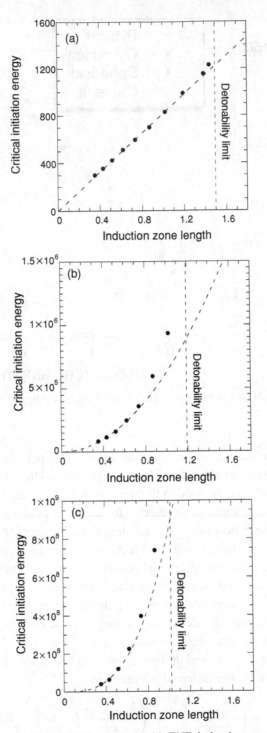

Figure 9.30. Variation of critical initiation energy with ZND induction zone length for (a) planar ($j = 0$), (b) cylindrical ($j = 1$), and (c) spherical ($j = 2$) geometries (Ng & Lee, 2003).

induction-zone length (representing the stability of the detonation) for different explosive mixtures. For the same mixture, we note that the critical explosion lengths for the different geometries are similar. The agreement is better for a stable detonation where the induction zone length is small (hence, the value of the stability parameter δ is small). For unstable detonations (i.e., large values of the induction zone length), instability plays a more significant role in the onset of detonation, and direct initiation may not be fully controlled by the decay of the blast generated by the energy source. Note that the explosion-length scaling applies only to the decay of the blast. Therefore, when other factors enter into the initiation process, the explosion-length scaling concept may not be applicable.

The results of the numerical simulation can also be used to verify the Zeldovich criterion of blast initiation. According to Zeldovich *et al.* (1957), the critical energy should depend on the induction zone length of the mixture, that is

$$E_c \approx \Delta_{\text{induction}}^{j+1},$$

where $j = 0, 1, 2$ for planar, cylindrical, and spherical geometry, respectively. Zeldovich's criterion is based on the argument that the blast energy must be such that, when the blast has decayed to the CJ state, the blast radius must be at least of the order of the induction-zone length. Figure 9.30 illustrates the numerical results of Ng for planar, cylindrical, and spherical geometries. Again, we can see that for small values of the induction-zone length, where δ is small and the detonation is stable, the agreement with Zeldovich's criterion is quite good. However, for unstable detonations where instability plays a more prominent role in controlling the initiation process, the numerical results deviate from the Zeldovich criterion $E_c \approx \Delta_{\text{induction}}^{j+1}$.

9.4. THE CRITICAL TUBE DIAMETER

Lafitte (1925) was perhaps the first to attempt to initiate a spherical detonation in a $CS_2 + 3O_2$ mixture using a planar detonation emerging from a 7-mm-diameter tube into the center of the spherical vessel. He failed to obtain direct initiation. In a later study, Zeldovich *et al.* (1957) used tubes of different diameters and found that, for a given mixture, there exists a critical value of the tube diameter for successful direct initiation. For tube diameters greater than the critical tube diameter, the planar detonation (upon emerging from the tube into an unconfined space) evolves successfully into a spherical detonation without failure. Below the critical tube diameter, the detonation exiting from the tube fails to transform into a spherical detonation.

Figure 9.31 shows frames from a schlieren movie of a planar detonation emerging from a tube into unconfined space. In this case, the tube diameter is below the critical diameter for the mixture. We can observe that the detonation fails, and a decoupled spherical deflagration is obtained. In Fig. 9.32, the tube diameter is greater than the critical value for the mixture, and direct initiation of a spherical detonation is observed. Note that the turbulent cellular structure of the spherical detonation is observed on the frontal half of the spherical detonation, whereas a smooth decoupled

μs

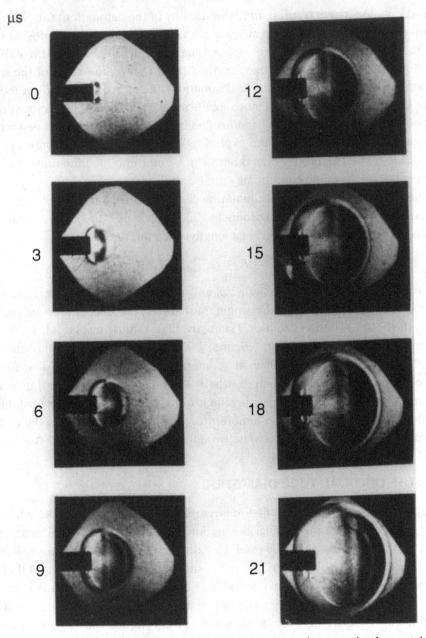

Figure 9.31. Sequence of schlieren photographs of a planar detonation emerging from a tube into unconfined space; subcritical case (courtesy of R.I. Soloukhin).

shock and reaction front is observed in the rear half of the diffracted detonation. This is due to the steep expansion waves that are generated at the corner of the tube exit that quench the detonation near the edge of the tube as it emerges into unconfined space. The expansion fan is less steep toward the tube axis, and the detonation may survive the quenching and reinitiate itself. Eventually, the reinitiated cellular detonation on the front half of the spherical wave near the tube axis will propagate

μs

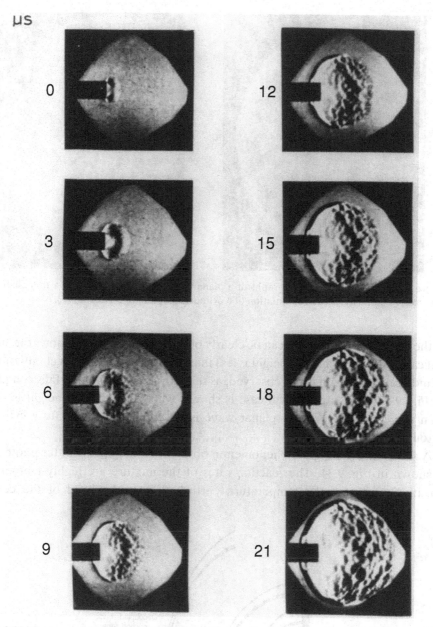

Figure 9.32. Sequence of schlieren photographs of a planar detonation emerging from a tube into unconfined space; supercritical case (courtesy of R.I. Soloukhin).

backward and reinitiate the quenched detonation. Eventually the entire diffracted spherical front becomes a self-sustained cellular detonation.

The same critical tube phenomenon also occurs in two dimensions, where a planar detonation in a channel emerges into a sudden opening and becomes a cylindrical detonation. Using a thin channel, open-shutter photographs can be taken. From the cellular structure, the transverse wave trajectories, the failure, and the reinitiation

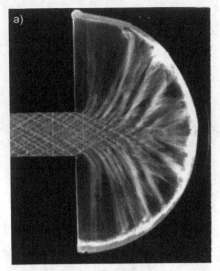 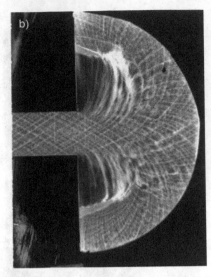

Figure 9.33. Open-shutter photographs of a planar detonation emerging into a thin channel: (a) unsuccessful and (b) successful initiation of a cylindrical detonation (Lee, 1995).

of the diffracted detonation can be clearly observed. Figure 9.33a shows the subcritical case where the channel height is less than the critical value. Direct initiation of a cylindrical detonation is not observed, as the diffracted detonation fails completely. In Fig. 9.33b, the supercritical case is shown, where the channel height is greater than the critical value. The planar wave is observed to evolve into a cylindrical detonation.

A sketch illustrating the phenomenon observed in the open-shutter photographs is shown in Fig. 9.34. The reaction rates of the mixture are highly temperature-sensitive such that the temperature perturbation at the head of the centered

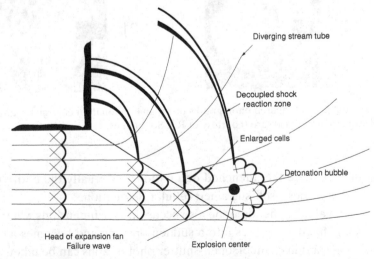

Figure 9.34. Sketch illustrating failure and reinitiation in unstable mixtures (Lee, 1995).

rarefaction fan that propagates toward the axis of the channel (as the detonation emerges from the channel exit) is sufficient to cause failure of the detonation, with the reaction zone decoupling from the shock front. Thus, the trajectory of the head of the rarefaction fan is essentially the trajectory of a *failure wave* behind which the reaction zone and the shock front are decoupled. In front of the failure wave, the detonation front is unperturbed, for information about the expansion propagates with the head of the rarefaction fan. As the rarefaction fan propagates toward the axis, the expansion gradient diminishes. Thus, if we consider a stream tube near the channel wall, the stream tube area will undergo a drastic increase due to the steepness of the rarefaction fan. However, closer to the axis of the channel, the gradient in the rarefaction fan decreases and the stream tube divergence is less. If we consider a stream tube as a detonation channel, then the area divergence will result in an attenuation of the detonation and a corresponding increase in the detonation cell size. We may evoke the Shchelkin criterion, and explain quantitatively that failure will result when the detonation cell size has increased to about twice its normal value. Thus, for the stream tubes near the corner, where the area increase is abrupt, the detonation fails. As the rate of area divergence decreases toward the tube axis, the area divergence is insufficient to cause failure of the detonation and only results in an increase in the cell size. If the rate of increase in the cell size is sufficiently small, instabilities grow, new cells can be generated within the enlarged cells, and reinitiation of the detonation occurs. This process is not unlike the propagation of a diverging self-sustained spherical detonation where the growth rate of new cells must be sufficiently fast to maintain a constant average number of cells per unit area as the surface of the detonation increases with the radius. In Fig. 9.33b, we can observe a few reinitiation sites near the head of the expansion fan as it propagates toward the channel axis. The reinitiation zone is indicated by the appearance of a finer transverse wave pattern, indicating that the detonation bubble is initially overdriven when formed.

A schlieren photograph of the detonation wave for the supercritical case in cylindrical geometry is shown in Fig. 9.35. Away from the axis, the failed portion of the diffracted detonation, where the shock and reaction zones are completely decoupled, can be seen. The reinitiated cylindrical detonation can be seen further downstream, closer to the channel axis.

When the tube diameter is near the critical value, the entire detonation front may fail when the head of the rarefaction fan has penetrated to the axis of the tube (or of the channel). However, local explosion centers can form at various locations of the decoupled shock and reaction front near the axis if the shock strength is still sufficiently high. Overdriven detonation bubbles then grow from the explosion centers and sweep across the decoupled shock–reaction-front surface to reinitiate the diffracted front and to form an asymmetrical spherical (or cylindrical) detonation. The sequence of shadowgraph photographs in Fig. 9.36 shows the growth of detonation bubbles as they engulf the decoupled shock–reaction-front surface to reinitiate the detonation near the critical regime.

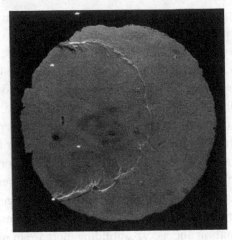

Figure 9.35. Schlieren photograph of supercritical transmission in cylindrical geometry (Lee, 1995).

Thus, we see that the direct initiation of an unconfined detonation by a planar detonation emerging from a tube near the critical condition is very similar to blast initiation. There exist analogous subcritical, critical, and supercritical regimes. The flow field depends on the initial conditions. For the blast initiation problem, the initial flow field is that of a decaying strong blast wave, whereas for the

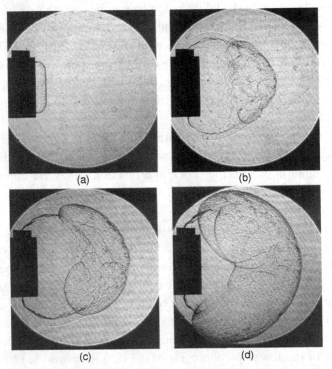

Figure 9.36. Sequence of schlieren photographs illustrating the critical transmission case in spherical geometry (Schultz & Shepherd, 2000).

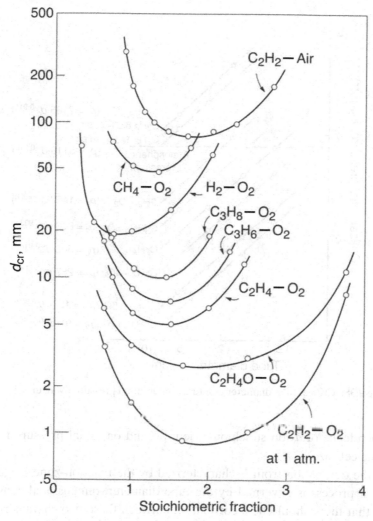

Figure 9.37. Critical tube diameter as a function of stoichiometry (Matsui & Lee, 1979).

critical-tube-diameter problem, the initial flow field is that of a diffracted planar detonation wave that emerges from a tube into unconfined space. Of particular importance to note is that, for the critical case, explosion centers form overdriven detonation bubbles that eventually engulf the entire shock surface and appear to be identical to both the critical-tube-diameter and the blast initiation phenomena.

The critical tube diameter for various fuel–oxygen mixtures as a function of stoichiometric ratio and initial pressure has been obtained by Matsui and Lee (1979). Figure 9.37 shows the dependence of the critical tube diameter on the stoichiometric fraction. Note that the minimum value of d_c occurs at about the stoichiometric composition. Figure 9.38 shows the dependence of the critical tube diameter on the initial pressure of the mixture. For pressures of the order of 1 atm and below, the critical tube diameter varies approximately inversely with the initial pressure.

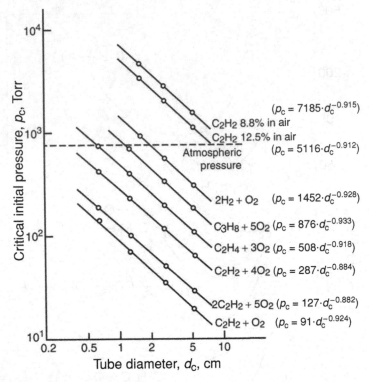

Figure 9.38. Critical tube diameter as a function of initial pressure (Matsui & Lee, 1979).

The dependence of d_c on stoichiometric ratio and on initial pressure is similar to that of the cell size.

Since the detonation front is characterized by the reaction-zone length, and the diffraction process is governed by the tube diameter, dimensional considerations indicate that there should be a correlation between these two characteristic length scales. For an unstable detonation, the characteristic length scale for the reaction zone is not well established. The cell size or the transverse wave spacing is often used as a characteristic length scale for unstable detonations. However, there is a spectrum of cell sizes in an unstable detonation, and, in general, we choose an average (or most probable) value as a representative value from smoked foils. It is not clear that a single average value can characterize the reinitiation process because the onset of detonation at critical conditions depends on the growth of instabilities. Nevertheless, an empirical correlation between the critical tube diameter and the cell size was first noted by Mitrofanov and Soloukhin (1965) for C_2H_2–O_2 mixtures (viz., $d_c \approx 13\lambda$). Later, Edwards et al. (1979) confirmed this observation for the same mixture and suggested that this empirical correlation should be applicable to other mixtures as well. A systematic study was then carried out by Knystautas et al. (1982) and confirmed that this correlation is indeed valid for a range of fuel–oxygen and fuel–air mixtures. Figure 9.39 shows the comparison between experimental values of the critical tube diameter and the 13λ correlation where the

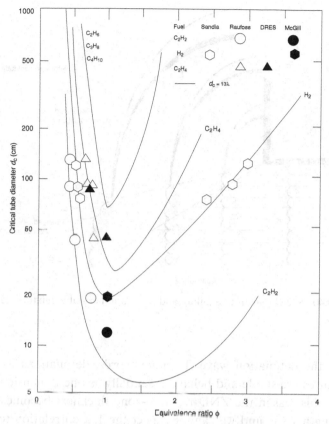

Figure 9.39. Comparison of experimental critical tube diameters and the 13λ correlation (Knystautas *et al.*, 1982).

cell size λ is determined from independent experiments by various investigators. In spite of the fact that no attempt was made to determine a precise limit between successful and unsuccessful initiation, the agreement with the 13λ correlation is quite adequate for the fuel–air mixtures investigated. It should also be noted that the experimental measurement of the detonation cell size is not very precise, particularly when the cell pattern is highly irregular for very unstable mixtures.

However, not all mixtures behave according to the 13λ correlation. It was found experimentally that, for mixtures highly diluted with argon, the $d_c \approx 13\lambda$ correlation does not apply. Instead, it was found that $d_c \approx 30\lambda$ or more for argon-diluted mixtures. In previous chapters, we have already discussed that for C_2H_2–O_2 mixtures with very high argon dilution (80% argon or greater), the detonation is stable. Even though a cellular pattern is still inscribed on a smoke foil, the transverse waves are weak and the cellular pattern is highly regular, indicating that the transverse waves are weak acoustic waves that play minor roles in the propagation of the detonation wave. Although two opposing sets of interacting weak transverse waves are observed, they are essentially analogous to Mach waves or characteristics in supersonic flow and do not play important roles in the processes within the

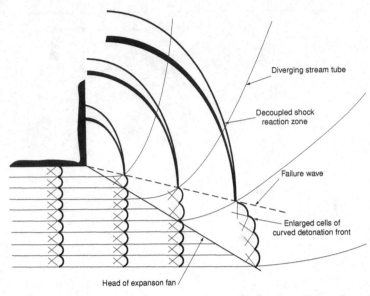

Figure 9.40. Sketch illustrating failure and reinitiation in stable mixtures (Lee, 1995).

structure of the detonation wave. In other words, detonations in highly argon-diluted mixtures are stable and behave essentially as one-dimensional ZND detonations. For this reason the ZND reaction-zone thickness becomes the more appropriate length scale, and we cannot expect the 13λ correlation to still apply. If the propagation mechanism of stable detonations in highly argon-diluted mixtures is different from cellular detonations in unstable mixtures, then the failure and the reinitiation mechanisms of a diffracting stable detonation emerging from a tube into unconfined space should also be different.

For stable detonations in mixtures highly diluted with argon, the reaction rates are less temperature sensitive and thus the small perturbations generated at the head of the rarefaction fan do not cause abrupt failure of the wave, resulting in complete decoupling of the reaction front from the leading shock. However, the expansion causes a decrease in the propagation velocity, resulting in a curved detonation front behind the head of the rarefaction fan. There exists a gradual increase in the front curvature from the head of the fan toward the rear. At some critical value when the curvature (due to stream tube divergence) becomes excessive, the detonation finally fails. For stable detonations, the failure wave no longer corresponds to the head of the rarefaction fan (as it does for unstable detonations). Between the head of the expansion fan and the failure wave, we see a curved attenuated detonation with a larger transverse wave spacing than a normal CJ detonation. This is illustrated in the sketch shown in Fig. 9.40. When the head of the rarefaction fan reaches the tube axis, we have an attenuated curved detonation front that extends from the axis to the failure wave. Global failure (in the subcritical case) occurs if the curvature of

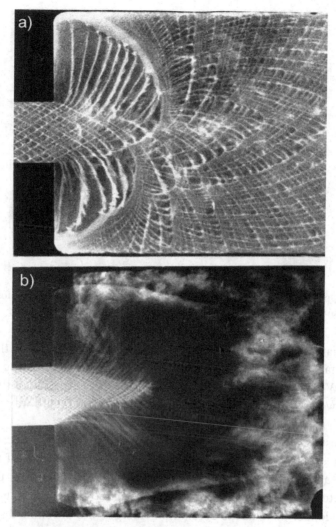

Figure 9.41. Open-shutter photograph of critical detonation transmission for the cylindrical geometry in (a) unstable and (b) stable mixtures (courtesy of A.A. Vasiliev).

the attenuated detonation wave exceeds a certain critical value when the head of the fan reaches the tube axis.

Figure 9.41 shows open-shutter photographs of the critical-tube-diameter phenomenon for the two cases of stable and unstable detonations. In Fig. 9.41a for an unstable detonation in an undiluted mixture of C_2H_2–O_2, we note that the failure wave propagates toward the axis of the channel, corresponding to the head of the rarefaction fan. Reinitiation is due to the formation of explosion centers in the enlarged cells of the attenuated detonation near the axis. Fine cell patterns corresponding to overdriven detonations can be identified as the detonation bubbles grow from local explosion centers to reinitiate the entire detonation front. In Fig. 9.41b, the detonation is stable in a mixture of C_2H_2–O_2 highly diluted with argon. For a

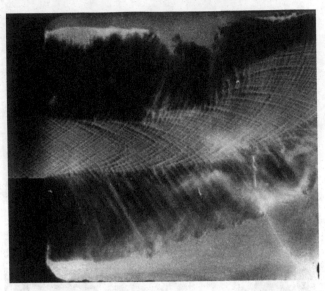

Figure 9.42. Open-shutter photograph of supercritical detonation transmission for cylindrical geometry in a stable mixture (courtesy of A.A. Vasiliev).

stable detonation, the failure wave trajectory does not correspond to the head of the rarefaction fan (which is just the transverse wave trajectory from the corner to the tube axis). We can also see the enlarged transverse wave spacing of the curved detonation between the head of the rarefaction fan and the failure wave as illustrated in the sketch of Fig. 9.40.

It is interesting to see that for stable detonations, the supercritical case is quite unlike that for unstable detonations, where detonation bubbles engulf the diffracted shock front and reinitiate the entire detonation. In the stable case, the curved attenuated detonation continues to propagate along the axis even though the channel walls are no longer there as the detonation emerges into unconfined space. Reinitiation is not due to overdriven detonation bubbles that grow to engulf the diffracted shock front. The curved detonation simply propagates along the axis as a curved detonation. Owing to the instability of this *detonation jet,* it tends to flip to either side of the axis as it propagates, and it cannot maintain its symmetry. This supercritical case for stable detonation is demonstrated in the open-shutter photograph in Fig. 9.42.

The existence of two failure mechanisms for unstable and stable detonations was first pointed out by Lee (1995). The difference between stable and unstable detonation is also manifested in the correlation between the critical diameter and the cell size in three and in two dimensions. For unstable detonations, the critical tube diameter for a three-dimensional spherical geometry follows the $d_c \approx 13\lambda$ correlation. If curvature were the failure mechanism, then for a two-dimensional cylindrical detonation formed by a planar detonation emerging from a slot of large aspect ratio, we would expect the correlation to differ by a factor of two so that $w_c \approx 6\lambda$,

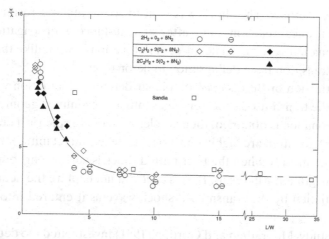

Figure 9.43. Critical channel height as a function of aspect ratio (Benedick *et al.*, 1985).

where w_c is the critical height of the slot. Experiments carried out by Benedick *et al.* (1985) for slots of different aspect ratios indicate that, as the aspect ratio approaches infinity, the critical slot height $w_c \to 3\lambda$ rather than 6λ. Figure 9.43 illustrates the results for the critical channel height as a function of the aspect ratio. For an aspect ratio of $L/w \approx 1$ (i.e., a square geometry where the phenomenon is essentially three-dimensional), the correlation of the critical height with λ is of the order of 13λ. When the aspect ratio increases, the critical channel height decreases and asymptotes to a value of the order of 3 rather than 6 as $L/w \to \infty$. This result suggests that for unstable detonations, failure and reinitiation are governed by the growth of instabilities rather than by curvature of the detonation front.

9.5. OTHER MEANS OF DIRECT INITIATION

A detonation is essentially a strong shock wave supported by chemical reactions, and early detonation researchers recognized that direct initiation can be achieved if a sufficiently strong shock wave is generated in the explosive mixture. A systematic study of the initiation of detonation waves by shock waves was carried out by Berets *et al.* (1950). They used a detonation driver section to generate a shock wave in an inert-gas buffer zone. The shock then propagated into the explosive test gas to effect ignition. In this manner, the free radical and product species from the detonation driver were isolated from the ignition process in the test gas, and only autoignition by shock compression was responsible for the initiation downstream. Berets *et al.* observed that, for a sufficiently strong shock wave, direct initiation occurred shortly after the shock entered the test gas. If the initiating shock was weak and the shocked temperature below the autoignition limit, the transmitted shock decayed similarly to its propagation in a non-reactive mixture. However, for a shock strong enough to cause autoignition but insufficient to effect direct initiation, the phenomenon became quite complex. The shock was seen to accelerate as a result of the energy released by chemical reactions, and, after some distance of propagation, the abrupt onset of detonation occurred. The onset of detonation is not unlike that which takes place in the transition from deflagration to detonation.

The investigation of Berets *et al.* on planar detonation initiation was carried out in a tube. If the transmitted shock emerges into an unconfined geometry and only a spherical detonation is obtained, then acceleration of the spherical deflagration and transition to detonation are highly improbable. Hence, direct initiation of a spherical detonation occurs only when the transmitted shock is very strong; otherwise, only a spherical deflagration will result. Berets *et al.* did not explore the detailed processes of direct initiation by the transmitted shock wave as it entered into the explosive mixture.

In a later study, Mooradian and Gordon (1951) investigated the detonation initiation process by a transmitted detonation rather than by a shock. No inert-gas buffer section was used, and the detonation from the initiator section transmitted directly into the explosive mixture in the test section. In the study of Berets *et al.*, the detonation in the driver section first generated a shock in the inert buffer gas when the high-pressure detonation products expanded into the buffer section. A normal shock was later transmitted into the test gas, and ignition and detonation development had to start downstream in the test gas, irrespective of the strength of the transmitted shock. On the other hand, if an established detonation was transmitted directly into the test gas, the detonation structure from the driver had only to readjust itself to suit the mixture downstream. In other words, the cell structure of the incident detonation in the initiator section has to evolve into the structure of the test gas. However, the instability in the structure is already present. Therefore, it is easier to initiate a detonation by a transmitted detonation than by a transmitted shock.

A good demonstration of the difference between a transmitted detonation and a transmitted shock was reported by Mooradian and Gordon. Figure 9.44a shows the detonation velocity in the test gas subsequent to the transmission of the detonation downstream into the test mixture. In this experiment, the diaphragm that initially separates the initiator section from the test section was quickly removed just before the detonation in the driver section was initiated. The mixture in the initiator section is a more sensitive $2 H_2 + O_2$ mixture, whereas the mixtures in the test section are much less sensitive, namely, 30% and 20% H_2–air mixtures. The detonation velocity in the driver section is about 2800 m/s, and, as can be observed, the transmitted detonation velocity drops almost instantaneously, settling at the CJ velocity of the test mixture. When the test mixture is near the detonation limits that correspond to the tube diameter being used, we note that the detonation initiated in the downstream section is unstable and undergoes slight oscillations.

To permit different initial pressures to be used in the driver and test section, a diaphragm has to be in place. Irrespective of how thin the diaphragm, its presence was found to have a drastic effect on the transmission process. Figure 9.44b

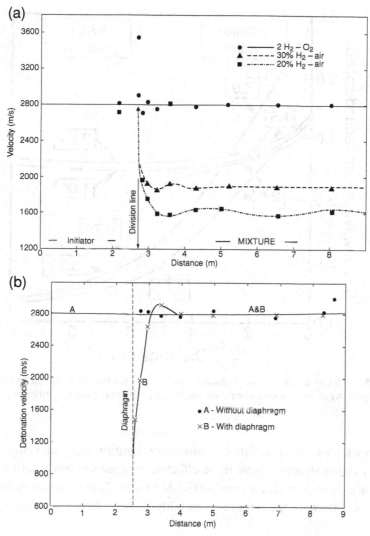

Figure 9.44. Velocity as a function of distance using 2 H$_2$–O$_2$ in both driver and test sections (Mooradian & Gordon, 1951).

shows the velocity variation with distance when the same gas (2 H$_2$–O$_2$) is used in both the driver and test sections. With the diaphragm quickly removed just before initiation of the detonation in the driver section, we note that the detonation simply transmits from one section to the other and no noticeable change of velocity is observed. However, with the diaphragm in place, the transmitted wave across the diaphragm is initially just a shock rather than a detonation. The transmitted shock velocity is slightly less than half the detonation velocity, and the wave rapidly reaccelerates to become an overdriven detonation (in about 5 or 6 diameters of propagation), which then decays back to a CJ detonation. The presence of the diaphragm essentially destroys the cellular structure of the incident detonation, and

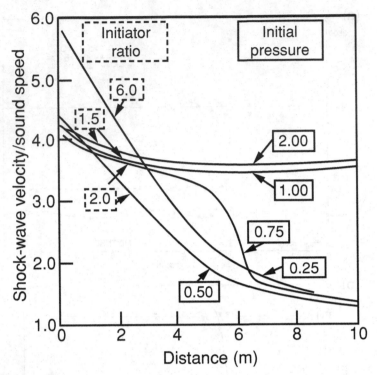

Figure 9.45. Velocity as a function of distance for a driver mixture of 2 H_2–O_2 and a test mixture of 2 H_2–O_2–26 Ar at various initiator ratios and initial pressures of the test mixture (Gordon *et al.*, 1958).

only a shock wave is transmitted downstream. Ignition and redevelopment of the instability in the structure have to be effected in order to form a detonation in the test mixture. It is clear that a transmitted detonation is more efficient than a transmitted shock, because the required instability in the detonation structure is already present.

Using a diaphragm such that the pressure in the initiator section can be different from that of the mixture in the downstream test section, Gordon *et al.* (1958) investigated the initiation phenomenon for different pressure ratios and combustible mixtures in the two sections for near-limit mixtures. Figure 9.45 shows typical velocity–distance plots for an initiator mixture of 2 H_2–O_2 and a test mixture of 2 H_2–O_2–26 Ar at various initiator pressure ratios and initial pressures of the test mixture. For the test mixture studied, the pressure limit in the 20-mm-diameter tube is about 1 atm (i.e., no stable detonation below an initial pressure of 1 atm). It was found that irrespective of the initiator pressure ratio, the detonation is observed to decay asymptotically with distance to a sonic wave. It is interesting to note that at an initial pressure of 0.75 atm, the detonation manages to propagate near the CJ velocity after initiation for almost 5 m before it abruptly decays to a weak shock. Thus, a metastable state can be maintained for quite a long distance when a sufficiently strong initiator is used to initiate an overdriven detonation wave. Above the

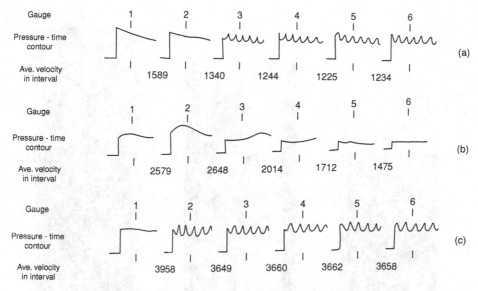

Figure 9.46. Pressure as a function of time for various initiator ratios and initial pressures of the test mixture (Mooradian & Gordon, 1951).

pressure limit (i.e., ≥ 1 atm), the initial overdriven detonation asymptotes to the CJ velocity of the mixture.

Using piezoelectric pressure gauges, the pressure–time profile at different locations along the test section was obtained by Mooradian and Gordon. Figure 9.46 shows the sequence of pressure–time profiles for various initiator ratios and test mixture pressures. In Fig. 9.46a, a high initiator ratio of 6:1 for $2\,H_2$–O_2 was used in the initiator section. The test gas $2\,H_2$–O_2–$17\,Ar$ was at an initial pressure of 0.28. The overdriven detonation decays to a spinning detonation in the test mixture. Note the absence of spin in the initially overdriven detonation. In Fig. 9.46b, a mixture of 7% O_2 in H_2 at an initial pressure of 2 atm was used in the test section. The initiator mixture was $2\,H_2$–O_2, and the initiator pressure ratio was 1.7:1. Direct initiation of detonation was not achieved, and we note the eventual decay to a sonic wave in the asymptotic limit. It is also interesting to note the pressure lump initially behind the incident shock wave when ignition occurs and chemical energy is released behind the shock. However, the reaction zone fails to couple with the shock wave and lags behind as the shock front propagates downstream. In Fig. 9.46c, the same test gas is used, but with a hydrogen-rich initiator mixture of $5\,H_2$–O_2. Even for a lower initiator pressure ratio of 1.5:1, direct initiation is successful, and the initial overdriven detonation decays to a steady spinning detonation downstream. It is well known in shock tube theory that a lighter driver gas results in a stronger driven shock wave in the test section. This is why the $5\,H_2$–O_2 mixture can successfully initiate a detonation, whereas the stoichiometric mixture of $2\,H_2$–O_2 fails to initiate the detonation even though the detonation pressure is higher in the stoichiometric mixture.

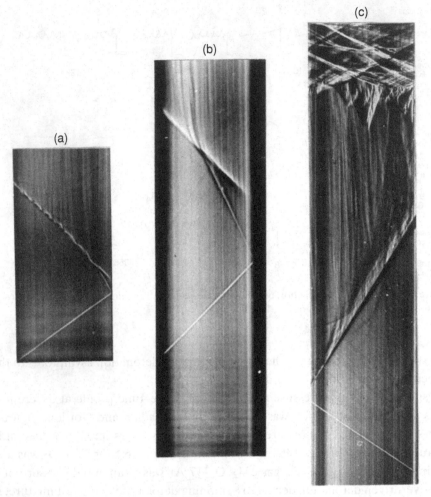

Figure 9.47. Streak schlieren photographs of direct initiation via shock waves (courtesy of R.I. Soloukhin).

In the preceding investigations of shock initiation, both Berets *et al.* and Mooradian and Gordon investigated direct initiation with incident shock waves. Direct initiation can also be achieved using reflected shock waves. Figure 9.47 shows three cases of direct initiation by shock waves reflected from the end wall of a shock tube. In Fig. 9.47a, a cellular detonation is observed to form shortly after the reflection of the incident shock from the end wall. In Fig. 9.47b, the initiation process is more resolved and one can observe the incident shock, the reflected shock, and then the detonation wave that forms after some induction delay time. The detonation catches up and merges with the reflected shock, and then continues to propagate as a detonation into the shock-heated mixture behind the incident shock. In certain unstable mixtures, ignition of the mixture behind the reflected shock after an induction period appears in discrete hot spots rather than uniformly at the end wall. The subsequent formation of the detonation behind the reflected shock is again similar

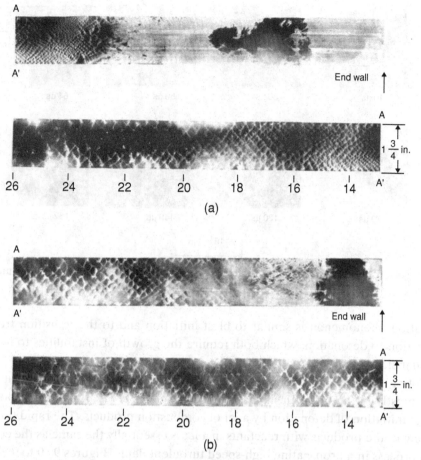

Figure 9.48. Smoked-foil records of detonation initiation due to shock reflection in a $2\,H_2-O_2-3\,N_2$ mixture at (a) $P_0 = 40$ Torr and (b) 30 Torr (courtesy of A.K. Oppenheim).

to the onset of detonation from local explosion centers under critical conditions observed in both blast initiation and the deflagration-to-detonation transition.

Perhaps the best illustration of the detonation structure during initiation by a reflected shock from an end wall of a shock tube is from a smoked foil. The development of an instability to form a cellular detonation structure is clearly indicated. The mixture is $2\,H_2-O_2-3\,N_2$, and in Fig. 9.48a, the initial pressure is 40 Torr. Note that there is no cellular structure near the end wall and that the onset of detonation occurs at about 9 inches from the end wall at a local explosion center. The detonation that is formed is initially overdriven (as indicated by the smaller cell sizes) and subsequently decays to the normal CJ velocity of the mixture. This is illustrated by the gradual increase in the cell size as shown in the continuation of the smoke foil below. In Fig. 9.48b, the initial pressure is lower, at 30 Torr. The incident shock and, thus, the reflected shock are stronger. The onset of detonation occurs closer to the end wall, indicating a shorter induction distance. Again the detonation is first overdriven and then relaxes to the normal CJ state. Figure 9.48 indicates that the

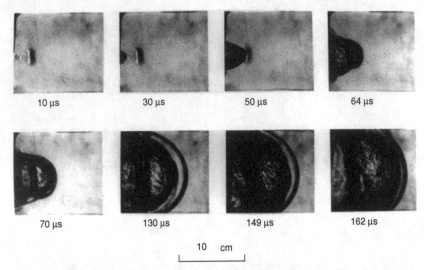

Figure 9.49. Unsuccessful detonation initiation via a jet of combustion products (Knystautas *et al.*, 1978).

initiation phenomenon is similar to blast initiation and to the transition from deflagration to detonation, which both require the growth of instabilities to form the cellular structure.

Noting that the onset of detonation in deflagration-to-detonation transition occurs mostly in the turbulent flame brush, Knystautas *et al.* (1978) investigated the direct initiation of detonation by a jet of combustion products. The rapid turbulent mixing of the products with reactants in a jet is essentially the same as the combustion process in a propagating high-speed turbulent flame. Figures 9.49 to 9.51 show

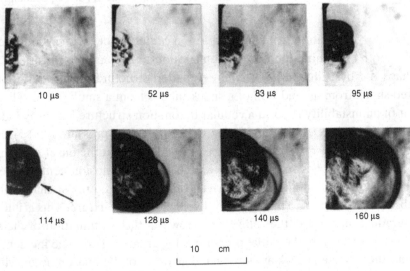

Figure 9.50. Successful detonation initiation via a jet of combustion products using a perforated plate (Knystautas *et al.*, 1978).

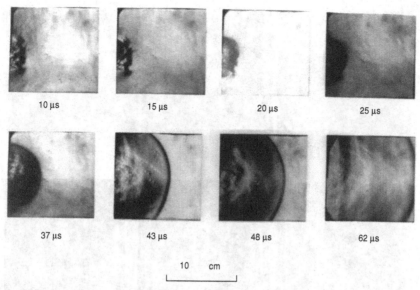

Figure 9.51. Successful detonation initiation via a jet of combustion products using a different perforated plate (Knystautas *et al.*, 1978).

sequences of schlieren photographs of turbulent jet initiation. The products were formed in a small prechamber where a constant volume explosion is effected in the mixture. The products then jet into the main combustion chamber, which contains the same mixture.

In Fig. 9.49, the starting vortex of the jet can first be observed. Because no diaphragm was used, a small amount of the mixture in the prechamber was displaced through the orifice during combustion in the prechamber. The starting vortex is essentially reactants that have been ejected from the prechamber. The products follow, and subsequent frames show the rapid combustion and formation of a spherical shock ahead of the turbulent deflagration. Direct initiation is not obtained in this case. In Fig. 9.50, a perforated plate is used to create more rapid turbulent mixing in the jet. The initial few frames illustrate the turbulent structure of the jet of ejected reactants and products from the prechamber. The onset of detonation is observed to originate from one or more local explosion centers within the turbulent mixing zone. The overdriven detonation bubbles then grow to engulf the jet to form a spherical detonation. Again, the initiation process is not unlike that of blast initiation at critical conditions or in the transition from deflagration to detonation. By using a strong turbulent jet, we have essentially bypassed the flame-acceleration process of the transition from deflagration to detonation. In Fig. 9.51, a different perforated plate is used, and the detonation is formed from a number of explosion centers in the mixing zone. A more symmetrical spherical detonation is formed, analogous to the supercritical energy regime of blast initiation.

Using a strong beam of ultraviolet radiation from a flash tube, direct initiation can also be obtained in the radical gradient field of the photon absorption zone

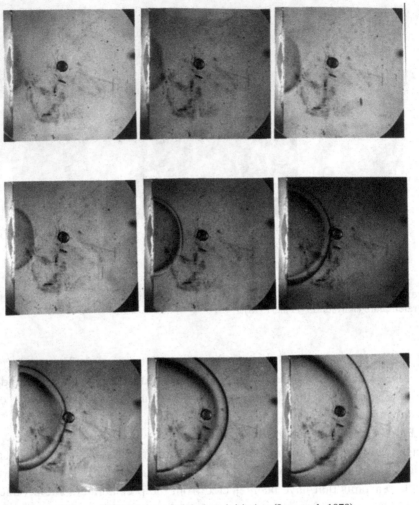

Figure 9.52. Unsuccessful photoinitiation (Lee *et al.*, 1978).

(Lee *et al.*, 1978). Figures 9.52 and 9.53 show the different regimes of photoinitiation. In Fig. 9.52, the mixture is H$_2$–Cl$_2$, and a circular quartz window is on the
left edge of the photograph where the intense ultraviolet beam from a flash lamp
enters the detonation chamber. Absorption of the ultraviolet light results in the dissociation of the chlorine molecules, and there exists a gradient in the chlorine atom
concentration decreasing exponentially away from the face of the window. After
a short induction period, rapid chemical reactions develop in the absorption zone,
and one can observe the formation of a shock wave and the subsequent spherical deflagration propagating away from the absorption zone. In Fig. 9.53, conditions are
such that an explosion center can be observed to form in the absorption zone. The
overdriven detonation bubble then grows to engulf the spherical deflagration at later
times. More than one explosion center can also form in the gradient field of the pho-
toabsorption zone. It is interesting to note that a gradient field is required for direct
initiation to occur. In experiments where the ultraviolet beam is extremely intense

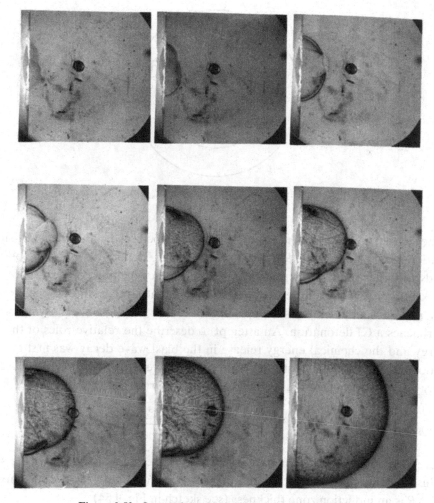

Figure 9.53. Successful photoinitiation (Lee *et al.*, 1978).

(thereby generating a more uniform distribution of free radicals), only a constant volume explosion of the irradiated mixture results.

It may be concluded that the various means of direct initiation of detonation refer to the different initial conditions that lead to the creation of the critical condition for the onset of detonation. Thus, it appears that all forms of initiation differ in the manner in which the critical conditions for the onset of detonation are achieved. However, the mechanisms responsible for the formation of a detonation appear to be common to all initiation methods, including the deflagration-to-detonation transition via flame acceleration.

9.6. THEORY OF BLAST INITIATION

Direct initiation of spherical detonation by an ideal point energy source is perhaps the simplest initiation mode and the most amenable to theoretical description. Close

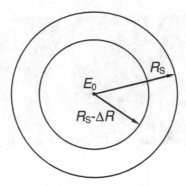

Figure 9.54. Sketch of a blast wave at radius R_s with blast energy E_0.

to the source, the decay of the strong blast is described by the analytical self-similar solution of von Neumann (1941), Sedov (1946), and Taylor (1950). As the blast expands to larger radii, the progressive release of chemical energy turns the strong shock into an overdriven detonation. Eventually, at large radii, the chemical energy release dominates the shock propagation, and the blast wave asymptotically approaches a CJ detonation. An attempt to describe the relative roles of the blast energy and the chemical energy release in the blast wave decay was first given by Zeldovich *et al.* (1957). The strength of the blast wave is proportional to the average energy density within the blast sphere, so we may write

$$M_s^2 = A \left\{ \frac{E_0 + \frac{4}{3}\pi(R_s - \Delta R)^3 \rho_0 Q}{\frac{4}{3}\pi R_s^3} \right\}, \tag{9.1}$$

where A is some proportionality constant, E_0 is the blast energy, R_s is the shock radius, Q is the chemical energy per unit mass, ρ_0 is the initial density of the mixture, and ΔR is an induction zone thickness (see sketch in Fig. 9.54).

Following Zeldovich, Eq. 9.1 can be rewritten as

$$n = \frac{m}{r^3} + \left(1 - \frac{1}{r}\right)^3, \tag{9.2}$$

where

$$n = \frac{M_s^2}{A\rho_0 Q}, \qquad m = \frac{E_0}{\rho_0 Q \frac{4}{3}\pi \, \Delta R^3}, \quad \text{and} \quad r = \frac{R_s}{\Delta R}.$$

The dimensionless parameters n, m, and r denote the shock strength, initiation energy, and shock radius, respectively. For $r \leq 1$, chemical energy does not play a role, and the blast decays as $M_s \propto 1/r^3$ as given by the similarity solution for strong point blast. For large radii, $M_s \rightarrow M_{CJ}$, which is a constant for a given mixture.

A plot of Eq. 9.2 is given in Fig. 9.55, where the shock strength n is plotted against the shock radius r for various values of the initiation energy m. It can be observed that the blast decays rapidly to a minimum below the CJ detonation strength before it reaccelerates asymptotically to a CJ detonation at large radii, irrespective of the

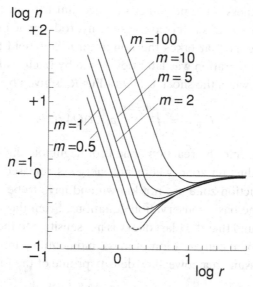

Figure 9.55. Shock strength as a function of shock radius (Zeldovich *et al.*, 1957).

value of the blast energy. Thus, a criterion is required to cut off the chemical energy contribution below a certain blast energy so that the shock does not reaccelerate to a CJ detonation. Zeldovich proposed that the duration of the shock, when its strength is above the CJ value, should be at least equal to the induction time of the mixture. This criterion then permits a minimum blast energy to be defined, and it was later referred to as the Zeldovich criterion.

Lee *et al.* (1966) attempted to formulate a quantitative theory based on Zeldovich's criterion. Using the similarity solution for strong blast decay, the shock radius is given by $R = At^{2/5}$, where A can be found from the energy integral as

$$A = \left(\frac{25E_0}{16\pi\rho_0 I} \right).$$

For $\gamma = 1.4$, the value for the integral from the similarity solution is $I = 0.423$. From the shock trajectory, the shock velocity (and hence the Mach number) can be obtained. Equating the shock Mach number to the CJ value and the time to the induction time τ_{CJ}, the initiation energy can be obtained as

$$E_0 = 62.5\gamma p_0 \pi M_{CJ}^5 (c_0\tau_{CJ})^3. \tag{9.3}$$

For a given mixture at given initial conditions, p_0, c_0, and M_{CJ} are known, and the induction time can also be computed if the dependence of the induction time on temperature (or, equivalently, the shock Mach number) obtained from shock tube data is given. Lee *et al.* (1966) found that Eq. 9.3 underestimates the initiation energy by at least three orders of magnitude. However, when an experimental value for the reaction-zone thickness (or reaction time) is used, the correct order of magnitude for the initiation energy is obtained. The experimentally measured reaction-zone thickness (or time) takes into account the three-dimensional turbulent cellular structure and is referred to as the hydrodynamic thickness.

An improved theory that takes the nonuniform flow field behind the decaying blast wave into consideration was later developed by Bach *et al.* (1971). The chemical energy released when the shock is at a radius R_s is given by the integral

$$E_{chem} = \int_0^{R_s - \Delta R} \rho(r, t) Q 4\pi r^2 \, dr,$$

where it is assumed that the reaction zone is of negligible thickness and ΔR is predominantly the induction zone where no energy is released. The density profile $\rho(r, t)$ and the induction zone are not known and have to be determined from the exact solution of the basic conservation equations. From the study of non-reactive blast waves, it is found that the blast decay is not sensitive to the flow profiles behind it as long as the global conservation of mass, momentum, and energy are satisfied. Thus, Bach *et al.* assumed a power-law density profile of the form

$$\frac{\rho}{\rho_0} = \frac{\rho_1}{\rho_0} \left(\frac{r}{R_s}\right)^q, \tag{9.4}$$

where the exponent q is obtained from the mass integral:

$$\rho_0 \frac{4}{3}\pi R_s^3 = \int_0^{R_s} \rho 4\pi r^2 \, dr = \frac{4\pi R_s^3 \rho_0}{q+3}\left(\frac{\rho_1}{\rho_0}\right);$$

thus,

$$q = 3\left(\frac{\rho_1}{\rho_0} - 1\right). \tag{9.5}$$

The exact density profile may not be of the power-law form given by Eq. 9.4, but with the exponent given by Eq. 9.5, global conservation of mass is satisfied. The density ratio at the front varies between $\rho_1/\rho_0 = \frac{\gamma+1}{\gamma-1}$ for a strong shock to $\rho_1/\rho_0 = \frac{\gamma+1}{\gamma}$ for a CJ detonation wave. Thus, $\frac{3}{\gamma} \le q \le \frac{6}{\gamma-1}$, and for $\gamma = 1.4$ we have $2.14 \le q \le 15$. In general, the density profile is very steep, with most of the mass concentrated near the shock front. Hence, even for a small induction-zone thickness ΔR, the amount of mass burned (and the amount of energy released) is relatively small when the blast is strong. As a result, the initial decay of the blast is mainly controlled by the blast energy, as described by the similarity solution of Taylor, Sedov, and von Neumann.

The conservation of total energy enclosed by the blast at any instant is given by the energy integral

$$E_0 + \int_0^{R_s - \Delta R} \rho Q 4\pi r^2 \, dr = \int_0^{R_s} \rho \left(e + \frac{u^2}{2}\right) 4\pi r^2 \, dr,$$

where we have neglected the initial internal energy of the mixture. Defining the nondimensional parameters

$$\xi = \frac{r}{R_s}, \qquad \psi = \frac{\rho}{\rho_0}, \qquad \phi = \frac{u}{\dot{R}_s}, \qquad \text{and} \qquad f = \frac{p}{\rho_0 \dot{R}_s^2},$$

the energy integral becomes

$$E_0 + 4\pi R_s^3 \rho_0 Q I_1 = 4\pi R_s^3 \rho_0 \dot{R}_s^2 I_2,$$

where

$$I_1 = \int_0^{1 - \frac{\Delta R}{R_s}} \psi \xi^2 \, d\xi,$$

$$I_2 = \int_0^1 \left(\frac{f}{\gamma - 1} + \frac{\psi \phi^2}{2} \right) \xi^2 \, d\xi,$$

and $\dot{R}_s = dR_s/dt$ is the shock velocity. Using Eq. 9.4 and 9.5, the integral I_1 can be obtained as

$$I_1 = \frac{1}{3} \left(1 - \frac{\Delta R}{R_s} \right)^{q+3},$$

where q is given by Eq. 9.5. The energy integral can be expressed as

$$M_s^2 = \frac{1}{4\pi\gamma I_2} \left(\frac{R_0}{R_s} \right)^3 + \frac{Q}{3c_0^2 I_2} \left(1 - \frac{\Delta R}{R_s} \right)^{q+3}, \tag{9.6}$$

where

$$M_s^2 = \frac{1}{4\pi\gamma I_2}, \qquad c_0^2 = \frac{\gamma p_0}{\rho_0} \quad \text{and} \quad R_0 = \left(\frac{E_0}{p_0} \right)^{\frac{1}{3}}.$$

R_0 is generally referred to as the explosion length.

Note that Eq. 9.6 is a better approximation than the original one derived by Zeldovich (i.e., Eq. 9.2). Qualitatively, however, Eqs. 9.2 and 9.6 are similar. For small radii (R_s/R_0), the first term on the right-hand side of Eq. 9.6 dominates and M_s^2 decays like $(R_s/R_0)^{-3}$ as given by the similarity solution for a strong point blast. For a large radius, the second term dominates because $\Delta R/R_s \to 0$ and $M_s \to M_{CJ}$.

The integral I_2 varies between the value given by the strong point–blast solution when the radius is small to $I_2 \to (1/3M_{CJ}^2) Q/c_0^2$ in the asymptotic limit when $M_s \to M_{CJ}$ and $R_s/R_0 \to \infty$. The asymptotic value of I_2 is $\frac{1}{6(\gamma^2-1)}$ because $M_{CJ}^2 = 2(\gamma^2 - 1) Q/c_0^2$; for $\gamma = 1.4$, we have $I_2 = 0.423$ in the strong-blast limit. $I_2 \to 0.176$ in the asymptotic limit at large radius when $R_s \to \infty$ and $M_s \to M_{CJ}$, so we have $0.176 \le I_2 \le 0.423$. Thus, the variation of I_2 is not large over the range from a strong shock to a CJ detonation. The energy integral given by Eq. 9.6 gives an equation for the variation of the shock strength with radius. However, the integral I_2 and the induction zone ΔR require additional equations for their description.

To evaluate I_2, the pressure and the particle velocity profiles behind the blast are required. Because the continuity equation involves only two dependent variables (the density and the particle velocity), knowledge of the density profile permits the particle velocity profile to be determined. The pressure distribution can then be obtained from the momentum equation. For the assumed power-law density profile of Eq. 9.4, the particle velocity and the pressure profiles were obtained by Sakurai (1959) and by Bach and Lee (1970) as

$$\frac{u}{\dot{R}_s} = \phi = \phi_1 \xi, \tag{9.7}$$

$$\frac{p}{\rho_0 \dot{R}_s^2} = f = f_1 - \frac{\psi_1 \left(1 - \xi^{q+2}\right)}{q+2} \left\{ 2\theta\eta \left(\frac{d\phi_1}{d\eta}\right) + \phi_1 \left(1 - \phi_1 - \theta\right) \right\}, \tag{9.8}$$

where

$$\theta = \frac{R_s \ddot{R}_s}{\dot{R}_s^2}. \tag{9.9}$$

The parameter θ is referred to as the shock decay coefficient, as it involves the time derivative of the shock velocity. With the profiles given by Eqs. 9.4, 9.7, and 9.8, the integral I_2 can be evaluated as

$$I_2 = \frac{f_1}{3(\gamma - 1)} + \frac{\psi_1 \phi_1^2}{2(q+5)} - \frac{\psi_1 \left(1 - \xi^{q+2}\right)}{q+2} \left\{ 2\theta\eta \left(\frac{d\phi_1}{d\eta}\right) + \phi_1 \left(1 - \phi_1 - \theta\right) \right\}, \tag{9.10}$$

where $\eta = 1/M_s^2$ is the shock strength parameter.

The boundary conditions for ψ_1, f_1, ϕ_1 at the shock front are given by the Rankine–Hugoniot equations and expressed as a function of the shock strength η (or M_s^2). The addition of the parameter θ in Eq. 9.10 requires another expression for θ. This can be obtained from the definition of θ itself, that is, Eq. 9.9 can be expressed in the form

$$\frac{d\eta}{dz} = \frac{-2\eta\theta}{z}, \tag{9.11}$$

where $z = R_s/R_0$ is the dimensionless shock radius (with respect to the explosion length R_0). The energy integral (Eq. 9.6) can be written in terms of z and η as

$$1 = Q\pi\gamma z^3 \left[\frac{I_2}{\eta} - \frac{Q}{3c_0^2} \left(1 - \frac{\epsilon}{z}\right)^{q+3} \right], \tag{9.12}$$

where I_2 is given by Eq. 9.10, and $\epsilon = \Delta R/R_0$ is the dimensionless induction zone thickness, which still requires an additional expression for its description when the shock strength changes. The temperature dependence of the induction-zone thickness can be found from shock tube data. In principle, the energy integral can be integrated simultaneously with the equation for I_2, Eq. 9.11 for θ, and an expression for the induction zone thickness. The derivative of ϕ_1 with respect to η can be obtained from the boundary conditions. The exponent $q(\eta) = 3[\psi_1(\eta) - 1]$ is also given by the boundary condition at the front. However, due to the highly implicit nature of the interdependence of the various parameters, numerical integration of the energy integral is quite difficult in practice. In the theory of Bach et al. (1971), they introduced further simplifications to facilitate the solution of the problem.

The strength of the blast at any instant depends on the average energy density of the shock sphere, leading Bach *et al.* to define an equivalent chemical energy

$$Q_e = Q \left(1 - \frac{\epsilon}{z}\right)^{q+3}. \tag{9.13}$$

This chemical energy is assumed to be released at the front and not in the reaction zone at some distance behind the front. The average energy density of the shock sphere remains the same. By assuming the chemical energy to be released at the shock, the front is made a reactive discontinuity and the boundary conditions for a reactive shock (or detonation) are given by the Rankine–Hugoniot equations as

$$\frac{\rho_1}{\rho_0} = \psi_1 = \frac{\gamma + 1}{\gamma - S + \eta}, \tag{9.14}$$

$$\frac{u_1}{\dot{R}_s} = \phi_1 = \frac{1 + S - \eta}{\gamma + 1}, \tag{9.15}$$

$$\frac{p_1}{\rho_0 \dot{R}_s^2} = f_1 = \frac{\gamma + \gamma S + \eta}{\gamma(\gamma + 1)}, \tag{9.16}$$

where

$$S = \left[(1 - \eta^2) - K\eta\right]^{\frac{1}{2}} \tag{9.17}$$

and

$$K = 2\left(\gamma^2 - 1\right)\frac{Q_e}{c_0^2}. \tag{9.18}$$

The preceding equations are essentially those for an overdriven detonation. When $M_s \gg 1$ (or $\eta \ll 1$), then $S \to 1$ and we have the strong shock conditions of

$$\psi_1 = \frac{\gamma + 1}{\gamma - 1} \quad \text{and} \quad \phi_1 = f_1 = \frac{2}{\gamma + 1}.$$

When $M_s \to M_{CJ}$ (or $\eta \to \eta_{CJ}$), then $S = 0$ and we have a CJ detonation where

$$\psi_1 \to \frac{\gamma + 1}{\gamma} \quad \text{and} \quad \phi_1 = f_1 \to \frac{1}{\gamma + 1}.$$

for $\eta_{CJ} \ll 1$.

From Eq. 9.13, we see that the equivalent energy depends on both the shock strength η and the shock radius (or curvature), $z = R_s/R_0$. The dependence on η is via the exponent q and also the induction zone thickness ΔR. Because $q = 3(\psi_1 - 1)$ and the density ratio, $\psi_1 = \rho_1/\rho_0$, is relatively insensitive to the shock strength when the shock is strong, Bach *et al.* took the limiting value for $\psi_1 = \frac{\gamma+1}{\gamma-1}$ and thus $q \approx \frac{6}{\gamma-1}$. The induction-zone thickness ΔR is also relatively insensitive to temperature (shock strength) when the temperature is high. Thus, Bach *et al.* considered that $\Delta R \approx \Delta R_{CJ}$ and defined a function $G(\eta)$ to describe the variation of the induction-zone thickness near the autoignition limit when ΔR increases exponentially with decreasing temperature (or, equivalently, the shock strength).

Accordingly, Eq. 9.13 is approximated by

$$Q_e = Q\left(1 - \frac{\Delta R_{CJ}/R_0}{R_s/R_0}\right)^{\frac{6}{\gamma-1}} G(\eta)$$

$$= Q\left(1 - \frac{\delta}{z}\right)^{\frac{6}{\gamma-1}} G(\eta), \tag{9.19}$$

where $G(\eta)$ is defined by

$$G(\eta) = \begin{cases} \dfrac{\left(1 - \frac{\eta}{\eta_c}\right)^2 \left(1 - \frac{3\eta_{CJ}}{\eta_c} + \frac{2\eta}{\eta_c}\right)}{\left(1 - \frac{\eta_{CJ}}{\eta_c}\right)^3}, & \eta_{CJ} \leq \eta \leq \eta_c, \\ 1, & 0 \leq \eta \leq \eta_{CJ}, \\ 0, & \eta \geq \eta_c. \end{cases} \tag{9.20}$$

The form for $G(\eta)$ given above is the lowest-order polynomial that gives a continuous derivative $\partial G(\eta)/\partial \eta$ and that satisfies the desired limiting conditions given by Eq. 9.20. Higher-order polynomials can equally be chosen, but they only have the effect of a sharper rise and cutoff as η approaches the two limits $\eta \to \eta_{CJ}$ and $\eta \to \eta_c$.

For the dependence of Q_e on the shock radius (i.e., curvature), Bach et al. specified that

$$Q_e = \begin{cases} 0, & z \leq \delta, \\ Q\left(1 - \frac{\delta}{z}\right)^{\frac{6}{\gamma-1}}, & z > \delta, \quad 0 \leq \eta \leq \eta_{CJ}, \\ Q\left(1 - \frac{\delta}{z}\right)^{\frac{6}{\gamma-1}} G(\eta), & \eta_{CJ} \leq \eta \leq \eta_c, \\ Q, & z \gg \delta, \quad \eta = \eta_{CJ}. \end{cases} \tag{9.21}$$

For $z > \delta$ and $\eta \geq \eta_c$, we have $G(\eta) = 0$. Hence, $Q_e = 0$ and chemical energy release is essentially cut off.

Thus, in the theory of Bach et al., the entire physics of the direct initiation process is modeled by Eqs. 9.19 to 9.21. Equations 9.20 and 9.21 are, in essence, the criteria whereby the specification of a cutoff shock strength, η_c, defines a critical value for the initiation energy. When the initiation energy is very small (hence, a small value of the explosion length R_0), the shock decays rapidly, $\eta \to \eta_c$ when $z < \delta$ and $Q_e = 0$, and the blast decays as a non-reactive blast. For larger values of the initiation energy and $\delta/z < 1$, curvature dominates, but chemical energy release progressively cuts in to support the shock as the shock radius increases and curvature decreases. The shock decay rate slows down. If the shock strength decays below the autoignition limit, then $G(\eta) \to 0$ as $\eta \to \eta_c$, and the chemical energy is effectively turned off. The shock decays to an acoustic wave in the limit. For successful initiation, the blast energy must be such that the blast decays sufficiently slowly for the chemical energy to slow the rate of decay of the shock. If η can be maintained above η_c as $z \gg 1$, then the chemical energy can support the shock, resulting in its progressive acceleration to a CJ detonation asymptotically as $z \to \infty$. Thus, Bach et al. essentially built upon

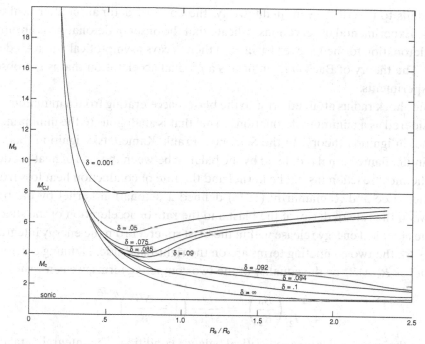

Figure 9.56. Decay of the blast wave as a function of radius (Bach *et al.*, 1971).

the idea of Zeldovich by incorporating the finite induction zone thickness and a cutoff autoignition limit directly into the blast decay process.

To obtain the actual decay of the blast with radius, Bach *et al.* solved for the energy integral directly from Eq. 9.6 together with the equations for I_2 (Eq. 9.10), θ (Eq. 9.11), and $d\phi_1/d\eta$ (from the boundary condition for ϕ_1, i.e., Eq. 9.15). For a chosen value of $\delta = \Delta R_{CJ}/R_0$, the shock decay curves are shown in Fig. 9.56. As can be observed, all the salient features of the initiation process are reproduced by the theory of Bach *et al.*, including the prediction of a critical energy. There exists a critical value of $\delta = \delta_c$ where the blast can reaccelerate after decaying to a minimum, and then eventually approach a CJ detonation in the limit of $z \to \infty$. For $\delta < \delta_c$, the blast fails to reaccelerate and decays until $M_s < M_c$ (or $\eta > \eta_c$), when chemical energy is cut off. The shock approaches an acoustic wave in the limit as $z \to \infty$. From the solution for the energy integral, the critical energy can be obtained from the value of δ_c:

$$\delta_c = \frac{\Delta_{CJ}}{R_0} = \Delta \left(\frac{p_0}{E_0}\right)^{\frac{1}{3}}, \qquad \text{or} \quad E_0 = p_0 \left(\frac{\Delta_{CJ}}{\delta_c}\right)^3. \qquad (9.22)$$

Thus, the cubic dependence of critical energy on the induction zone thickness of the Zeldovich criterion is recovered.

Using the induction zone thickness of a ZND detonation computed from shock tube data, the theory of Bach *et al.* predicts a critical energy three orders of magnitude less than the experimental value. However, if experimental values for the thickness of a turbulent cellular detonation are used for Δ_{CJ}, the theory of Bach *et al.* gives values in good agreement with experiments. The theory of Bach *et al.*

also fails to describe, even qualitatively, the onset of detonation at critical conditions. Experimental observations indicate that the onset of detonation is abrupt, and the detonation formed is overdriven and then decays asymptotically to a CJ detonation. The theory of Bach *et al.* indicates a gradual acceleration that is not observed in experiments.

The shock radius attained prior to the blast reaccelerating from a minimum can be considered as a minimum detonation kernel that is analogous to the minimum flame kernel in ignition theory. In the Semenov–Frank–Kamenetsky ignition theory, the minimum flame kernel is defined by the balance between the rate of heat production by chemical reaction inside the kernel and the rate of conduction heat loss from the surface. Lee and Ramamurthi (1976) defined a detonation kernel by the balance between the rate of decay of the blast and the rate of acceleration of the shock due to the chemical energy release within the blast sphere. From the energy integral (i.e., Eq. 9.6), the two competing terms are on the right-hand side. Defining a detonation kernel as $R_s = R_s^*$ by the equality of the two competing terms, we can write

$$\frac{1}{4\pi\gamma I_2^*}\left(\frac{R_0}{R_s^*}\right)^3 = \frac{Q}{3I_2^*c_0^2}\left(1 - \frac{\Delta R^*}{R_s^*}\right)^{q^*+3}, \tag{9.23}$$

where the asterisk denotes the critical balance condition. The integral I_2 takes on a value from the similarity solution for strong point blast in the limit when $M_s \gg 1$ or $\eta \ll 0$. For $\gamma = 1.4$, this limiting value of strong blast gives $I_2 = 0.423$. In the asymptotic limit when $M_s \to M_{CJ}$ and $R_s \to \infty$, a value for I_2 can also be obtained from the energy integral and Eq. 9.23 as

$$I_2 \to \frac{Q}{3c_0^2 M_{CJ}^2}.$$

Because $M_{CJ}^2 = 2(\gamma - 1)Q/c_0^2$, the asymptotic value of I_2 can be obtained as

$$I_2 = \frac{1}{6(\gamma^2 - 1)},$$

which for $\gamma = 1.4$ is equal to 0.173. Thus, I_2 is not a strongly varying function of the shock strength, and a mean value may be chosen for I_2^*.

From the equality of the two competing terms (Eq. 9.23), the detonation kernel size R_s^* can be obtained from the energy integral as

$$R_s^* = \frac{\Delta R^* - M_s^*}{1 - \left[\frac{1}{2}\left(\frac{M_s^*}{M_{CJ}}\right)^2\right]^{\frac{1}{q+3}}}. \tag{9.24}$$

A value for M_s^* can be chosen for the autoignition limit, and ΔR^* can be computed from shock tube data. Thus, the detonation kernel radius R_s^* can be found from Eq. 9.24. From the energy integral, equality of the competing terms gives

$$M_s^* = \frac{2}{4\pi\gamma I_2^*}\left(\frac{R_0^*}{R_s^*}\right)^3 = \frac{2Q}{c_0^2 I_2^*}\left(1 - \frac{\Delta R^*}{R_s^*}\right)^{q^*+3},$$

and solving for the initiation energy gives

$$E_0 = 2\pi\gamma p_0 I_2^* M_s^{*2} R_s^{*3}. \tag{9.25}$$

Since the kernel size, R_s^*, is proportional to the induction-zone length, the above expression also gives the cubic dependence of the initiation energy on the induction-zone thickness.

Thus, we see that the essence of all blast initiation theories requires the specification of some characteristic length scale that corresponds to a critical shock strength. The reaction-zone thickness of a one-dimensional, steady ZND detonation does not represent the real three-dimensional, transient cellular detonation structure, but the detonation cell size is difficult to measure accurately. With the cubic dependence of the critical initiation energy on the characteristic reaction-zone length scale, the error in the critical energy can be quite large. On the other hand, the critical tube diameter can generally be determined more precisely. Lee and Matsui (1977) proposed an initiation theory using the critical tube diameter as the characteristic length scale. They pointed out that the critical tube diameter gives a minimum surface area for which a detonation can survive the expansion waves coming in from the sides to quench it. In blast initiation, Lee and Matsui argued that the critical initiation energy must be such that when the blast has decayed to the CJ velocity, the surface area of the blast must be of a certain minimum size to survive the quenching due to nonsteady expansion and curvature. Thus, they equated the minimum surface area of the blast sphere to the area of the critical tube (i.e., $4\pi R_*^2 = \pi d_c^2/4$) and obtained the critical blast radius as $R_* = d_c/4$. From strong blast theory, the shock strength–radius relationship can be obtained, and the critical energy from this theory can be expressed as

$$E_c = \frac{\gamma\pi p_0 I_2 M_{CJ}^2 d_c^3}{16}. \tag{9.26}$$

The value of I_2 is found from the similarity solution for a strong point blast (e.g., for $\gamma = 1.4$, $I_2 = 0.423$), the CJ Mach number is found from equilibrium calculations, and thus, with an experimental value for the critical tube diameter, the critical energy can be determined from Eq. 9.26. Alternatively, if the empirical correlation $d_c = 13\lambda$ is used, Eq. 9.26 can be written in terms of the cell size λ as

$$E_c = \frac{2197\gamma\pi p_0 I_2 M_{CJ}^2 \lambda^3}{16}. \tag{9.27}$$

A comparison between the critical initiation energy predicted by Eq. 9.27 and experimental values for a range of fuel–air mixtures is shown in Fig. 9.57. The cell size λ has been measured independently from smoked foils. As can be observed, the agreement is quite good in spite of the simplicity of the model. For explosive mixtures, where the detonation is relatively stable, λ may not be an appropriate length scale to characterize the detonation structure. The critical tube diameter and Eq. 9.26 should be used in these cases.

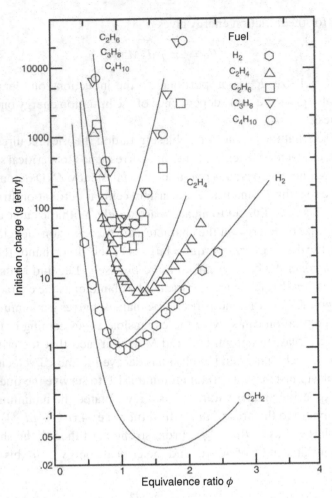

Figure 9.57. Comparison of critical initiation energy with experimental values for a range of fuel–air mixtures (Knystautas *et al.*, 1985).

9.7. THE SWACER MECHANISM

Since most self-sustained detonations are unstable, the formation of a detonation requires the growth of instabilities to form an unstable cellular detonation front. The growth of instabilities is essentially the coupling of gasdynamic perturbations with the chemical energy release, requiring the proper phase relationship. The SWACER concept is, in essence, Rayleigh's stability criterion applied to a traveling pulse. In the numerical simulation of direct initiation of one-dimensional detonations, the onset of detonation is effected by a pressure pulse originating from the reaction zone that surges rapidly to an amplitude much greater than the CJ pressure as it propagates toward the shock front. When the pressure pulse merges with the precursor shock, a highly overdriven detonation is formed that relaxes to a CJ detonation, as is observed experimentally. Figure 9.58 illustrates the process of the onset of

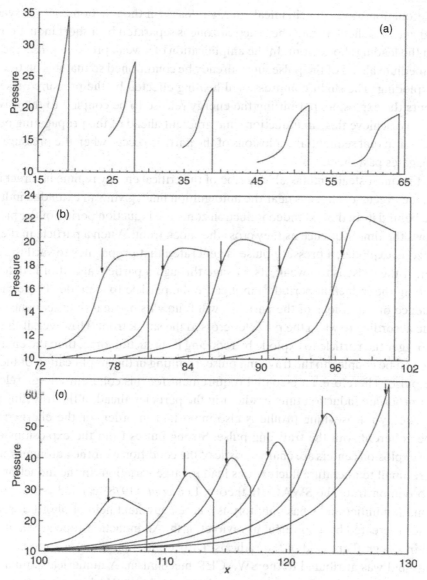

Figure 9.58. Pressure amplification leading to the onset of detonation in the critical energy regime; the arrow denotes the location of the reaction zone (Lee & Higgins, 1999).

detonation in the critical regime of blast initiation. The rapid amplification of the pressure pulse from the reaction zone as it propagates toward and eventually merges with the leading shock front to form an overdriven detonation can be observed in Fig. 9.58c.

The rapid amplification of the traveling compression pulse requires the proper synchronization of the chemical energy release with the propagating wave. This is referred to as *coherent energy release* in the SWACER mechanism. In a propagating

CJ detonation wave, the chemical energy release (in the reaction zone) is synchronized with the shock front. The reaction zone is separated by a short induction zone from the leading shock front. In the amplification of a weak pressure pulse, the reactive medium ahead of the pulse must already be conditioned so that it is on the verge of exploding. The slight compressional heating effected by the pressure pulse then triggers the explosion, permitting the energy release to be coupled to the traveling pulse. To achieve this, an induction time gradient ahead of the propagating pulse is required so that sequential explosions of the particles occur when the pressure pulse propagates past them.

In the quasi-steady metastable period of the critical energy regime in direct initiation, the precursor shock is near the autoignition limit, giving an extended induction zone behind it. In this extended induction zone, the induction period of the particles follows the time sequence as they cross the shock front. When a particle in the reaction zone explodes, a pressure pulse is generated that propagates toward the shock front. If the pulse is too weak, its passage through a particle ahead of it that is undergoing the induction period cannot cause the particle to explode. Therefore, the sequence of explosions of the particles will follow its original time sequence to explode according to when the particles crossed the shock front. However, if the pulse can trigger the particle to explode by reducing its induction time, then the energy release can be coupled to the traveling pulse, resulting in the amplification of the pressure pulse. Therefore, an essential requirement for this coherent energy release is an appropriate induction time gradient in the particles ahead of the traveling pulse. The energy-release–time profile is also important in order for the energy release to be coherent with the traveling pulse. Strong pulses from the explosions of discrete explosion centers are characteristic of the condition near the autoignition limit, where small temperature fluctuations lead to large variations in the induction time.

To demonstrate the SWACER theory, Lee *et al.* (1978) carried out a study on detonation initiation by flash photolysis where a gradient field of photo-dissociated species is created by a powerful ultraviolet flash. An induction-time gradient is created in the gradient field of the radicals. Direct initiation of detonation in the gradient field was attributed to the SWACER mechanism. A numerical simulation of the detonation formation process was also carried out (Yoshikawa & Lee, 1992). Figure 9.59 shows the results of the numerical simulation of the SWACER mechanism in an induction-time gradient field due to photodissociation in an H_2–Cl_2 mixture.

Figure 9.59a shows the supercritical case where the irradiation intensity is $I_0 = 2 \, \text{kW/cm}^2$. The pressure profiles of the shock wave as it propagates in the induction-time gradient field are shown. It can be seen that the shock builds up to an overdriven detonation in a distance of about 10 cm and then decays to the CJ detonation state of the mixture. When the intensity of the irradiation is reduced, we arrive at a critical value $I_0 \approx 1 \, \text{kW/cm}^2$ where the shock accelerates to the detonation pressure, but rapidly decays past the CJ state to some quasi-steady condition of about half the CJ detonation state. If detonation results eventually from this metastable state, it

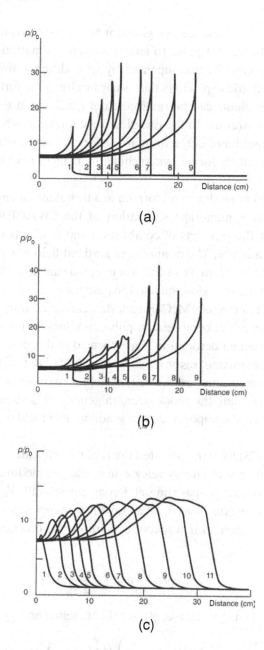

Figure 9.59. Numerical simulation of SWACER in an induction time gradient (Yoshikawa & Lee, 1992).

is due to instability mechanisms rather than that of the SWACER mechanism. Of particular interest is the case when a much more intense radiation pulse is used, thus destroying the steepness of the gradient field.

Figure 9.59c illustrates the pressure profiles when no amplification is observed. The pressure remains at about the constant volume explosion value, and no detonation is initiated. These numerical simulations, together with the experimental

studies, confirm that the presence of a gradient field is essential for the rapid amplification of an initially sub-CJ pulse to an overdriven detonation. It seems that the
onset of detonation must be accompanied by an additional amplification process
other than the volumetric explosions that occur locally in the turbulent zone.

Induction-time gradients can be realized from gradients in temperature or free
radical species concentration. In the turbulent reaction zone, where local explosion
centers and the formation of detonation are observed experimentally, the required
gradient field is most likely formed via turbulent mixing of reacting gases and products with the reactants.

Nonuniform gradient fields can be formed in a turbulent mixing zone. Figure 9.60
shows the results of a numerical simulation of the SWACER mechanism in a
nonuniform field of the products of combustion and reactants consisting of a stoichiometric H_2–air mixture. The nonuniform gradient field was prescribed by varying the products fraction from 0 to 1 linearly over a distance of 30 cm. Both detailed
chemical kinetics of the H_2 oxidation and the gasdynamics of the transient flow field
are described by a Lagrangian McCormack flux-corrected transport code. The results indicate that an initial compression pulse amplifies and steepens to an amplitude of 10 bars. Although detonation is not formed in this particular computation,
the results for a larger mixing region do show the onset of detonation in the gradient
field presented. If we assume just a constant volume explosion of a hot spot in an
inert surrounding medium, the shock strength generated is about an order of magnitude less. Therefore, the importance of a gradient field for the onset of detonation
is evident.

Thibault *et al.* (1978) have investigated the rapid amplification of a pressure pulse
from the timed sequence of energy release in a reactive medium. They considered
a traveling energy source propagating at a constant velocity V_0, depositing energy
Q_0 per unit mass at a rate of $\omega(t)$. The traveling energy source simulates a constant induction-time gradient in a reactive mixture. The functional form for $\omega(t)$ is
assumed to be

$$\omega(t) = Q_0 \frac{t}{\tau_R^2} \exp\left(-\frac{1}{2}\frac{t^2}{\tau_R^2}\right). \tag{9.28}$$

The rate of chemical energy release, $\dot{q}(t)$, of the traveling energy source is written as

$$\dot{q}(t) = \omega\left(t - \frac{x}{V_0}\right) H\left(t - \frac{x}{V_0}\right), \tag{9.29}$$

where $\omega(t)$ satisfies the condition

$$\int_0^\infty \omega(t)\, dt = Q_0, \tag{9.30}$$

and $H(y)$ denotes the Heaviside function

$$H(y) = \begin{cases} 0, & y < 0, \\ 1, & y \geq 0. \end{cases}$$

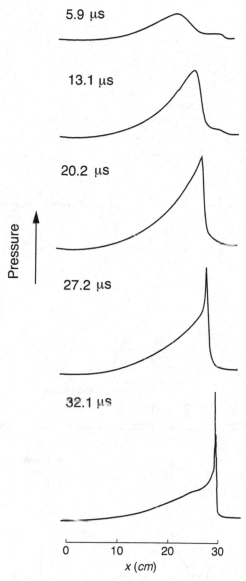

5.9 μs

13.1 μs

20.2 μs

27.2 μs

32.1 μs

Pressure

0   10   20   30

x (cm)

Figure 9.60. Numerical simulation of SWACER in a gradient field of products and reactants of H_2–air (courtesy of P. Thibault).

In Eq. 9.28, τ_R represents the time required for $\omega(t)$ to reach its peak value and is referred to as the reaction time. The development of pressure waves in the medium due to the traveling energy source is computed from the numerical integration of the reactive Euler equations. Figure 9.61 shows the pressure profiles at different instances of time. The dimensionless distance and time are defined by $t^* = t/\tau_R$ and $x^* = x/c_0\tau_R$, respectively, and $Q = Q_0/c_0^2$, where $c_0^2 = \gamma p_0/\rho_0$ is the sound speed. In Fig. 9.61a, where the source Mach number $A = V_0/c_0 = 0.5$, we note that the pressure pulse runs away from the source. For a supersonic source where $A = 3$, a

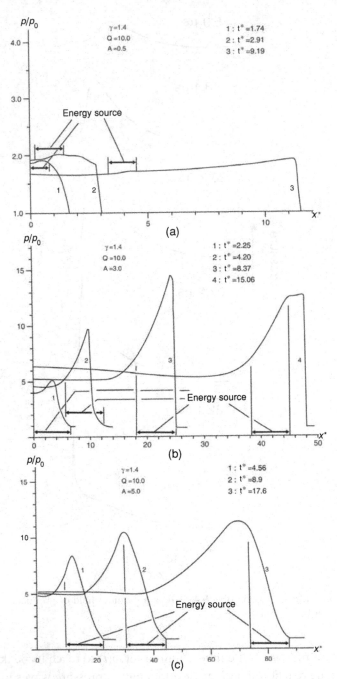

Figure 9.61. Pressure profile as a function of time for different traveling energy sources (Thibault *et al.*, 1978).

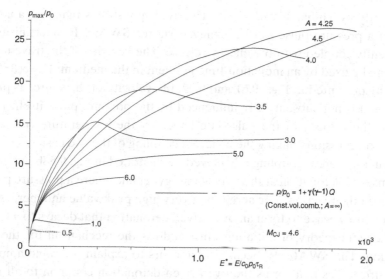

Figure 9.62. Peak pressure as a function of total energy released for various source Mach numbers (Thibault *et al.*, 1978).

very high peak pressure pulse is developed initially, which steepens to form a shock wave. However, at later times the shock propagates ahead and decouples from the energy source, and the amplitude drops (Fig. 9.61b). It is interesting to note that, for a highly supersonic source where $A = 5.0$ (the CJ Mach number corresponding to the chemical energy release Q is $M_{CJ} = 4.6$), a shock wave is not formed, because the rate of energy release that pressurizes the medium is more rapid than the rate of steepening of the front of the pressure pulse from nonlinear convective effects (Fig. 9.61c). The peak pressure developed is less than that for the case of a lower source Mach number where the surge generates a shock. The shock then decouples itself and propagates ahead of the source. In the limit where the source travels at infinite velocity (i.e., $A = \infty$), the entire volume of the mixture explodes simultaneously and the pressure developed corresponds to the constant volume explosion pressure where $p/p_0 = 1 + \gamma(\gamma - 1)Q$.

Figure 9.62 shows the peak pressure plotted against the total energy released. The different curves correspond to the different Mach numbers of the source, $A = V_0/c_0$. For $A < 1$ (i.e., $V_0 < c_0$), the peak pressure of the pulse generated by the source is very low, as the front of the pressure wave generated runs ahead of the energy source and disperses the pressure. For $A \gg 1$, the rate of pressure rise is controlled by the rate of the energy release, and in the limit of $A \to \infty$ we have a constant volume explosion where the pressure builds up to the $p/p_0 = 1 + \gamma(\gamma - 1)Q$. In the range of $1 < A < M_{CJ}$, the pressure pulse surges to a high value before it asymptotes to a lower steady value as the source outruns the shock wave that is generated and the optimal (coherent) coupling of the energy release with the pressure wave is destroyed.

The study by Thibault *et al.* (1978) clearly demonstrates that the rapid amplification of a pressure pulse can be achieved via the SWACER mechanism when it is coherently coupled with the energy release. The velocity of the traveling energy source can be fixed by an induction-time gradient in the medium: $V_0 = (\partial\tau/\partial x)^{-1}$. In the results presented in Figs. 9.61 and 9.62, the velocity of the source is prescribed initially and is not subsequently influenced by the pressure pulse itself. In reality, however, the velocity of the pulse is effected by the temperature rise associated with the compression, making the coherent coupling of the two easier. The pressure surge, when coherent coupling is achieved as illustrated in Fig. 9.58b, is very similar to the onset of detonation at the critical energy regime where a pressure pulse emanating from the reaction zone surges to a very high peak value and merges with the precursor shock wave to form an overdriven detonation that decays to a CJ wave.

A successful theory of initiation must address the mechanism for the onset of detonation. The SWACER mechanism appears to explain the rapid amplification of pressure waves that forms the overdriven detonation common to all modes of detonation formation.

9.8. CLOSING REMARKS

Direct initiation simply refers to the generation of a detonation wave without going through the process of accelerating a slow laminar flame to some critical state for the spontaneous onset of detonation to occur. Therefore, a direct initiation source is required to produce this critical state for the direct onset of detonation. The critical state for the onset of detonation has not been clearly defined. However, if one considers the propagation mechanism of a self-sustained detonation, it is clear that cellular instability plays an important role in most explosive mixtures. Thus, it may be said that onset of detonation is achieved by providing the necessary conditions for the spontaneous development of instabilities. The growth of instabilities requires pressure perturbations to be coupled to the chemical energy release so as to be amplified. Rayleigh's criterion, or the SWACER concept, which requires the proper phase synchronization of the pressure oscillations (or the traveling pressure pulse) with the chemical energy release, appears to be the required condition for the onset of detonation.

In general, the onset of detonation is observed to originate from local explosion centers. The explosion of a volume of explosive mixture generates a pressure (shock) pulse that radiates away from the local explosion site. The subsequent rapid amplification of the pressure pulse requires the medium ahead of it to be *preconditioned* in such a manner that the energy release can be synchronized with the traveling pulse through the preconditioned reactants. Direct initiation is essentially the production of this critical state by the initiation source. In blast initiation in the critical energy regime, the blast decays to some quasi-steady metastable state where some pressure perturbations traverse the reaction zone and amplify rapidly to catch up with the leading shock, forming an overdriven detonation (e.g., Figs. 9.22

and 9.55). In direct initiation by flash pyrolysis, an induction-time gradient field is generated by photodissociation in the absorption path of the light pulse. The propagation of the pressure pulse through this gradient field is amplified via the SWACER mechanism. One can observe the similarity of the pressure profiles from the numerical simulation of the photoinitiation in Figs. 9.59 and 9.58.

A detonation is essentially a strong shock coupled to the energy release from the chemical reactions it initiates, and it is evident that a strong shock wave ($M_S > M_{CJ}$) of sufficiently long duration ($t_S > t_{CJ}$) will form a detonation when it propagates into an explosive mixture. However, because the shock wave does not have any instability associated with it, the formation of a self-sustained cellular detonation in the explosive mixture will require transverse instabilities to be developed. On the other hand, if a detonation is initiated in a driver section and then is transmitted into a driven section containing the test explosive mixture, then a cellular self-sustained detonation can be easily formed in the driven section, because the instabilities are already present in the driven detonation. The instabilities (cellular structure) only have to adjust to the cellular structure of the detonation in the driven section.

If the initiation shock is weaker (i.e., $M_S < M_{CJ}$) but still sufficient to induce auto-ignition in the explosive gas of the driven section, then the onset of detonation occurs through the formation of local explosion centers. This is similar to the onset of detonation under critical conditions for the different initiation sources as well as in the transition from deflagration to detonation, indicating that the physics responsible for detonation initiation are universal.

Bibliography

Bach, G.G., and J.H.S. Lee. 1970. *AIAA J.* 8:271.

Bach, G.G., R. Knystautas, and J.H.S. Lee. 1969. In *12th Int. Symp. on Combustion*, 855.

Bach, G.G., R. Knystautas, and J.H.S. Lee. 1971. In *13th Int. Symp. on Combustion*, 1097–1110.

Benedick, W.R., R. Knystautas, and J.H.S. Lee. 1985. Dynamics of shock waves, explosion and detonations. In *Progress in astronautics and aeronautics*, Vol. 94, ed. J.R. Bowen, N. Manson, A.K. Oppenheim, and R.I. Soloukhin, 546–555.

Berets, D.J., E.F. Greene, and G.J. Kistiakowsky. 1950. *J. Am. Chem. Soc.* 7(2):1080.

Edwards, H., G.O. Thomas, and M.A. Nettleton. 1979. *J. Fluid Mech.* 95:79.

Fay, J. 1953. In *4th Int. Symp. on Combustion*, 507.

Gordon, W.E, A.J. Mooradian, and S.A. Harper. 1959. In *7th Int. Symp. on Combustion*, 752.

Knystautas, R., C. Guirao, J.H.S. Lee, and A. Sulmistras. 1985. Dynamics of shock waves, explosions, and detonations. In *Progress in astronautics and aeronautics*, Vol. 94, ed. J.R. Bowen, N. Manson, A.K. Oppenheim, and R.I. Soloukhin, 23–37. AIAA.

Knystautas, R., J.H.S. Lee, and C. Guirao. 1982. *Combust. Flame* 48:63–82.

Knystautas, R., J.H.S. Lee, I. Moen, and H.G. Wagner. 1978. In *17th Int. Symp. on Combustion*, 1235.

Korobeinikov, V.P. 1969. *Astronaut.* Acta 14(5):411–420. See also Lee, J.H.S. 1965. Hypersonic flow of a detonation gas. Rept. 65-1 Mech. Eng., McGill University, Montreal, Canada.

Lafitte, P. 1925. *C.R. Acad. Sci. Paris* 177:178. See also 1925. *Ann Phys. Ser.* 4:587.

Lee, J.H.S. 1965. The propagation of shocks and blast waves in a detonating gas. Ph.D. dissertation, McGill University.

Lee, J.H.S. 1977. *Am. Rev. Phys. Chem.* 28:75–104.

Lee, J.H.S. 1995. On the critical diameter problem. In *Dynamics of Exothermicity*, 321–335. Gordon and Breach.

Lee, J.H.S., and A.J. Higgins. 1999. *Phil. Trans. R. Soc. Lond. A.* 357:3503–3521.

Lee, J.H.S., R. Knystautas, and N. Yoshikawa. 1978. *Acta Astronaut.* 5:971.

Lee, J.H., R. Knystautas, C. Guirao, A. Bekesy, and S. Sabbagh. 1972. *Combust. Flame* 18:321–325.

Lee, J.H.S., B.H.K. Lee, and R. Knystautas. 1966. *Phys. Fluids* 9:221–222.

Lee, J.H., and H. Matsui. 1977. *Combust. Flame* 28:61–66.

Lee, J.H.S. and K. Ramamurthi. 1976. *Combust. Flame* 27:331–340.

Lin, S.C. 1954. *J. Appl. Phys.* 25(1).

Matsui, H., and J.H.S. Lee. 1979. In *17th Int. Symp. on Combustion*, 1269.

McBride, B.J., and S. Gordon. 1996. Computer program for calculation of complex chemical equilibrium compositions and applications II. Users manual and program description. NASA Rep. NASA RP-1311-P2.

Mitrofanov, V.V., and R.I. Soloukhin. 1965. *Sov. Phys. Dokl.* 9(12):1055.

Mooradian, A.J., and W.E. Gordon. 1951. *J. Chem. Phys.* 19:1166.

Murray, S.B., I. Moen, P., Thibault, R. Knystautas, and J.H.S. Lee. 1991. *Dynamics of detonations and explosions*, ed. A. Kuhl, J.C. Reyer, A.A. Borisov, and W. Siriguano, 91. AIAA.

Ng, H.D. 2005. The effect of chemical kinetics on the structure of gaseous detonations. Ph.D. dissertation, McGill University, Montreal.

Ng, H.D., and J.H.S. Lee. 2003. *J. Fluid Mech.* 476:179–211.

Radulescu, M., A. Higgins, J.H.S. Lee, and S. Murray. 2003. *J. Fluid Mech.* 480:1–24.

Sakurai, A. 1953, 1954. *J. Phys. Soc. Jpn.* 8(5), 9(2).

Sakurai, A. 1959. In *Exploding wires*, ed. W.G. Chase and H.K. Moore, 264. Plenum Press.

Schultz, E., and J. Shepherd. 2000. Detonation diffraction through a mixture gradient. Explosion Dynamics Laboratory Report FM00-1.

Sedov, L.I. 1946. Propagation of strong blast waves. *Prikl. Mat. Mekh.* 10:244–250.

Shepherd, W.C.F. 1949. In *3rd Int. Symp. on Combustion*, 301.

Short, M., and J. Quirk. 1997. *J. Fluid Mech.* 339:89–119.

Taylor, G.I. 1950. *Proc. Roy. Soc. Lond. A.* 201:159–174.

Thibault, P., N. Yoshikawa, and J.H.S. Lee. 1978. Shock wave amplification through coherent energy release. Presented at the 1978 Fall Technical Meeting of the Eastern Section of the Combustion Institute, Miami Beach.

von Neumann, J. 1941. The point source solution. Nat. Defense Res. Comm. Div. B Rept. AM-9.

Wadsworth, J. 1961. *Nature* 190:623–624.

Yoshikawa, N., and J.H.S. Lee. 1992. *Prog. Astronaut. Aeronaut.* 153:99–123.

Zeldovich, Ya.B., S.M. Kogarko, and N.N. Simonov. 1957. *Sov. Phys. Tech. Phys.* 1:1689–1713.

Epilogue

The Chapman–Jouguet (CJ) theory, formulated over a century ago, provides a simple method for determining the detonation velocity using the conservation equations and the equilibrium thermodynamic properties of the reactants and products. The Zeldovich–von Neumann–Döring (ZND) theory for the detonation structure, developed in the 1940s, permits the variation of the state in the reaction zone to be computed by integrating the chemical-kinetic rate equations simultaneously with the flow equations. The CJ criterion, which chooses the minimum velocity (or tangency) solution on the equilibrium Hugoniot curve, had been shown by von Neumann to be invalid for certain explosives that have a temperature overshoot (or intersecting partially reacted Hugoniot curves). For these explosives, the detonation velocity is higher than the CJ value and corresponds to weak detonation solutions on the equilibrium Hugoniot curve. Experimental evidence from the past 50 years has also confirmed that detonations are intrinsically unstable and have a transient three-dimensional structure. This throws further doubt on the general validity of the steady one-dimensional CJ theory. However, in spite of all the lack of support for the CJ theory, the CJ detonation velocity is found to agree remarkably well with experiments. Even for near-limit mixtures where three-dimensional transient effects are significant, the averaged velocity still generally agrees with the CJ value to within 10%. A possible explanation for this is perhaps the insensitivity of the CJ velocity to the chemical energy release ($V_{CJ} \approx \sqrt{Q}$). The energy that goes into the three-dimensional fluctuations, and hence is not available to drive the detonation in the direction of propagation, has only a small effect on the detonation velocity. Therefore, the CJ theory remains as the cornerstone of the foundation of detonation theory.

The steady one-dimensional ZND theory cannot describe the three-dimensional nonsteady cellular structure of detonations, in general. The one-dimensional ZND reaction length is generally two to three orders of magnitude smaller than the characteristic length scale of unstable detonations (e.g., the cell size). It is of critical importance to develop a theory that can determine an effective thickness of unstable detonations. The prediction of all the important non-equilibrium dynamic detonation parameters (limits, direct initiation energy, critical tube diameter, etc.) requires the knowledge of a length scale that characterizes the effective reaction-zone thickness of a cellular detonation. The effective thickness of an unstable detonation is often referred to as the *hydrodynamic thickness*. To formulate a theory for the hydrodynamic thickness, one can follow the approach used in turbulence theory. One can consider a steady mean flow with three-dimensional fluctuations superimposed onto it, analogous to the Reynolds-averaged Navier–Stokes equations in turbulent flow. Additional terms (analogous to the Reynolds stress, Reynolds flux, etc.) will appear, and the development of models for these terms would be a formidable task. The "turbulent" reaction zone of a cellular detonation consists of a variety of complex nonlinear processes of shock waves, vortices, shear layers, and density interface interactions. The modeling of these nonlinear processes would be far more difficult than the modeling of incompressible, non-reacting turbulent flows that still remain unresolved. Thus, the description of the turbulent structure of cellular detonation is not forthcoming in the foreseeable future. Satisfactory methods for the experimental measurement of the hydrodynamic thickness have yet to be developed.

A few reviewers of this manuscript have commented on the lack of any discussion on the practical applications of detonations. In particular, they pointed out the recent worldwide interest in the application of detonation to propulsion (i.e., the ram accelerator and pulse detonation engines). There are some fundamental difficulties in the use of detonation in propulsion systems. One is the control of the detonation so that it is coupled to the device itself. Detonations are self-propagating waves that respond to variations in initial and boundary conditions on microsecond time scales. On the other hand, devices of finite mass have response time scales larger by orders of magnitude or more. Thus, it is extremely difficult to control the detonation so that it can be coupled to an accelerating mass. Fuel–air mixtures are also relatively insensitive to detonation. Direct blast initiation requires energy of the order of megajoules, in contrast with the millijoules required for igniting a deflagration in the mixture. As yet, no practical ram accelerator or pulse detonation engine has been developed.

Perhaps the most successful application of detonation has been harnessing its enormous power for destructive purposes. Major advances in detonation theory were made by top scientists recruited to work on explosives during the two world wars. The largest concentration of detonation researchers is still to be found in government weapons laboratories. Detonations also have important civilian applications in construction work. The power of detonations literally enables man to move mountains and permit large-scale construction projects to be realized. However, it could be argued that such major defacing of the earth's surface morphology cannot be beneficial to mankind in the long run. More environmentally acceptable alternatives to these projects can perhaps be found.

It should also be noted that knowledge of the detonation phenomenon can be used to prevent and investigate detonations in accidental explosions. In fact, the discovery of detonations by Mallard and Le Châtelier (1883) was a result of their effort to understand explosions in coal mines.* Accidental explosions in mines and the chemical industries have claimed countless lives and cost billions of dollars in property damage annually. Their prevention is of great importance to the economic development of our modern society.

A scientist seeks to understand the phenomenon being studied. Its application by society – for good or evil – should not dissuade one in one's search for the truth.

* Mallard, E., and H. Le Châtelier. 1883. Recherches experimentales et theoriques sur la combustion des mélanges gazeux explosifs. *Ann. Mines Ser. 4.* 8:274–618.

Index

Printed in the United States
By Bookmasters